Classification and Identification of Lie Algebras

Volume 33

CRM MONOGRAPH SERIES

Centre de Recherches Mathématiques
Montréal

Classification and Identification of Lie Algebras

Libor Šnobl
Pavel Winternitz

American Mathematical Society
Providence, Rhode Island USA

The production of this volume was supported in part by the Fonds de recherche du Québec–Nature et technologies (FRQNT) and the Natural Sciences and Engineering Research Council of Canada (NSERC).

2010 *Mathematics Subject Classification.* Primary 17Bxx, 17B05 81Rxx, 81R05, 70Hxx, 37J15; Secondary 17B20, 17B30, 17B40, 70Sxx, 37Jxx.

For additional information and updates on this book, visit
www.ams.org/bookpages/crmm-33

Library of Congress Cataloging-in-Publication Data

Šnobl, Libor, 1976- author.
Classification and identification of Lie algebras / Libor Šnobl, Pavel Winternitz.
pages cm. – (CRM monograph series ; volume 33)
Includes bibliographical references and index.
ISBN 978-0-8218-4355-0 (alk. paper)
1. Lie algebras. 2. Lie superalgebras. I. Winternitz, Pavel, author. II. Title.

QA252.3.S66 2014
512′.482–dc23 2013034225

∞ The paper used in this book is acid-free and falls within the guidelines
established to ensure permanence and durability.
Visit the AMS home page at http://www.ams.org/

10 9 8 7 6 5 4 3 2 1 19 18 17 16 15 14

Contents

Preface

The purpose of this book is to serve as a tool for practitioners of Lie algebra and Lie group theory, i.e., for those who apply Lie algebras and Lie groups to solve problems arising in science and engineering. It is not intended to be a textbook on Lie theory, nor is it oriented towards one specific application, for instance the analysis of symmetries of differential equations. We restrict our attention to finite-dimensional Lie algebras over the fields of complex and real numbers.

In any application Lie algebras typically arise as sets of linear operators that commute with a given operator, say the Hamiltonian of a physical system. Alternatively, Lie groups arise as groups of (local) transformations leaving some object invariant; the corresponding Lie algebra then consists of vector fields generating 1-parameter subgroups. The object may be for instance the set of all solutions of a system of equations. The equations can be differential, difference, algebraic or integral ones, or some combination of such equations. They may be linear or nonlinear. In any case, the Lie algebra is realized by some operators in a basis that is usually not the standard one and that depends crucially on the manner in which it was obtained. The structure constants of Lie algebras can be calculated in any basis, but they in turn are basis dependent and reveal very little about the actual structure of the given Lie algebra.

After the Lie algebra $\mathfrak{g}$ associated with a studied problem is found, the next task that faces the researcher is to identify the Lie algebra as an abstract Lie algebra. In some cases $\mathfrak{g}$ may be isomorphic to a known algebra given in some accessible list. This is certainly the case for semisimple Lie algebras in view of Cartan's classification of all simple Lie algebras over the complex numbers, and subsequent classification of their real forms.

The fundamental Levi theorem, stating that every finite dimensional Lie algebra is isomorphic to a semidirect sum of a semisimple Lie algebra and the maximal solvable ideal (the radical) greatly simplifies the task of identifying a given Lie algebra. The weak link is that no complete classification of solvable Lie algebras exists, nor can one be expected to be produced in the future.

The problem addressed in this book is that of transforming a randomly obtained basis of a Lie algebra into a "canonical basis" in which all basis independent features of the Lie algebra are directly visible. For low dimensional Lie algebras (of dimension less or equal six) this makes it possible to identify the Lie algebra completely. In this book we give a representative list of all such Lie algebras. As stated above, in any dimension a complete identification can be performed for semisimple Lie algebras. We also describe some classes of nilpotent and solvable Lie algebras of arbitrary finite dimensions for which a complete classification exists and hence an exact identification is possible.

The book has four parts. The first presents some general results and concepts that are used in the subsequent chapters. In particular such invariant notions as the dimension of ideals in the characteristic series, and the invariants of the coadjoint representation are introduced.

In Part 2 we present algorithms that accomplish the following tasks:

(1) An algorithm for determining whether the algebra $\mathfrak{g}$ can be decomposed into a direct sum. If $\mathfrak{g}$ is decomposable the algorithm provides a basis in which $\mathfrak{g}$ is explicitly decomposed into a direct sum of indecomposable Lie subalgebras.

(2) A further algorithm is presented to find the radical $R(\mathfrak{g})$ and the Levi factor, i.e., the semisimple component of $\mathfrak{g}$.

(3) If the Lie algebra is solvable, for instance if it is the radical of a larger algebra, then it is necessary to identify its nilradical, i.e., the maximal nilpotent ideal. A rational (i.e., avoiding calculation of eigenvalues) algorithm for performing this is presented.

The text includes many examples illustrating various situations that may arise in such computations. All these algorithms have been implemented on computers.

Part 3 is devoted to solvable and nilpotent Lie algebras. While a complete classification of such algebras seems not to be feasible, it is possible to take a class of nilpotent Lie algebras and construct all extensions of these algebras to solvable ones. Finite-dimensional solvable Lie algebras with Abelian, Heisenberg, Borel, filiform and quasifiliform nilradicals are presented in Part 3.

Part 4 of the book consists of tables of all indecomposable Lie algebras of dimension n where $1 \leq n \leq 6$. They are ordered in such a way as to make the identification of any given low-dimensional Lie algebra written in an arbitrary basis as simple as possible. Any Lie algebra up to dimension 6 is isomorphic to precisely one entry in the tables. Essential characteristics of each algebra including its Casimir invariants are also provided.

The book is based on material that was previously dispersed in journal articles, many of them written by one or both of the authors of this book together with collaborators. The tables in Part 4 are based on older results and have been independently verified, in some cases corrected, unified and ordered by structural properties of the algebras (rather than by the way they were originally obtained).

Libor Šnobl and Pavel Winternitz

Acknowledgements

Our research was funded by multiple sources during the years it took us to write this book. Research of L. Šnobl was supported by the postdoctoral fellowship of the Centre de recherches mathématiques, Université de Montréal in 2004 – 2006. Next, his research at Czech Technical University in Prague was funded mainly by the research plans MSM210000018 and MSM6840770039 of the Ministry of Education of the Czech Republic. The research of P. Winternitz was partly supported by research grants from NSERC of Canada. LŠ thanks the Centre de recherches mathématiques, Université de Montréal for hospitality during numerous visits there while working on the manuscript. PW thanks the Faculty of Nuclear Sciences and Physical Engineering, Czech Technical University in Prague for hospitality during his visits there.

We thank Professors I. Anderson, A. G. Elashvili, J. Patera, Dr. A. Bihlo and D. Karásek for interesting and helpful discussions. We also thank all our colleagues and students with whom we collaborated on the results that are presented in this book.

We are particularly indebted to A. Montpetit for his efficient help in transforming our manuscript into a publishable book. We also thank Ms. I. Mette of AMS Book Acquisitions for keeping us on track.

L. Šnobl dedicates this book to his parents Libuše and Zdeněk. P. Winternitz dedicates this book to his wife Milada and his sons Peter and Michael. We both thank them for their support and encouragement.

Part 1

General Theory

CHAPTER 1

Introduction and Motivation

Lie groups and Lie algebras appear in science in many different guises. They may be a priori parts of the theory, like Lorentz or Galilei invariance of most physical theories, or the (semi)simple Lie groups of the Standard model in particle theory.

Alternatively, specific Lie groups may appear as consequences of specific dynamics. Consider any physical system with dynamics described by a system of ordinary or partial differential equations. This system of equations will be invariant under some local Lie group of local point transformations, taking solutions into solutions. This symmetry group G and its Lie algebra $\mathfrak{g}$ can be determined in an algorithmic manner [86]. The Lie algebra $\mathfrak{g}$ is obtained as an algebra of vector fields, usually in some nonstandard basis, depending on the way in which the algorithm is applied.

An immediate task is to identify the algebra found as being isomorphic to some known abstract Lie algebra. To do this we must transform it to a canonical basis in which all basis independent properties are manifest. Thus, if $\mathfrak{g}$ is decomposable into a direct sum, it should be explicitly decomposed into components that are further indecomposable

$$\mathfrak{g} = \mathfrak{g}_1 \oplus \mathfrak{g}_2 \oplus \cdots \oplus \mathfrak{g}_k. \tag{1.1}$$

Each indecomposable component must be further identified. Let $\mathfrak{g}$ now denote such an indecomposable Lie algebra. A fundamental theorem due to E. E. Levi [59, 71] tells us that any finite-dimensional Lie algebra can be represented as the semidirect sum

$$\mathfrak{g} = \mathfrak{p} \niplus \mathfrak{r}, \quad [\mathfrak{p}, \mathfrak{p}] = \mathfrak{p}, \quad [\mathfrak{r}, \mathfrak{r}] \subset \mathfrak{r}, \quad [\mathfrak{p}, \mathfrak{r}] \subseteq \mathfrak{r}, \tag{1.2}$$

where the subalgebra $\mathfrak{p}$ is semisimple and $\mathfrak{r}$ is the *radical* of $\mathfrak{g}$, i.e., its maximal solvable ideal. If $\mathfrak{g}$ is simple, we have $\mathfrak{r} = 0$ (an indecomposable semisimple algebra is actually simple). If $\mathfrak{g}$ is solvable, we have $\mathfrak{p} = 0$. Algorithms realizing decompositions (1.1), (1.2) exist [102] and are presented below.

In view of the Levi theorem (1.2) the classification of all finite-dimensional Lie algebras can be reduced to three steps:

(1) Classification of all simple Lie algebras. This also provides a classification of all semisimple ones.
(2) Classification of all solvable Lie algebras.
(3) Classification of all possible linear actions of the semisimple algebra $\mathfrak{p}$ on the radical $\mathfrak{r}$.

Semisimple Lie algebras over the field of complex numbers $\mathbb{C}$ have been completely classified by W. Killing and É. Cartan [22, 60], over the field of real numbers $\mathbb{R}$ by É. Cartan in [20, 21] (the analysis in the real case was later simplified by F. Gantmacher in [50]).

The third step is basically a task of the representation theory of semisimple Lie algebras. The main unsolved problem is the classification of all finite dimensional solvable Lie algebras, a task that does not have a realistic solution. Even all nilpotent Lie algebras are impossible to classify. (For instance in [127] the authors claim that they have found 24 168 nonisomorphic 9-dimensional nilpotent algebras Lie algebras with a maximal Abelian ideal of dimension 7 alone.) A realistic partial classification problem is to classify all solvable Lie algebras with a given nilradical of an arbitrary finite dimension n. So far this has been done for certain series of nilpotent Lie algebras, namely Abelian, Heisenberg and Borel nilradicals, as well as certain filiform and quasifiliform algebras. The results of the original articles [83, 84, 106, 115–118, 123, 124] are presented in a unified manner in Part 3 of this book.

An interesting physical application of the classification of low-dimensional Lie algebras is in general relativity. Indeed, the classification of Einstein spaces according to their isometry groups [95] is based on the work of Bianchi and his successors [6, 65]. The Petrov classification concerns Einstein spaces of dimension 4 and hence involves isometry groups of relatively low dimensions [95, 120]. String theory, brane cosmology and some other elementary particle theories going beyond the standard model require the use of higher-dimensional spaces. Any attempt at a Lie group classification of such spaces will require knowledge of higher-dimensional Lie groups, including solvable ones.

One of the very useful applications of Lie group analysis in science is the identification of seemingly different problems that are mathematically equivalent. Indeed, let us consider two systems, say (A) and (B), of differential equations

$$
\begin{gathered}
E_a^A(\vec{X},\vec{U};\partial_{X_i}U_\alpha,\partial^2_{X_iX_j}U_\alpha,\dots)=0,\\
E_a^B(\vec{x},\vec{u};\partial_{x_i}u_\alpha,\partial^2_{x_ix_j}u_\alpha,\dots)=0,\\
\vec{X}=(X_1,\dots,X_p),\qquad \vec{U}=(U_1,\dots,U_q),\\
\vec{x}=(x_1,\dots,x_p),\qquad \vec{u}=(u_1,\dots,u_q),\\
1\le a\le N,\quad 1\le\alpha\le q,\quad 1\le i,j,\dots\le p
\end{gathered}
\tag{1.3}
$$

of the same order describing different physical processes. From the mathematical viewpoint, we may consider these two systems equivalent if there exists a local invertible transformation of the independent and dependent variables

$$
X_i=\Lambda_i(\vec{x},\vec{u}),\qquad U_\alpha=\Omega_\alpha(\vec{x},\vec{u})
\tag{1.4}
$$

transforming the systems (A) and (B) into each other. The transformation (1.4) together with the mapping between the spaces of functions $\vec{u}(\vec{x})$ and $\vec{U}(\vec{X})$ induced by it is called a *point transformation.*

A necessary condition for the equivalence of the systems (A) and (B) is that they have isomorphic Lie algebras of infinitesimal point symmetries $\mathfrak{g}_A$ and $\mathfrak{g}_B$. The algebras $\mathfrak{g}_A$ and $\mathfrak{g}_B$ are by construction realized by vector fields in variables $(\vec{X},\vec{U})$ and $(\vec{x},\vec{u})$, respectively.

If $\mathfrak{g}_A$ and $\mathfrak{g}_B$ are isomorphic then a local point transformation (1.4) may exist such that it transforms the vector fields of $\mathfrak{g}_A$ into those of $\mathfrak{g}_B$ and vice versa. The transformation is unique up to point transformations leaving $\mathfrak{g}_A$ or $\mathfrak{g}_B$ invariant. In any case, the transformation taking $\mathfrak{g}_A$ into $\mathfrak{g}_B$ will also take the system (A) into a class of systems with the symmetry algebra $\mathfrak{g}_B$, which includes (B). If the symmetry algebra $\mathfrak{g}_B$ is large enough, the family of systems with the symmetry algebra $\mathfrak{g}_B$ is significantly restricted and often contains only the system (B). In this case the systems (A) and (B) are necessarily equivalent.

Let us consider an example of this procedure which will also demonstrate the usefulness of the identification algorithms presented in this book. The system (A) in our example will be the shallow water equations

$$\begin{aligned} U_T + UU_X + VU_Y + H_X &= 0, \\ V_T + UV_X + VV_Y + H_Y &= 0, \\ H_T + (UH)_X + (VH)_Y &= 0 \end{aligned} \tag{1.5}$$

describing the motion of an incompressible fluid in shallow water without boundaries in a gravitational field perpendicular to the flat bottom of the infinite basin [92]. The dependent variables are U, V, the two components of the fluid velocity and H, the depth. The independent variables are the time T and the Cartesian spatial coordinates X, Y. In our mathematical analysis we have set the gravitational constant $g = 1$ in (1.5).

The system (B) consists of the rotating shallow water equations for a fluid in a finite curved basin subject to the force of gravity (perpendicular to the surface at rest), and to a Coriolis force due to the rotation of the fluid. The equations are [68]

$$\begin{aligned} u_t + uu_x + vu_y + (Z+h)_x &= fv, \\ v_t + uv_x + vv_y + (Z+h)_y &= -fu, \\ h_t + (uh)_x + (vh)_y &= 0. \end{aligned} \tag{1.6}$$

The dependent and independent variables have the same interpretation as their uppercase counterparts in system (1.5), f corresponds to the Coriolis force,

$$Z(x, y) = A(x^2 + y^2) \tag{1.7}$$

describes the form of the bottom of the basin containing the fluid. The basin is chosen as a circular paraboloid.

Instead of the parameter A it is convenient to introduce an angular frequency ω such that

$$A = \frac{\omega^2 - f^2}{8}$$

and denote

$$R_1 = \tfrac{1}{2}(\omega + f), \quad R_2 = \tfrac{1}{2}(\omega - f).$$

By an explicit algorithmic computation one arrives at the generators of the point symmetries of these two sets of equations (1.5) and (1.6). For the free liquid (1.5) one finds the generators [27]

$$\begin{gathered} P_T = \partial_T, \quad P_X = \partial_X, \quad P_Y = \partial_Y, \quad G_X = T\partial_X + \partial_U, \quad G_Y = T\partial_Y + \partial_V, \\ D_1 = T\partial_T + X\partial_X + Y\partial_Y, \quad D_2 = -T\partial_T + U\partial_U + V\partial_V + 2H\partial_H, \\ L_1 = -Y\partial_X + X\partial_Y - V\partial_U + U\partial_V, \\ \Pi = T^2\partial_T + TX\partial_X + TY\partial_Y + (X - TU)\partial_U + (Y - TV)\partial_V - 2TH\partial_H. \end{gathered} \tag{1.8}$$

Their Lie brackets define the following structure of a Lie algebra $\mathfrak{g}_A$

	P_T	P_X	P_Y	G_X	G_Y	D_1	D_2	L_1	Π
P_T	0	0	0	P_X	P_Y	P_T	$-P_T$	0	$D_1 - D_2$
P_X	0	0	0	0	0	P_X	0	P_Y	G_X
P_Y	0	0	0	0	0	P_Y	0	$-P_X$	G_Y
G_X	$-P_X$	0	0	0	0	0	G_X	G_Y	0
G_Y	$-P_Y$	0	0	0	0	0	G_Y	$-G_X$	0
D_1	$-P_T$	$-P_X$	$-P_Y$	0	0	0	0	0	Π
D_2	P_T	0	0	$-G_X$	$-G_Y$	0	0	0	$-\Pi$
L_1	0	$-P_Y$	P_X	$-G_Y$	G_X	0	0	0	0
Π	$-D_1 + D_2$	$-G_X$	$-G_Y$	0	0	$-\Pi$	Π	0	0

The radical of the Lie algebra $\mathfrak{g}_A$ is spanned by $L_1, D_1 + D_2, G_X, G_Y, P_X, P_Y$, the nilradical by P_X, P_Y, G_X, G_Y and the Levi factor by $P_T, D_1 - D_2, \Pi$. Thus the choice of basis of the vector space spanned by the generators (1.8) is not the most natural one from the point of view of the structure of the algebra $\mathfrak{g}_A$. When we introduce a new basis

$$\begin{aligned} e_1 = P_T, \quad e_2 = D_1 - D_2, \quad e_3 = -\Pi, \quad e_4 = -(D_1 + D_2), \\ e_5 = L_1, \quad e_6 = P_Y, \quad e_7 = P_X, \quad e_8 = G_Y, \quad e_9 = G_X, \end{aligned} \tag{1.9}$$

the structure of the Lie algebra $\mathfrak{g}_A$, i.e., the Levi decomposition, the nilradical and the properties of the action of the radical and of the Levi factor on the nilradical, become immediately obvious. Namely, the Lie brackets read

	e_1	e_2	e_3	e_4	e_5	e_6	e_7	e_8	e_9
e_1	0	$2e_1$	$-e_2$	0	0	0	0	e_6	e_7
e_2	$-2e_1$	0	$2e_3$	0	0	$-e_6$	$-e_7$	e_8	e_9
e_3	e_2	$-2e_3$	0	0	0	e_8	e_9	0	0
e_4	0	0	0	0	0	e_6	e_7	e_8	e_9
e_5	0	0	0	0	0	e_7	$-e_6$	e_9	$-e_8$
e_6	0	e_6	$-e_8$	$-e_6$	$-e_7$	0	0	0	0
e_7	0	e_7	$-e_9$	$-e_7$	e_6	0	0	0	0
e_8	$-e_6$	$-e_8$	0	$-e_8$	$-e_9$	0	0	0	0
e_9	$-e_7$	$-e_9$	0	$-e_9$	e_8	0	0	0	0

The Levi factor is spanned by e_1, e_2, e_3 with the standard Lie brackets of the algebra $\mathfrak{sl}(2, \mathbb{R})$. The radical is spanned by $e_4, e_5, e_6, e_7, e_8, e_9$, the nilradical by e_6, e_7, e_8, e_9, the basis elements $e_4, e_5 \in \mathrm{R}(\mathfrak{g}_A) \setminus \mathrm{NR}(\mathfrak{g}_A)$ commute with the Levi factor and act on the nilradical as a dilation and a rotation, respectively.

Similarly, one finds the Lie algebra of generators of point symmetries also for the rotating fluid in a circular paraboloidal basin (1.6). The generators read [68]

$$
(1.10)\quad
\begin{aligned}
P_0 &= \partial_t, \quad D = x\partial_x + y\partial_y + u\partial_u + v\partial_v + 2h\partial_h,\\
Y_1 &= \cos(R_1 t)\partial_x - \sin(R_1 t)\partial_y - R_1\sin(R_1 t)\partial_u - R_1\cos(R_1 t)\partial_v,\\
Y_2 &= \sin(R_1 t)\partial_x + \cos(R_1 t)\partial_y + R_1\cos(R_1 t)\partial_u - R_1\sin(R_1 t)\partial_v,\\
Y_3 &= \cos(R_2 t)\partial_x + \sin(R_2 t)\partial_y - R_2\sin(R_2 t)\partial_u + R_2\cos(R_2 t)\partial_v,\\
Y_4 &= \sin(R_2 t)\partial_x - \cos(R_2 t)\partial_y + R_2\cos(R_2 t)\partial_u + R_2\sin(R_2 t)\partial_v,\\
R &= y\partial_x - x\partial_y + v\partial_u - u\partial_v,\\
K_1 &= \frac{1}{2}\cos(\omega t)(x\partial_x + y\partial_y - u\partial_u - v\partial_v + f(y\partial_u - x\partial_v) - 2h\partial_h)\\
&\quad + \frac{1}{2\omega}\sin(\omega t)(f(y\partial_x - x\partial_y + v\partial_u - u\partial_v) - \omega^2(x\partial_u + y\partial_v) + 2\partial_t),\\
K_2 &= -\frac{1}{2}\sin(\omega t)(x\partial_x + y\partial_y - u\partial_u - v\partial_v + f(y\partial_u - x\partial_v) - 2h\partial_h)\\
&\quad + \frac{1}{2\omega}\cos(\omega t)(f(y\partial_x - x\partial_y + v\partial_u - u\partial_v) - \omega^2(x\partial_u + y\partial_v) + 2\partial_t).
\end{aligned}
$$

Their Lie brackets define the Lie algebra $\mathfrak{g}_B$ (shown in Table 1.1) with the radical

$$\mathrm{R}(\mathfrak{g}_B) = \operatorname{span}\{D, R, Y_1, Y_2, Y_3, Y_4\},$$

the nilradical

$$\mathrm{NR}(\mathfrak{g}_B) = \operatorname{span}\{Y_1, Y_2, Y_3, Y_4\},$$

and the Levi factor

$$\mathfrak{p} = \operatorname{span}\left\{P_0 + \frac{f}{2}R, K_1, K_2\right\}.$$

From the indefinite signature of the Killing form of the Levi factor $\mathfrak{p}$ one may conclude that $\mathfrak{p}$ is isomorphic to the simple algebra $\mathfrak{sl}(2,\mathbb{R})$.

Thus the structure of the algebra $\mathfrak{g}_B$ is rather similar to that of the algebra $\mathfrak{g}_A$. Their relation is more transparent when $\mathfrak{g}_B$ is written in a basis resembling the basis $(e_1, \dots, e_9)$ introduced above, i.e., with the standard structure of the Levi factor and nonnilpotent basis elements of the radical chosen commuting with the Levi factor. A (nonunique) change of basis fulfilling these requirements is

$$
(1.11)\quad
\begin{aligned}
\tilde{e}_1 = -\frac{1}{\omega}\left(P_0 + \frac{f}{2}R\right) + K_2, \quad \tilde{e}_2 = -2K_1, \quad \tilde{e}_3 = \frac{1}{\omega}\left(P_0 + \frac{f}{2}R\right) + K_2,\\
\tilde{e}_4 = -D, \quad \tilde{e}_5 = R, \quad \tilde{e}_6 = Y_1 - Y_3,\\
\tilde{e}_7 = Y_2 + Y_4, \quad \tilde{e}_8 = -Y_2 + Y_4, \quad \tilde{e}_9 = Y_1 + Y_3.
\end{aligned}
$$

When the Lie brackets of the basis elements $(\tilde{e}_1, \dots, \tilde{e}_9)$ are computed they turn out to be the same as the Lie brackets of the algebra $\mathfrak{g}_A$ expressed in the basis $(e_1, \dots, e_9)$. That means that the symmetry algebras $\mathfrak{g}_A$ and $\mathfrak{g}_B$ are isomorphic.

The isomorphism between the symmetry algebras is a necessary condition for the existence of a point transformation mapping the system (1.5) to the system (1.6). However, it is far from sufficient.

The next step is to look for a point transformation that maps the algebra $\mathfrak{g}_B$ realized as the algebra of vector fields on a manifold $M \subset \mathbb{R}^6$ with the coordinates t, x, y, u, v, h to the algebra $\mathfrak{g}_A$ of the vector fields on the same manifold with the coordinates T, X, Y, U, V, H. If such a transformation exists, then the two vector realizations $\mathfrak{g}_A$, $\mathfrak{g}_B$ of the same abstract Lie algebra are equivalent. (Both the

TABLE 1.1

	P_0	K_1	K_2	D	R	Y_1	Y_2	Y_3	Y_4
P_0	0	ωK_2	$-\omega K_1$	0	0	$-\frac{f+\omega}{2}Y_2$	$\frac{f+\omega}{2}Y_1$	$\frac{f-\omega}{2}Y_4$	$\frac{\omega-f}{2}Y_3$
K_1	$-\omega K_2$	0	$\frac{-1}{\omega}\left(P_0+\frac{f}{2}R\right)$	0	0	$-\frac{1}{2}Y_3$	$\frac{1}{2}Y_4$	$-\frac{1}{2}Y_1$	$\frac{1}{2}Y_2$
K_2	ωK_1	$\frac{1}{\omega}\left(P_0+\frac{f}{2}R\right)$	0	0	0	$\frac{1}{2}Y_4$	$\frac{1}{2}Y_3$	$\frac{1}{2}Y_2$	$\frac{1}{2}Y_1$
D	0	0	0	0	0	$-Y_1$	$-Y_2$	$-Y_3$	$-Y_4$
R	0	0	0	0	0	Y_2	$-Y_1$	$-Y_4$	Y_3
Y_1	$\frac{f+\omega}{2}Y_2$	$\frac{1}{2}Y_3$	$-\frac{1}{2}Y_4$	Y_1	$-Y_2$	0	0	0	0
Y_2	$-\frac{f+\omega}{2}Y_1$	$-\frac{1}{2}Y_4$	$-\frac{1}{2}Y_3$	Y_2	Y_1	0	0	0	0
Y_3	$\frac{\omega-f}{2}Y_4$	$\frac{1}{2}Y_1$	$-\frac{1}{2}Y_2$	Y_3	Y_4	0	0	0	0
Y_4	$\frac{f-\omega}{2}Y_3$	$-\frac{1}{2}Y_2$	$-\frac{1}{2}Y_1$	Y_4	$-Y_3$	0	0	0	0

transformation and the equivalence can be local, defined in a neighborhood of a generic point of M only).

Using computer algebra we find a locally invertible map ϕ

$$\begin{aligned} X &= \phi_X(t,x,y,u,v,h), \quad Y = \phi_Y(t,x,y,u,v,h), \quad T = \phi_T(t,x,y,u,v,h), \\ U &= \phi_U(t,x,y,u,v,h), \quad V = \phi_V(t,x,y,u,v,h), \quad H = \phi_H(t,x,y,u,v,h) \end{aligned} \tag{1.12}$$

such that $e_j = \phi_*(\tilde{e}_j)$. Explicitly it reads

$$\begin{aligned} T &= \cot\left(\frac{\omega}{2}t\right), \\ X &= \frac{1}{2\sin((\omega/2)t)}\left(\cos\left(\frac{f}{2}t\right)x - \sin\left(\frac{f}{2}t\right)y\right), \\ Y &= -\frac{1}{2\sin((\omega/2)t)}\left(\sin\left(\frac{f}{2}t\right)x + \cos\left(\frac{f}{2}t\right)y\right), \\ U &= \frac{1}{2\omega}\Bigg(-2\sin\left(\frac{\omega}{2}t\right)\cos\left(\frac{f}{2}t\right)u + 2\sin\left(\frac{\omega}{2}t\right)\sin\left(\frac{f}{2}t\right)v \\ &\quad + \left(\sin\left(\frac{\omega}{2}t\right)\sin\left(\frac{f}{2}t\right)f + \cos\left(\frac{f}{2}t\right)\cos\left(\frac{\omega}{2}t\right)\omega\right)x \\ &\quad + \left(\sin\left(\frac{\omega}{2}t\right)\cos\left(\frac{f}{2}t\right)f - \sin\left(\frac{f}{2}t\right)\cos\left(\frac{\omega}{2}t\right)\omega\right)y\Bigg), \\ V &= \frac{1}{2\omega}\Bigg(2\sin\left(\frac{\omega}{2}t\right)\sin\left(\frac{f}{2}t\right)u + 2\sin\left(\frac{\omega}{2}t\right)\cos\left(\frac{f}{2}t\right)v \\ &\quad + \left(\sin\left(\frac{\omega}{2}t\right)\cos\left(\frac{f}{2}t\right)f - \sin\left(\frac{f}{2}t\right)\cos\left(\frac{\omega}{2}t\right)\omega\right)x \\ &\quad - \left(\sin\left(\frac{\omega}{2}t\right)\sin\left(\frac{f}{2}t\right)f + \cos\left(\frac{f}{2}t\right)\cos\left(\frac{\omega}{2}t\right)\omega\right)y\Bigg), \\ H &= Ch\sin\left(\frac{\omega}{2}t\right)^2, \end{aligned} \tag{1.13}$$

where C is an arbitrary nonvanishing integration constant.

Since the symmetry algebras of the systems of equations (1.5) and (1.6) are equivalent, it is possible that the two models are related by a point transformation. In order to verify that we perform an explicit change of independent and dependent variables (1.13) in the PDE system (1.5) and we find that we indeed recover the system of PDEs (1.6) provided that we set the integration constant C in (1.13) to

$$C = \frac{1}{\omega^2}.$$

Thus, the two dynamical systems (1.5) and (1.6) can be (locally) transformed one into the other by a point transformation and any solution of (1.5) gives rise to a (local) solution of (1.6) and vice versa. The point transformation (1.13) is highly nontrivial and its existence and explicit form would be very difficult to obtain without first noticing the isomorphism between the symmetry algebras $\mathfrak{g}_A$ and $\mathfrak{g}_B$. We mention that the physical interpretations of these two mathematically isomorphic algebras are quite different. For instance the nilradical of $\mathfrak{g}_A$ corresponds to translations and Galilei transformations in the x and y directions. The nilradical of $\mathfrak{g}_B$ corresponds to transformations related to the Coriolis force and the presence of the boundary (1.7). The physical interpretations of the $\mathfrak{sl}(2, \mathbb{R})$ Levi factors of $\mathfrak{g}_A$ and $\mathfrak{g}_B$ are also different.

CHAPTER 2

Basic Concepts

The purpose of this chapter is to establish the notation and clarify the definitions used in the following text. Our aim here is not a comprehensive introduction to the theory of Lie algebras which can be found in numerous textbooks together with detailed proofs of theorems and propositions, e.g., [41, 46, 59, 62, 72, 108, 109].

2.1. Definitions

A *Lie algebra* $\mathfrak{g}$ is a vector space over a field $\mathbb{F}$ equipped with a multiplication (also called a Lie bracket), i.e., a bilinear map $[\ ,\ ]\colon \mathfrak{g}\times\mathfrak{g}\to\mathfrak{g}$ such that

$$[y,x] = -[x,y] \qquad \text{(antisymmetry)} \tag{2.1}$$

$$\big[x,[y,z]\big] + \big[y,[z,x]\big] + \big[z,[x,y]\big] = 0 \qquad \text{(Jacobi identity)} \tag{2.2}$$

for all elements $x, y, z \in \mathfrak{g}$. We shall consider the fields $\mathbb{F} = \mathbb{R}$, $\mathbb{C}$ only.

A *subalgebra* $\mathfrak{h}$ of the Lie algebra $\mathfrak{g}$ is a vector subspace of $\mathfrak{g}$ which is closed under the bracket,

$$[\mathfrak{h},\mathfrak{h}] \subseteq \mathfrak{h}. \tag{2.3}$$

An *ideal* $\mathfrak{i}$ of the Lie algebra $\mathfrak{g}$ is a subalgebra such that

$$[\mathfrak{i},\mathfrak{g}] \subseteq \mathfrak{i}. \tag{2.4}$$

The Lie algebra $\mathfrak{g}$ itself and $\{0\}$ are *trivial* ideals. A Lie algebra which does not possess any nontrivial ideal is called *simple.*

Lemma 2.1. *If* $\mathfrak{i},\mathfrak{j} \subset \mathfrak{g}$ *are ideals, so is* $[\mathfrak{i},\mathfrak{j}] = \operatorname{span}\{[a,b] \mid a\in\mathfrak{i}, b\in\mathfrak{j}\}$.

PROOF. Using

$$\big[x,[a,b]\big] = \big[[x,a],b\big] + \big[a,[x,b]\big] \in [\mathfrak{i},\mathfrak{j}]$$

for all $x\in\mathfrak{g}$, $a\in\mathfrak{i}$, $b\in\mathfrak{j}$. Thus $[x,y]\in[\mathfrak{i},\mathfrak{j}]$ holds also for any $y\in\operatorname{span}\{[a,b] \mid a\in\mathfrak{i}, b\in\mathfrak{j}\}$. □

Three different series of ideals can be associated with any given Lie algebra. The dimensions of the ideals in each of these series are important characteristic features of the given Lie algebra.

The *derived series* $\mathfrak{g} = \mathfrak{g}^{(0)} \supseteq \mathfrak{g}^{(1)} \supseteq \cdots \supseteq \mathfrak{g}^{(k)} \supseteq \cdots$ is defined recursively

$$\mathfrak{g}^{(k)} = [\mathfrak{g}^{(k-1)},\mathfrak{g}^{(k-1)}], \quad \mathfrak{g}^{(0)} = \mathfrak{g}. \tag{2.5}$$

The second term in the series, namely $\mathfrak{g}^{(1)} = [\mathfrak{g},\mathfrak{g}]$, is called the *derived algebra* of $\mathfrak{g}$ and may be also denoted $D(\mathfrak{g})$ or $\mathfrak{g}^2$. If the derived series terminates, i.e., there exists $k\in\mathbb{N}$ such that $\mathfrak{g}^{(k)} = 0$, then $\mathfrak{g}$ is called a *solvable Lie algebra.* If the derived series contains only the algebra $\mathfrak{g}$ itself, i.e., $\mathfrak{g}^{(1)} = \mathfrak{g}$, then the Lie algebra $\mathfrak{g}$ is called *perfect.*

The *lower central series* $\mathfrak{g} = \mathfrak{g}^1 \supseteq \cdots \supseteq \mathfrak{g}^k \supseteq \cdots$ is again defined recursively

$$\mathfrak{g}^k = [\mathfrak{g}^{k-1}, \mathfrak{g}], \quad \mathfrak{g}^1 = \mathfrak{g}. \tag{2.6}$$

If the lower central series terminates, i.e., there exists $k \in \mathbb{N}$ such that $\mathfrak{g}^k = 0$, then $\mathfrak{g}$ is called a *nilpotent Lie algebra*. The highest value of k for which we have $\mathfrak{g}^k \neq 0$ is the *degree of nilpotency* of the nilpotent Lie algebra $\mathfrak{g}$.

Obviously, a nilpotent Lie algebra is also solvable. An Abelian Lie algebra is nilpotent of degree 1.

The *upper central series* is $\mathfrak{z}_1(\mathfrak{g}) \subseteq \cdots \subseteq \mathfrak{z}_k(\mathfrak{g}) \subseteq \cdots \subseteq \mathfrak{g}$. In this series $\mathfrak{z}_1$ is the *center* of $\mathfrak{g}$

$$\mathfrak{z}_1(\mathfrak{g}) = C(\mathfrak{g}) = \{x \in \mathfrak{g} \mid [x, y] = 0, \ \forall y \in \mathfrak{g}\}. \tag{2.7}$$

Now let us consider the factor algebra $\mathfrak{f}_1 \simeq \mathfrak{g}/\mathfrak{z}_1(\mathfrak{g})$. Its center is $C(\mathfrak{f}_1) = C\big(\mathfrak{g}/\mathfrak{z}_1(\mathfrak{g})\big)$. We define the *second center* $\mathfrak{z}_2(\mathfrak{g})$ of $\mathfrak{g}$ to be the unique ideal in $\mathfrak{g}$ such that

$$\mathfrak{z}_2(\mathfrak{g})/\mathfrak{z}_1(\mathfrak{g}) = C\big(\mathfrak{g}/\mathfrak{z}_1(\mathfrak{g})\big). \tag{2.8}$$

Recursively we define *kth-center* $\mathfrak{z}_k$ as the unique ideal in $\mathfrak{g}$ such that its image under factorization by $\mathfrak{z}_{k-1}(\mathfrak{g})$ is the center of $\mathfrak{g}/\mathfrak{z}_{k-1}(\mathfrak{g})$, i.e.,

$$\mathfrak{z}_k(\mathfrak{g})/\mathfrak{z}_{k-1}(\mathfrak{g}) = C\big(\mathfrak{g}/\mathfrak{z}_{k-1}(\mathfrak{g})\big). \tag{2.9}$$

The union of the upper central series is the *hypercenter*

$$\mathfrak{z}_\infty(\mathfrak{g}) = \bigcup_{i=1}^{\infty} \mathfrak{z}_i(\mathfrak{g}). \tag{2.10}$$

Since the higher centers $\mathfrak{z}_i(\mathfrak{g})$ are ordered by inclusion, the hypercenter can be also viewed as the largest ideal in the sequence and coincides with $\mathfrak{z}_k(\mathfrak{g})$ for all k larger than certain $k_0 \in \mathbb{N}$. The upper central series terminates, i.e., the hypercenter is equal to the whole algebra $\mathfrak{g}$, if and only if $\mathfrak{g}$ is nilpotent [59]. The hypercenter is always a nilpotent ideal in $\mathfrak{g}$ because we have

$$\big(\mathfrak{z}_k(\mathfrak{g})\big)^{k+1} = \underbrace{[\mathfrak{z}_k(\mathfrak{g}), [\mathfrak{z}_k(\mathfrak{g}), \ldots, \mathfrak{z}_k(\mathfrak{g})]]}_{k+1 \text{ times}} \subseteq \underbrace{[\mathfrak{g}, [\mathfrak{g}, \ldots, [\mathfrak{g}}_{k \text{ times}}, \mathfrak{z}_k(\mathfrak{g})] \ldots] = 0$$

for any $k \in \mathbb{N}$ by construction.

We shall call these three series of ideals the *characteristic series* of the algebra $\mathfrak{g}$. We shall use the notations DS, CS and US for sets of integers denoting the dimensions of subalgebras in the derived, lower central and upper central series, respectively.

Corollary 2.2 (of Lemma 2.1). *Let $\mathfrak{i} \subset \mathfrak{g}$ be an ideal, so are $\mathfrak{i}^k$ and $\mathfrak{i}^{(k)}$, $k \in \mathbb{N}$.*

Lemma 2.3. *For any Lie algebra $\mathfrak{g}$,*

- *the sum of any two solvable ideals of $\mathfrak{g}$ is a solvable ideal of $\mathfrak{g}$,*
- *the sum of any two nilpotent ideals of $\mathfrak{g}$ is a nilpotent ideal of $\mathfrak{g}$.*

PROOF. See [59, 62, 139]. □

Thus in any given Lie algebra we have two unique ideals, the radical and the nilradical. The *radical* is the maximal solvable ideal and we shall denote it by $\mathrm{R}(\mathfrak{g})$.

It is unique because it is a sum of all solvable ideals. Similarly, the *nilradical* $\mathrm{NR}(\mathfrak{g})$ is the maximal nilpotent ideal. The following relation between them holds

$$(\mathrm{R}(\mathfrak{g}))^2 \subseteq \mathrm{NR}(\mathfrak{g}) \subseteq \mathrm{R}(\mathfrak{g}). \tag{2.11}$$

Both of them can vanish, then the algebra is *semisimple*. When the radical of $\mathfrak{g}$ is nonvanishing, also the nilradical is nonvanishing — either $(\mathrm{R}(\mathfrak{g}))^2 = 0$, i.e., both $\mathrm{NR}(\mathfrak{g})$ and $\mathrm{R}(\mathfrak{g})$ are Abelian and therefore coincide, or $0 \neq (\mathrm{R}(\mathfrak{g}))^2 \subseteq \mathrm{NR}(\mathfrak{g})$.

Lemma 2.4. *For any subalgebra* $\mathfrak{h}$ *of the Lie algebra* $\mathfrak{g}$ *we have*

$$\mathrm{NR}(\mathfrak{g}) \cap \mathfrak{h} \subseteq \mathrm{NR}(\mathfrak{h}), \tag{2.12}$$

$$\mathrm{R}(\mathfrak{g}) \cap \mathfrak{h} \subseteq \mathrm{R}(\mathfrak{h}). \tag{2.13}$$

Notice that the equality sign in (2.12), instead of the $\subseteq$ sign would, in general, not be appropriate. Indeed, consider the example of the two-dimensional non-Abelian Lie algebra $\mathrm{span}\{e_1, e_2\}$ with commutation relation

$$[e_1, e_2] = e_1. \tag{2.14}$$

The nilradical is $\mathrm{NR}(\mathfrak{g}) = \mathrm{span}\{e_1\}$. Consider the subalgebra $\mathfrak{s} = \mathrm{span}\{e_2\}$. We have

$$\mathrm{NR}(\mathfrak{g}) \cap \mathfrak{s} = \{0\} \subsetneq \mathrm{NR}(\mathfrak{s}) = \mathfrak{s}.$$

In relation to the nilradical the hypercenter satisfies

$$\mathfrak{z}_\infty(\mathfrak{g}) \subseteq \mathrm{NR}(\mathfrak{g}). \tag{2.15}$$

The radical of a Lie algebra $\mathfrak{g}$ is characterized by means of repeated nilradical formation as follows [139]:

$$\begin{gathered}\mathrm{NR}(\mathfrak{g}) \equiv (\mathrm{NR})_1(\mathfrak{g}) \subseteq (\mathrm{NR})_2(\mathfrak{g}) \subseteq (\mathrm{NR})_3(\mathfrak{g}) \subseteq \cdots \subseteq (\mathrm{NR})_\infty(\mathfrak{g}),\\ (\mathrm{NR})_{i+1}(\mathfrak{g})/(\mathrm{NR})_i(\mathfrak{g}) = \mathrm{NR}\big(\mathfrak{g}/(\mathrm{NR})_i(\mathfrak{g})\big) \quad (i \in \mathbb{N}),\\ \bigcup_{i\in\mathbb{N}} \mathrm{NR}_i(\mathfrak{g}) \equiv (\mathrm{NR})_\infty(\mathfrak{g}) = \mathrm{R}(\mathfrak{g}).\end{gathered}$$

We use lower indices for this repeated nilradical formation to avoid confusion with the lower central series of equation (2.6). There is always a first index $j = j(\mathfrak{g})$ for which

$$(\mathrm{NR})_j(\mathfrak{g}) = \mathrm{R}(\mathfrak{g}) \tag{2.16}$$

and either $j = 1$ or else $j > 1$ with $(\mathrm{NR})_{j-1}(\mathfrak{g}) \subsetneq \mathrm{R}(\mathfrak{g})$.

The *centralizer* $\mathrm{cent}_\mathfrak{g}(\mathfrak{h})$ of a given subalgebra $\mathfrak{h} \subseteq \mathfrak{g}$ in $\mathfrak{g}$ is the set of all elements in $\mathfrak{g}$ commuting with all elements in $\mathfrak{h}$, i.e.,

$$\mathrm{cent}_\mathfrak{g}(\mathfrak{h}) = \{x \in \mathfrak{g} \mid [x, y] = 0, \ \forall y \in \mathfrak{h}\}. \tag{2.17}$$

The *normalizer* $\mathrm{norm}_\mathfrak{g}(\mathfrak{h})$ of a given subspace $\mathfrak{h} \subseteq \mathfrak{g}$ in $\mathfrak{g}$ is the set of all elements x in $\mathfrak{g}$ such that $[x, h]$ is in the subspace $\mathfrak{h}$ for any $h \in \mathfrak{h}$, i.e.,

$$\mathrm{norm}_\mathfrak{g}(\mathfrak{h}) = \{x \in \mathfrak{g} \mid [x, y] \in \mathfrak{h}, \ \forall y \in \mathfrak{h}\}. \tag{2.18}$$

When $\mathfrak{h}$ is a subalgebra then necessarily $\mathfrak{h} \subseteq \mathrm{norm}_\mathfrak{g}(\mathfrak{h})$. The normalizer of an ideal in $\mathfrak{g}$ is the whole algebra $\mathfrak{g}$.

Both the centralizer and the normalizer can be also taken with respect to a subset $\mathfrak{f} \subset \mathfrak{g}$,

$$\mathrm{cent}_\mathfrak{f}(\mathfrak{h}) = \mathrm{cent}_\mathfrak{g}(\mathfrak{h}) \cap \mathfrak{f}, \quad \mathrm{norm}_\mathfrak{f}(\mathfrak{h}) = \mathrm{norm}_\mathfrak{g}(\mathfrak{h}) \cap \mathfrak{f}.$$

A *representation* ρ of a given Lie algebra $\mathfrak{g}$ on a vector space V is a linear map of $\mathfrak{g}$ into the space $\mathscr{L}(V)$ of linear operators acting on V

$$\rho\colon \mathfrak{g} \to \mathscr{L}(V)\colon x \to \rho(x)$$

such that for any pair x, y of elements of $\mathfrak{g}$

$$\rho([x, y]) = \rho(x) \circ \rho(y) - \rho(y) \circ \rho(x) \tag{2.19}$$

holds. When the map ρ is injective, the representation is called *faithful*.

A subspace W of V is called *invariant* if

$$\rho(\mathfrak{g})W = \{\rho(x)w \,|\, x \in \mathfrak{g}, w \in W\} \subseteq W.$$

A representation ρ of $\mathfrak{g}$ on V is

- *reducible* if a proper nonvanishing invariant subspace W of V exists,
- *irreducible* if no nontrivial invariant subspace of V exists,
- *fully reducible* when every invariant subspace W of V has an invariant complement $\widetilde{W}$, i.e.,

$$V = W \oplus \widetilde{W}, \quad \rho(\mathfrak{g})\widetilde{W} \subseteq \widetilde{W}. \tag{2.20}$$

In particular, any irreducible representation is also fully reducible.

An important criterion for irreducibility of a given representation is

Theorem 2.5 (Schur lemma). *Let $\mathfrak{g}$ be a complex Lie algebra and ρ its representation on a finite-dimensional vector space V.*

(1) *Let ρ be irreducible. Then any operator A on V which commutes with all $\rho(x)$,*

$$[A, \rho(x)] = 0, \quad \forall x \in \mathfrak{g},$$

has the form $A = \lambda \mathbf{1}$ for some complex number λ.

(2) *Let ρ be fully reducible and such that every operator A on V which commutes with all $\rho(x)$ has the form $A = \lambda \mathbf{1}$ for some complex number λ. Then ρ is irreducible.*

For future reference we recall that the Schur lemma holds also in a more general situation where the representation of a Lie algebra $\mathfrak{g}$ is replaced by an arbitrary set of operators S acting on the complex vector space V and by (ir)reducibility we understand the (non)existence of a nontrivial subspace invariant under all operators $B \in S$, respectively. In such a context the following corollary to the Schur lemma is useful:

Corollary 2.6. *Let S be a set of operators acting on a complex vector space V such that every traceless operator A on V which commutes with all $B \in S$ is nilpotent. Then the vector space V cannot be written as a direct sum of two nontrivial subspaces invariant under S.*

A particular representation is defined for any Lie algebra $\mathfrak{g}$, namely the *adjoint representation* of a given Lie algebra $\mathfrak{g}$ is a linear map of $\mathfrak{g}$ into the space of linear operators acting on $\mathfrak{g}$

$$\mathrm{ad}\colon \mathfrak{g} \to \mathscr{L}(\mathfrak{g})\colon x \to \mathrm{ad}(x)$$

defined for any pair x, y of elements of $\mathfrak{g}$ via

$$\mathrm{ad}(x)\ y = [x, y]. \tag{2.21}$$

The image of ad is denoted by $\mathrm{ad}\,\mathfrak{g}$.

A well-known theorem allows us to express nilpotency of a given algebra in terms of operators in the adjoint representation.

Theorem 2.7 (Engel theorem). *A Lie algebra $\mathfrak{g}$ is nilpotent if and only if $k \in \mathbb{N}$ exists such that $\big(\mathrm{ad}(x)\big)^k = 0$ for all $x \in \mathfrak{g}$, i.e., if all $\mathrm{ad}(x)$ are nilpotent operators on $\mathfrak{g}$.*

Similarly, the solvability of a given complex algebra can be formulated in terms of operators representing it in any faithful representation.

Theorem 2.8 (Lie theorem). *Let $\mathfrak{g}$ be a Lie algebra and ρ its faithful representation on a complex vector space V, $n = \dim V$. The algebra $\mathfrak{g}$ is solvable if and only if a filtration of V by codimension 1 subspaces $(V_k)_{k=1}^n$ invariant with respect to the representation ρ exists, i.e.,*

$$0 \subsetneq V_1 \subsetneq V_2 \subsetneq \cdots \subsetneq V_n \equiv V, \tag{2.22}$$

$$\rho(\mathfrak{g}) V_k \subset V_k. \tag{2.23}$$

Remark 2.9. When the representation ρ is chosen to be the adjoint representation, the statement of the Lie theorem holds even when it is not faithful.

As a direct consequence of the Lie and Engel theorems 2.8 and 2.7 we find that we have

$$D\big(\mathrm{R}(\mathfrak{g})\big) = [\mathrm{R}(\mathfrak{g}), \mathrm{R}(\mathfrak{g})] \subseteq \mathrm{NR}(\mathfrak{g}). \tag{2.24}$$

and moreover

$$[\mathrm{R}(\mathfrak{g}), \mathfrak{g}] \subseteq \mathrm{NR}(\mathfrak{g}).$$

(For proofs of Theorems 2.7, 2.8 and their consequences see e.g. [59, p. 36–37, 48–51; 62, p. 42–49].)

An *automorphism* Φ of a given Lie algebra $\mathfrak{g}$ is a linear map

$$\Phi \colon \mathfrak{g} \to \mathfrak{g}$$

such that for any pair x, y of elements of $\mathfrak{g}$

$$\Phi([x, y]) = [\Phi(x), \Phi(y)]. \tag{2.25}$$

A *derivation* D of a given Lie algebra $\mathfrak{g}$ is a linear map

$$D \colon \mathfrak{g} \to \mathfrak{g}$$

such that for any pair x, y of elements of $\mathfrak{g}$

$$D([x, y]) = [D(x), y] + [x, D(y)]. \tag{2.26}$$

If an element $z \in \mathfrak{g}$ exists, such that

$$D = \mathrm{ad}(z), \quad \text{i.e., } D(x) = [z, x],\ \forall x \in \mathfrak{g}$$

the derivation D is called an *inner derivation*, any other one is an *outer derivation.*

The set of all automorphisms of $\mathfrak{g}$ with composition as the group law forms a Lie group $\mathrm{Aut}(\mathfrak{g})$ and its Lie algebra coincides with the algebra $\mathfrak{Der}(\mathfrak{g})$ of all derivations of $\mathfrak{g}$.

An ideal $\mathfrak{i}$ in $\mathfrak{g}$ is called a *characteristic ideal* if it is invariant with respect to every derivation $D \in \mathfrak{Der}(\mathfrak{g})$.

Theorem 2.10. *The radical, nilradical as well as any ideal belonging to the characteristic series of a finite-dimensional Lie algebra $\mathfrak{g}$ is invariant under any derivation of $\mathfrak{g}$.*

Proof. We notice that any of the mentioned ideals is by construction invariant under every automorphism of $\mathfrak{g}$. E.g., if the solvable ideals $R(\mathfrak{g})$ and $\Phi\big(R(\mathfrak{g})\big)$ differed for some automorphism $\Phi \in \mathrm{Aut}(\mathfrak{g})$, their sum would be a larger solvable ideal in $\mathfrak{g}$, thus contradicting the maximality of $R(\mathfrak{g})$. For any derivation $D \in \mathfrak{Der}(\mathfrak{g})$ and any ideal $\mathfrak{i}$ invariant under every automorphism we consider the one-parameter family of automorphisms $\Phi(t) = \exp(tD)$. We have

$$\exp(tD)(\mathfrak{i}) = \mathfrak{i}, \quad \forall t \in \mathbb{F}\,(= \mathbb{R}, \mathbb{C}),$$

i.e., also

$$D(\mathfrak{i}) \subseteq \mathfrak{i}. \tag{2.27}$$

□

An even stronger result than Theorem 2.10 holds. Namely, any derivation of a solvable Lie algebra maps the algebra into its nilradical (see, e.g., [62, Proposition 1.40] for a proof).

A symmetric bilinear form B on a given Lie algebra $\mathfrak{g}$ such that

$$B(\mathrm{ad}(x)y, z) + B(y, \mathrm{ad}(x)z) = 0 \tag{2.28}$$

for any triple $x, y, z \in \mathfrak{g}$ is called ad-*invariant* or *invariant*. A symmetric bilinear form B is *invariant with respect to automorphisms* if

$$B\big(\Phi(x), \Phi(y)\big) = B(x, y) \tag{2.29}$$

for any automorphism Φ of $\mathfrak{g}$ and any pair $x, y \in \mathfrak{g}$.

We recall that any form B invariant with respect to automorphisms is also ad-invariant. This statement follows from the fact that (2.29) implies upon differentiation

$$B(D(x), y) + B\big(x, D(y)\big) = 0 \tag{2.30}$$

for any derivation D, in particular for all inner derivations. The converse is not true in general because outer derivations of $\mathfrak{g}$ may exist or there may be discrete transformations in $\mathrm{Aut}(\mathfrak{g})$ which are not obtained by exponentiation of elements in $\mathfrak{Der}(\mathfrak{g})$.

The *Killing form* K of a given Lie algebra $\mathfrak{g}$ is a symmetric bilinear form on $\mathfrak{g}$ defined by

$$K(x, y) = \mathrm{Tr}\big(\mathrm{ad}(x) \cdot \mathrm{ad}(y)\big). \tag{2.31}$$

The Killing form is invariant with respect to automorphisms. In the particular case of complex simple Lie algebras, any invariant symmetric bilinear form is a multiple of the Killing form.

The Killing form provides the following important criteria for semisimplicity and solvability.

Theorem 2.11 (Cartan's criteria). *A Lie algebra $\mathfrak{g}$ is*

- *semisimple if and only if its Killing form is nondegenerate;*
- *solvable if and only if the restriction of its Killing form to the derived algebra vanishes.*

2.2. Levi theorem

In any Lie algebra $\mathfrak{g}$ there exists a unique maximal solvable ideal $\mathrm{R}(\mathfrak{g})$ called the *radical* as was already mentioned. The radical satisfies $\mathrm{R}(\mathfrak{g}) = 0$ if and only if $\mathfrak{g}$ is semisimple. On the other hand the radical $\mathrm{R}(\mathfrak{g})$ coincides with $\mathfrak{g}$ if and only if $\mathfrak{g}$ is solvable.

Theorem 2.12 (Levi theorem). *Any finite-dimensional Lie algebra $\mathfrak{g}$ can be decomposed into a semidirect sum*

$$\mathfrak{g} = \mathfrak{p} \ni\!\!\!\!+ \mathrm{R}(\mathfrak{g}) \tag{2.32}$$

where the complement $\mathfrak{p}$ of the radical $\mathrm{R}(\mathfrak{g})$ in $\mathfrak{g}$ is a semisimple Lie algebra, isomorphic to the factor algebra $\mathfrak{g}/\mathrm{R}(\mathfrak{g})$. The semisimple Lie algebra $\mathfrak{p}$ is called the Levi factor *or* Levi subalgebra *of $\mathfrak{g}$.*

A sequel to the Levi theorem is the result of Mal′cev

Theorem 2.13 (Mal′cev theorem). *Any two Levi factors $\mathfrak{p}_1$ and $\mathfrak{p}_2$ of the Lie algebra $\mathfrak{g}$ are isomorphically mapped one into the other by some inner automorphism Φ of the form*

$$\Phi = \exp(\operatorname{ad} z) \tag{2.33}$$

where $z \in \mathrm{NR}(\mathfrak{g})$.

(For proofs see, e.g., [59, 62, 71, 73, 139]).

We notice that the derived algebra $\mathfrak{g}^2$ is necessarily contained in $\mathfrak{p} \ni\!\!\!\!+ \mathrm{NR}(\mathfrak{g})$ where $\mathrm{NR}(\mathfrak{g})$ is the nilradical of $\mathfrak{g}$. The reason is that for any $x \in \mathfrak{p}$ its adjoint representation $\operatorname{ad}|_{\mathrm{R}(\mathfrak{g})}(x)$ is a derivation of the radical $\mathrm{R}(\mathfrak{g})$, i.e., maps $\mathrm{R}(\mathfrak{g})$ to $\mathrm{NR}(\mathfrak{g})$.

2.3. Classification of complex simple Lie algebras

Let us briefly review classical results concerning the classification of simple and semisimple Lie algebras. This classification belongs to the greatest achievements in the theory of Lie algebras. These results were originally obtained by W. Killing [60] and É. Cartan [22].

Let $\mathfrak{g}$ be a complex Lie algebra. Any nilpotent subalgebra $\mathfrak{g}_0$ of $\mathfrak{g}$ coinciding with its normalizer $\operatorname{norm}_{\mathfrak{g}}(\mathfrak{g}_0)$ is called a *Cartan subalgebra.* It can be constructed in the following way.

Let $x \in \mathfrak{g}$. Consider the linear operator $\operatorname{ad}(x) \in \mathscr{L}(\mathfrak{g})$ and find its generalized nullspace

$$\mathfrak{g}_0(x) = \lim_{k\to\infty} \ker\big(\operatorname{ad}(x)\big)^k.$$

When $\dim \mathfrak{g}_0(x)$ is minimal, i.e.,

$$\dim \mathfrak{g}_0(x) = \min_{y\in\mathfrak{g}} \dim \mathfrak{g}_0(y),$$

we call the element $x \in \mathfrak{g}$ *regular.*

Proposition 2.14. *Let $x \in \mathfrak{g}$ be a regular element of the complex Lie algebra $\mathfrak{g}$. Then $\mathfrak{g}_0(x)$ is a Cartan subalgebra of $\mathfrak{g}$. Any other Cartan subalgebra of $\mathfrak{g}$ is related to $\mathfrak{g}_0(x)$ by an automorphism of $\mathfrak{g}$.*

Consequently, the dimension of the Cartan subalgebra $\mathfrak{g}_0(x)$ is independent of the choice of the regular element x and is called the *rank* of the Lie algebra $\mathfrak{g}$. We point out that the proposition holds whether or not $\mathfrak{g}$ is semisimple, i.e., any complex Lie algebra has a Cartan subalgebra unique up to automorphisms. The uniqueness is lost for real algebras; e.g., a finite number of distinct Cartan subalgebras exists for each finite-dimensional semisimple Lie algebra over $\mathbb{R}$, see [63, 121].

Cartan subalgebras of semisimple algebras have special properties. The Cartan subalgebra $\mathfrak{g}_0$ of a semisimple Lie algebra is Abelian. In addition, all elements of the Cartan subalgebra are ad-*diagonalizable* or *semisimple*, meaning that $\mathrm{ad}(h) \in \mathfrak{gl}(\mathfrak{g})$ is diagonalizable for every $h \in \mathfrak{g}_0$. Therefore, there exist common eigenspaces $\mathfrak{g}_\lambda \subset \mathfrak{g}$ of all operators $\mathrm{ad}(h)$, $h \in \mathfrak{g}_0$ and nonvanishing functionals $\lambda \in \mathfrak{g}_0^*$ such that

$$\mathrm{ad}(h)e_\lambda = \lambda(h) \cdot e_\lambda, \quad h \in \mathfrak{g}_0, e_\lambda \in \mathfrak{g}_\lambda$$

(where $\mathfrak{g}_0^*$ is the dual space of the vector space $\mathfrak{g}_0$.) These functionals λ are called *roots* of the semisimple Lie algebra $\mathfrak{g}$. The collection of all roots is called the *root system* of the algebra $\mathfrak{g}$ and denoted by Δ. The diagonalizability of $\mathrm{ad}(h)$ implies that

$$\mathfrak{g} = \mathfrak{g}_0 \dotplus \left(\dotplus\{\mathfrak{g}_\lambda \mid \lambda \in \Delta\}\right)$$

where $\dotplus$ stands for a direct sum of vector spaces.

It is always possible to introduce an ordering among the roots via a choice of $h_0 \in \mathfrak{g}_0$ such that $\lambda(h_0) \neq 0$ and $\lambda(h_0) \in \mathbb{R}$ for all roots λ. This ordering is not unique but different choices give results equivalent up to automorphism of $\mathfrak{g}$. For any pair of roots λ, κ one writes $\lambda > \kappa$ if and only if $\lambda(h_0) > \kappa(h_0)$. Similarly one defines *positive roots* $\lambda > 0$, i.e., $\lambda(h_0) > 0$ and *negative roots* $\lambda < 0$, i.e., $\lambda(h_0) < 0$. The set of all positive roots is denoted Δ^+, the set of negative roots Δ^-. We have $\Delta = \Delta^+ \cup \Delta^-$. *Simple roots* are positive roots which cannot be written as a sum of two positive roots. We denote the set of all simple roots by Δ^S.

We list the most important properties of the root system Δ and *root subspaces* $\mathfrak{g}_\lambda$ of a semisimple complex Lie algebra $\mathfrak{g}$:

(1) The Killing form K of $\mathfrak{g}$ when restricted to $\mathfrak{g}_0 \times \mathfrak{g}_0$ is nondegenerate.

(2) To any functional $\lambda \in \mathfrak{g}_0^*$ we can associate a unique element $h_\lambda \in \mathfrak{g}_0$ such that

$$\lambda(h) = K(h_\lambda, h), \quad \forall h \in \mathfrak{g}_0 \tag{2.34}$$

and we can define a nondegenerate bilinear symmetric form $\langle\ ,\ \rangle$ on $\mathfrak{g}_0^*$ so that

$$\langle \lambda, \kappa \rangle = K(h_\lambda, h_\kappa), \quad \forall \lambda, \kappa \in \mathfrak{g}_0^*.$$

(3) If λ is a root then so is $-\lambda$ and no other multiple of λ is a root.

(4) All root subspaces $\mathfrak{g}_\lambda$ are 1-dimensional.

(5) $[\mathfrak{g}_\lambda, \mathfrak{g}_\kappa] = \mathfrak{g}_{\lambda+\kappa}$ whenever λ, κ and $\lambda + \kappa$ are roots.

(6) When $\lambda + \kappa$ is neither 0 nor a root we have $[\mathfrak{g}_\lambda, \mathfrak{g}_\kappa] = 0$.

(7) $[\mathfrak{g}_\lambda, \mathfrak{g}_{-\lambda}] \subset \mathfrak{g}_0$.

(8) There is a basis of $\mathfrak{g}$ consisting of elements of the Cartan subalgebra $\mathfrak{g}_0$ and of the root subspaces $\mathfrak{g}_\lambda$ such that the structure constants of $\mathfrak{g}$ in this basis are integers; such a basis is called the *Weyl – Chevalley basis* of $\mathfrak{g}$ and the real form of the Lie algebra $\mathfrak{g}$ corresponding to this choice of basis is called the *split real form* of $\mathfrak{g}$.

(9) Simple roots are linearly independent.

(10) Any positive root is a linear combination of simple roots with nonnegative integer coefficients; therefore, the root system Δ is contained in the real subspace of $\mathfrak{g}_0^*$ spanned by the simple roots, we denote this subspace by $\mathfrak{h}^*$.

(11) $\langle\ ,\ \rangle$ defines a real scalar product on $\mathfrak{h}^*$.

(12) The whole Lie algebra $\mathfrak{g}$ is obtained by multiple Lie brackets of root vectors e_α where $\alpha \in \Delta^S$ or $-\alpha \in \Delta^S$.

(13) The root system Δ is invariant under all reflections S_λ of the form

$$S_\lambda(\alpha) = \alpha - 2\frac{\langle \alpha, \lambda \rangle}{\langle \lambda, \lambda \rangle}\lambda, \qquad \lambda, \alpha \in \Delta;$$

all such reflections generate a finite group called the *Weyl group* of the root system Δ.

(14) Any root is an image of some simple root under the action of some element of the Weyl group; in particular, it has the same length.

It turns out that the structure of a semisimple complex Lie algebra is fully determined up to isomorphism by angles and relative lengths of its simple roots in the Euclidean space $\mathfrak{h}^*$. This information is usually encoded either in the *Cartan matrix* $A = (a_{\kappa\lambda})$

$$a_{\kappa\lambda} = 2\frac{\langle \kappa, \lambda \rangle}{\langle \lambda, \lambda \rangle}, \qquad \kappa, \lambda \in \Delta^S$$

or equivalently in Dynkin diagrams. The Cartan matrix has only integer entries: 2 on the diagonal, $0, -1, -2, -3$ off the diagonal. The *Dynkin diagram* associated with the Cartan matrix is a graph with vertices corresponding to the simple roots where the number of edges connecting the vertices labelled by κ and λ is equal to $a_{\kappa\lambda}a_{\lambda\kappa} \in \{0, 1, 2, 3\}$. Further, one distinguishes graphically between shorter and longer roots either by different symbols for vertices or different types of arrows connecting vertices. We shall use a convention that the arrow goes from the longer root to the shorter one, e.g., a subdiagram of the form

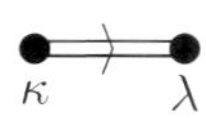

implies the following values of the elements of the Cartan matrix

$$a_{\kappa\lambda} = 2\frac{\langle \kappa, \lambda \rangle}{\langle \lambda, \lambda \rangle} = -2, \quad a_{\lambda\kappa} = 2\frac{\langle \lambda, \kappa \rangle}{\langle \kappa, \kappa \rangle} = -1.$$

The structure of any root system can be shown to be such that:

(1) Simple components $\mathfrak{g}_k$ of a semisimple Lie algebra $\mathfrak{g}$ correspond to connected subdiagrams of the Dynkin diagram of $\mathfrak{g}$.
(2) There are no closed loops in Dynkin diagrams.
(3) A connected Dynkin diagram is either *simply laced* meaning that it contains only simple edges and consequently all roots are of the same length, or the corresponding root system contains roots of precisely two different lengths.

The fundamental classification result is due to W. Killing [60] and É. Cartan [22] whose computations were later significantly simplified by E. Dynkin [43, 44]. It states that a finite-dimensional complex simple Lie algebra $\mathfrak{g}$ either belongs to one of the classical series of simple Lie algebras, or is one of the exceptional simple Lie algebras. The classical Lie algebras are

TABLE 2.1. Dynkin diagrams of simple Lie algebras.

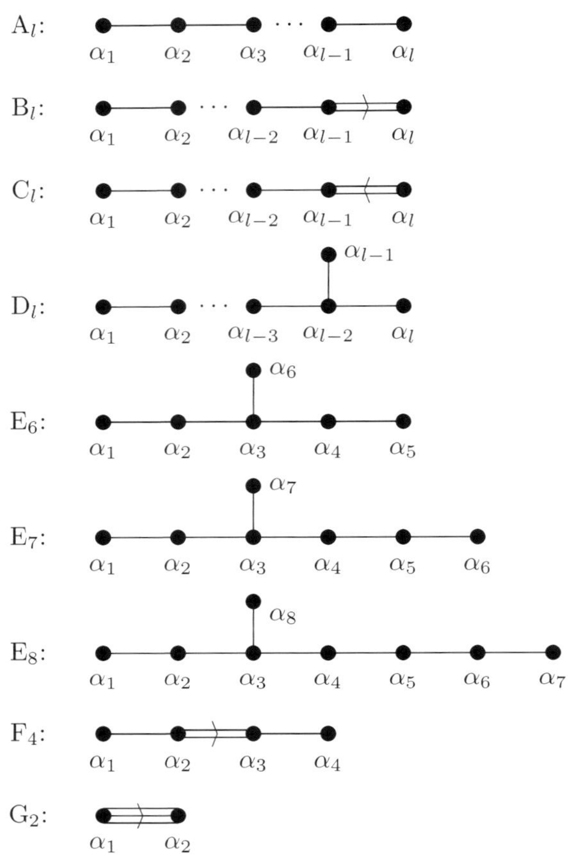

- $\mathfrak{sl}(l+1,\mathbb{C})$ of rank $l \geq 1$, also denoted A_l, the algebra of traceless $(l+1) \times (l+1)$ matrices,
- $\mathfrak{so}(2l+1,\mathbb{C})$ of rank $l \geq 2$, also denoted B_l, the algebra of skew-symmetric $(2l+1) \times (2l+1)$ matrices,
- $\mathfrak{sp}(2l,\mathbb{C})$ of rank $l \geq 3$, also denoted C_l, the algebra of $2l \times 2l$ matrices skew-symmetric with respect to a nondegenerate antisymmetric form on $\mathbb{C}^{2l}$,
- $\mathfrak{so}(2l,\mathbb{C})$ of rank $l \geq 4$, also denoted D_l, the algebra of skew-symmetric $2l \times 2l$ matrices.

The five exceptional algebras are denoted by E_6, E_7, E_8, F_4, G_2. Out of these algebras, the algebras A_l, D_l, E_6, E_7, E_8 are simply laced. The corresponding Dynkin diagrams are given in Table 2.1.

2.4. Chevalley cohomology of Lie algebras

Certain properties of Lie algebras are best described in terms of Chevalley (or Chevalley – Eilenberg) cohomology [29]. Therefore, let us review its definition here.

Let us assume that a Lie algebra $\mathfrak{g}$ and its representation ρ on a vector space V are given. The *k-cochain* c is a totally antisymmetric linear map from $\mathfrak{g}^{\times k} =$

$\mathfrak{g} \times \mathfrak{g} \times \cdots \times \mathfrak{g}$ to V. The vector space of all k-cochains is denoted by $C^k(\mathfrak{g}, V; \rho)$. The direct sum of all $C^k(\mathfrak{g}, V; \rho)$, $k = 0, \dots, \dim \mathfrak{g}$ is called the *cochain complex* and denoted by $C^\bullet(\mathfrak{g}, V; \rho)$. On the cochain complex the *cohomology operator* d is defined by its action on an arbitrary k-cochain

$$\text{(2.35)} \quad \mathrm{d}\, c(x_1, \dots, x_{k+1}) = \sum_{i=1}^{k+1} (-1)^{i+1} \rho(x_i) c(x_1, \dots, \hat{x}_i, \dots, x_{k+1}) + \sum_{1 \le i < j \le k+1} (-1)^{i+j} c([x_i, x_j], x_1, \dots, \hat{x}_i, \dots, \hat{x}_j, \dots, x_{k+1})$$

for any $(k+1)$-tuple of vectors $x_1, \dots, x_{k+1} \in \mathfrak{g}$ ($\hat{x}_i$ indicates that the term is omitted) and linearly extended to the whole of $C^\bullet(\mathfrak{g}, V; \rho)$. It is only in (2.35) that the representation ρ enters the structure of $C^\bullet(\mathfrak{g}, V; \rho)$. When the choice of the representation ρ is obvious from the context we may omit it in the notation and write $C^k(\mathfrak{g}, V)$, $C^\bullet(\mathfrak{g}, V)$.

The cohomology operator d satisfies the fundamental relation

$$\text{(2.36)} \quad \mathrm{d} \circ \mathrm{d} = 0$$

which can be demonstrated, e.g., by a straightforward, if tedious, explicit calculation using the definition (2.35) and the definition of a representation. Equation (2.36) implies that $\operatorname{Im} \mathrm{d} \subseteq \operatorname{Ker} \mathrm{d}$.

A k-cochain c is called a *k-cocycle* if

$$\text{(2.37)} \quad \mathrm{d}\, c = 0.$$

Explicitly, the condition reads

$$\text{(2.38)} \quad 0 = \sum_{i=1}^{k+1} (-1)^{i+1} \rho(x_i) c(x_1, \dots, \hat{x}_i, \dots, x_{k+1}) + \sum_{1 \le i < j \le k+1} (-1)^{i+j} c([x_i, x_j], x_1, \dots, \hat{x}_i, \dots, \hat{x}_j, \dots, x_{k+1}).$$

We denote the vector space of k-cocycles by $Z^k(\mathfrak{g}, V; \rho)$, i.e.,

$$Z^k(\mathfrak{g}, V; \rho) = \operatorname{Ker} \mathrm{d} \cap C^k(\mathfrak{g}, V; \rho).$$

A k-cochain c is called a *coboundary* if a $(k-1)$-cochain c' exists such that

$$\text{(2.39)} \quad c = \mathrm{d}\, c'.$$

The vector space of k-coboundaries is denoted by $B^k(\mathfrak{g}, V; \rho)$, i.e.,

$$B^k(\mathfrak{g}, V; \rho) = \operatorname{Im} \mathrm{d} \cap C^k(\mathfrak{g}, V; \rho).$$

Due to (2.36) any coboundary is a cocycle,

$$\text{(2.40)} \quad B^k(\mathfrak{g}, V; \rho) \subseteq Z^k(\mathfrak{g}, V; \rho).$$

Therefore, one may construct the factor vector space

$$\text{(2.41)} \quad H^k(\mathfrak{g}, V; \rho) = Z^k(\mathfrak{g}, V; \rho) / B^k(\mathfrak{g}, V; \rho)$$

of classes of cocycles modulo coboundaries. $H^k(\mathfrak{g}, V; \rho)$ is called the kth *cohomology space* or kth *cohomology group* (with the group structure being the vector addition). The direct sum of all $H^k(\mathfrak{g}, V; \rho)$ is called the *cohomology complex* and denoted by $H^\bullet(\mathfrak{g}, V; \rho)$.

The Chevalley cohomology finds numerous applications in the theory of Lie algebras. For example, let us consider the adjoint representation ad of the given Lie algebra $\mathfrak{g}$ and its 1st cohomology space $H^1(\mathfrak{g}, \mathfrak{g}; \mathrm{ad})$. From the definition of a cocycle we find that in this case 1-cocycles are precisely the derivations of $\mathfrak{g}$ because (2.38) reduces to

$$0 = \mathrm{ad}(x_1)c(x_2) - \mathrm{ad}(x_2)c(x_1) - c([x_1, x_2]) \tag{2.42}$$

which is just another formulation of the definition of a derivation (2.26)

$$c([x_1, x_2]) = [x_1, c(x_2)] + [c(x_1), x_2].$$

Similarly, 1-coboundaries in $B^1(\mathfrak{g}, \mathfrak{g}; \mathrm{ad})$ are inner derivations since by application of d to a 0-cochain $x \in \mathfrak{g}$ we obtain

$$\mathrm{d}\,x \in B^1(\mathfrak{g}, \mathfrak{g}; \mathrm{ad}) : \ (\mathrm{d}\,x)y = \mathrm{ad}(y)x = -\,\mathrm{ad}(x)y, \tag{2.43}$$

i.e., we have $\mathrm{d}\,x = -\,\mathrm{ad}(x)$. Altogether, we conclude that

$$H^1(\mathfrak{g}, \mathfrak{g}; \mathrm{ad}) = \mathfrak{Der}(\mathfrak{g}) / \mathfrak{Inn}(\mathfrak{g}). \tag{2.44}$$

That means that $H^1(\mathfrak{g}, \mathfrak{g}; \mathrm{ad})$ provides a convenient mathematical description of the equivalence classes of outer derivations modulo inner derivations.

Similarly, the 2nd cohomology group becomes useful in the construction of central extensions, as will be indicated in Chapter 8.

CHAPTER 3

Invariants of the Coadjoint Representation of a Lie Algebra

3.1. Casimir operators and generalized Casimir invariants

The term *Casimir operator*, or *Casimir invariant*, of a Lie algebra $\mathfrak{g}$ is usually reserved for elements of the center of the enveloping algebra of the Lie algebra $\mathfrak{g}$ [25, 98]. Casimir operators play a crucial role in the representation theory of Lie algebras (see, e.g., [26]). These elements should be algebraically independent so that all other elements of the center of the enveloping algebra are appropriately symmetrized polynomials in the given ones. The Casimir operators are in one-to-one correspondence with polynomial invariants characterizing orbits of the coadjoint representation of $\mathfrak{g}$ [61].

In general, invariants of the coadjoint representation of a Lie algebra are not necessarily polynomials and we shall call these nonpolynomial invariants *generalized Casimir invariants.* On the other hand, for certain classes of Lie algebras it is known that their invariants can be expressed in a particular form. For example, for perfect Lie algebras including semisimple ones and for nilpotent Lie algebras, all invariants of the coadjoint representation are functions of polynomial ones. More generally, for the so-called algebraic Lie algebras [28] a basis of independent invariants of the coadjoint representation can be chosen in the form of ratios of polynomial functions on $\mathfrak{g}^*$, i.e., as elements of the corresponding quotient field (see [1]).

Casimir invariants are of primordial importance in physics. They represent such important quantities as angular momentum, elementary particle mass and spin, Hamiltonians of various physical systems, etc. The same can be said of generalized Casimir invariants. Indeed, Hamiltonians and integrals of motion for physical systems are not necessarily polynomials in the momenta [55, 99], though typically they are invariants of some group action.

Essentially two methods of calculating Casimir and generalized Casimir invariants exist and both of them are based on the calculation of invariants of the coadjoint representation. The first method is an infinitesimal one and amounts to solving a system of first order linear partial differential equations. The second method is more global in nature; it uses the (local) action of the Lie group G corresponding to the Lie algebra $\mathfrak{g}$. It is an application of Cartan's method of moving frames [23, 24] and its modern formulation is due to M. Fels and P. Olver [48].

We shall discuss the two methods separately. Both of them simultaneously yield Casimir and generalized Casimir invariants and have been applied to low dimensional Lie algebras [1, 2, 13, 15, 87] and to certain classes of Lie algebras of arbitrary dimensions [7–11, 83, 94, 106, 115–117, 124]. Below the term "generalized Casimir invariant" will also include the polynomial ones.

3.2. Calculation of generalized Casimir invariants using the infinitesimal method

3.2.1. Formulation of the problem and number of functionally independent solutions. In order to calculate the (generalized) Casimir invariants we consider some basis $(x_1, \ldots, x_n)$ of $\mathfrak{g}$, in which the structure constants are ${c_{ij}}^k$. A basis for the coadjoint representation is given by the vector fields,

$$\widehat{X}_k = \sum_{a,b=1}^{n} x_b {c_{ka}}^b \frac{\partial}{\partial x_a}, \quad 1 \leq k \leq n. \tag{3.1}$$

In (3.1) the quantities x_a are commuting independent variables which can be identified with coordinates in the dual basis of the space $\mathfrak{g}^*$, dual to the algebra $\mathfrak{g}$. (Recall that there is a canonical isomorphism ι between any finite dimensional vector space V and its double dual $(V^*)^*$, defined through the relation

$$\iota(v)(\alpha) = \alpha(v)$$

for any $v \in V$ and all $\alpha \in V^*$).

The invariants of the coadjoint representation, i.e., the *generalized Casimir invariants*, are solutions of the following system of partial differential equations

$$\widehat{X}_k I(x_1, \ldots, x_n) = 0, \quad k = 1, \ldots, n. \tag{3.2}$$

Let us first determine the number of functionally independent solutions of the system (3.2). We can rewrite this system as

$$C \cdot \nabla I = 0 \tag{3.3}$$

where C is the antisymmetric matrix

$$C = \begin{pmatrix} 0 & {c_{12}}^b x_b & \ldots & {c_{1n}}^b x_b \\ -{c_{12}}^b x_b & 0 & \ldots & {c_{2n}}^b x_b \\ \vdots & & & \vdots \\ -{c_{1,n-1}}^b x_b & \ldots & 0 & {c_{n-1,n}}^b x_b \\ -{c_{1n}}^b x_b & \ldots & -{c_{n-1,n}}^b x_b & 0 \end{pmatrix} \tag{3.4}$$

in which summation over the repeated index b is to be understood in each term and ∇ is the gradient operator $\nabla = (\partial_{x_1}, \ldots, \partial_{x_n})^T$. The number of independent equations in the system (3.2) is $r(C)$, the generic rank of the matrix C. The number of functionally independent solutions of the system (3.2) is hence

$$n_I = n - r(C). \tag{3.5}$$

Since C is antisymmetric, its rank is even. It follows that n_I has the same parity as n. Equation (3.5) gives the number of functionally independent generalized Casimir invariants. The individual equations in the system of partial differential equations (PDEs) (3.2) can be solved by the method of characteristics, or equivalently by integration of the vector fields (3.1).

3.2.2. Method of characteristics for a single PDE. The *method of characteristics* is applicable to a linear homogeneous first-order PDE

$$\sum_{k=0}^{n} f^k(x_1, \ldots, x_n) \frac{\partial}{\partial x_k} u(x_1, \ldots, x_n) = 0 \tag{3.6}$$

for an unknown function u. (It may be generalized to inhomogeneous and quasi-linear PDEs [135, 140] but we shall not need that here.) Instead of attempting to solve (3.6) directly, we can consider an associated systems of ODEs

$$\frac{\mathrm{d}\tilde{x}_k(t)}{\mathrm{d}t} = f^k\big(\tilde{x}_1(t), \ldots, \tilde{x}_n(t)\big), \quad 1 \le k \le n \tag{3.7}$$

and find its solution satisfying a generic initial condition

$$\tilde{x}_k(0) = x_k, \quad 1 \le k \le n. \tag{3.8}$$

In the language of differential geometry that means that we are constructing the flow, i.e., the collection of all integral curves, of the vector field

$$\widehat{F} = \sum_{j=0}^{n} f^a(x_1, \ldots, x_n)\frac{\partial}{\partial x_a}. \tag{3.9}$$

Once the integral curves, i.e., solutions of (3.7), are known, we construct functionally independent functions which are constant along the integral curves in the following way. We choose a hypersurface in $\mathbb{R}^n$ such that it is transversal to all integral curves (this is often done only locally). We associate to every integral curve its intersection with the chosen hypersurface. The coordinates of that point of intersection are invariants of the vector field $\widehat{F}$, i.e., solutions of (3.6), because they are by construction the same for any pair of points connected by an integral curve of $\widehat{F}$ and consequently are annihilated by the vector field.

For the sake of the argument let us assume that the hypersurface is expressed in our coordinates as the hyperplane $x_1 = 1$. Let us take $(x_1, \ldots, x_n)$ as the initial condition (3.8). We determine the value of the curve parameter $t(x_1, \ldots, x_n)$ such that $\tilde{x}\big(t(x_1, \ldots, x_n)\big)$ lies on the hyperplane $x_1 = 1$, i.e., $\tilde{x}_1\big(t(x_1, \ldots, x_n)\big) = 1$. The remaining $n-1$ coordinates $\tilde{x}_k\big(t(x_1, \ldots, x_n)\big)$, $2 \le k \le n$ of the intersection of the integral curve with the hyperplane $x_1 = 1$ are invariants of the vector field (3.9)

$$I_k(x_1, \ldots, x_n) = \tilde{x}_{k+1}\big(t(x_1, \ldots, x_n)\big), \quad 1 \le k \le n-1.$$

They are by construction functionally independent.

We remark that any invariant of the vector field $\widehat{F}$ is obviously also an invariant of the vector field $\widehat{G} = f\widehat{F}$ for any smooth function $f : \mathbb{R}^n \to \mathbb{R}$. On the other hand, the integral curves of $\widehat{G}$ differ from those of $\widehat{F}$ by a reparameterization, i.e., the differential equations (3.7) are different for $\widehat{G}$ and for $\widehat{F}$. Consequently, the solution of the system of ODEs (3.7) can be often significantly simplified through a suitable choice of the function f. This independence of the invariants on the reparameterization of the integral curves is symbolically depicted by rewriting of the system (3.7) in the form

$$\frac{\mathrm{d}x_1}{f^1(x_1, \ldots, x_n)} = \frac{\mathrm{d}x_2}{f^2(x_1, \ldots, x_n)} = \cdots = \frac{\mathrm{d}x_n}{f^n(x_1, \ldots, x_n)}. \tag{3.10}$$

The method of characteristics relies on our ability to find the integral curves of the vector field $\widehat{F}$, i.e., to solve the system of ODEs (3.7). This can be done explicitly only for certain classes of functions $f^1, \ldots, f^n$. One particular case when we can integrate the system (3.7) is when all functions f^k are linear in the coordinates x_j. Luckily, this is the case for the vector fields $\widehat{X}_a$ encountered in (3.1). We shall now study this case in some detail.

Let the vector field $\widehat{F}$ take the form

$$\widehat{F} = \sum_{k=1}^{n}\sum_{j=1}^{n} \alpha_k^j x_j \frac{\partial}{\partial x_k}. \tag{3.11}$$

Let x denote the column vector of the coordinates $(x_1, \ldots, x_n)^T$. To the vector field (3.11) we associate its flow determined by the equations

$$\frac{\mathrm{d}\tilde{x}_k(t)}{\mathrm{d}t} = \sum_{j=1}^{n} \alpha_k^j \tilde{x}_j, \quad 1 \leq k \leq n$$

or, in vector notation,

$$\frac{\mathrm{d}\tilde{x}(t)}{\mathrm{d}t} = F \cdot \tilde{x}(t), \quad F = \begin{pmatrix} \alpha_1^1 & \alpha_1^2 & \ldots & \alpha_1^n \\ \alpha_2^1 & \alpha_2^2 & \ldots & \alpha_2^n \\ \vdots & \vdots & & \vdots \\ \alpha_n^1 & \alpha_n^2 & \ldots & \alpha_n^n \end{pmatrix} \tag{3.12}$$

subject to the initial conditions $\tilde{x}(0) = x$.

We can perform a linear change of coordinates $(x_1, \ldots, x_n)$ to simplify the ODE system (3.12). Putting

$$y = S \cdot x \tag{3.13}$$

where $S = (s_k^j)$ is a constant invertible matrix, we transform (3.12) to

$$\frac{\mathrm{d}\tilde{y}(t)}{\mathrm{d}t} = \widetilde{F} \cdot \tilde{y}(t), \quad \widetilde{F} = S \cdot F \cdot S^{-1} \tag{3.14}$$

with the initial condition $\tilde{y}(0) = y$.

If F is diagonalizable then we can choose S^{-1} to be a matrix of eigenvectors of F and completely decouple the system (3.12). The solution of the system (3.14) is

$$\tilde{y}_k(t) = y_k \mathrm{e}^{\alpha_k t}, \quad k = 1, \ldots, n \tag{3.15}$$

where α_k are eigenvalues of the matrix F ordered so that we have $\alpha_1 \neq 0$. We choose our hypersurface as $y_1 = 1$ and compute $t(y_1, \ldots, y_n)$ such that $\tilde{y}_1\big(t(y_1, \ldots, y_n)\big) = 1$. The result is

$$t(y_1, \ldots, y_n) = -\frac{1}{\alpha_1} \ln(y_1).$$

The remaining coordinates $\tilde{y}_k(t)$ of the intersection of the integral curve and the hyperplane $y_1 = 1$ are then the invariants of the vector field $\widehat{F}$. We obtain $n - 1$ invariants

$$I_{k-1} = \frac{y_k}{y_1^{\alpha_k/\alpha_1}}, \quad k = 2, \ldots, n. \tag{3.16}$$

In terms of the original coordinates $(x_1, \ldots, x_n)$ our invariants read

$$I_{k-1} = \frac{\sum_{j=1}^{n} s_k^j x_j}{(\sum_{j=1}^{n} s_1^j x_j)^{\alpha_k/\alpha_1}}, \quad k = 2, \ldots, n. \tag{3.17}$$

The situation becomes more complicated when the matrix F is not diagonalizable. If that is the case we reduce it to its Jordan canonical form $\widetilde{F}$ (over the field of complex numbers) as in (3.14). Let us assume that F already is in its Jordan form,

$$F = \begin{pmatrix} J_{q_1}(\alpha_1) & & & & & \\ & \ddots & & & & \\ & & J_{q_M}(\alpha_M) & & & \\ & & & J_{p_1}(0) & & \\ & & & & \ddots & \\ & & & & & J_{p_K}(0) \end{pmatrix}, \tag{3.18}$$

where

$$J_q(\alpha) = \begin{pmatrix} \alpha & & & & \\ 1 & \alpha & & & \\ & 1 & \ddots & & \\ & & \ddots & \alpha & \\ & & & 1 & \alpha \end{pmatrix} \in \mathbb{C}^{q\times q},$$

$$\alpha_j \neq 0, \quad q_j \geq 1, \quad j = 1, \ldots, M, \qquad p_i \geq 1, \quad i = 1, \ldots, K, \tag{3.19}$$

$$M \geq 1, \quad K \geq 0, \quad \sum_{j=1}^{M} q_j + \sum_{i=1}^{K} p_i = n.$$

We shall assume that by a suitable rescaling of the vector field $\widehat{F}$ we have set $\alpha_1 = 1$.

The operator $\widehat{F}$ can be read off the matrix F. We have

$$\widehat{F} = \sum_{j=1}^{M} \left\{ \alpha_j \sum_{k=1}^{q_j} x_{s_{j-1}+k} \frac{\partial}{\partial_{x_{s_{j-1}+k}}} + \sum_{k=2}^{q_j} x_{s_{j-1}+k-1} \frac{\partial}{\partial_{x_{s_{j-1}+k}}} \right\} + \sum_{k=1}^{K} \sum_{m=2}^{p_j} x_{t_{k-1}+m-1} \frac{\partial}{\partial_{x_{t_{k-1}+m}}} \tag{3.20}$$

where we have set

$$s_0 = 0, \qquad s_j = \sum_{a=1}^{j} q_a, \qquad j = 1, \ldots, M,$$

$$t_0 = s_M, \qquad t_k = s_M + \sum_{b=1}^{k} p_b, \quad k = 1, \ldots, K.$$

Before treating the general case with F as in (3.18), let us first present two auxiliary results, for cases when F is indecomposable.

Lemma 3.1. *Consider the vector field $\widehat{F}_0$ defined by the matrix*

$$F_0 = \begin{pmatrix} 0 & & & & \\ 1 & 0 & & & \\ & 1 & \ddots & & \\ & & \ddots & 0 & \\ & & & 1 & 0 \end{pmatrix} \in \mathbb{C}^{n\times n}.$$

There are $n-1$ invariants of the vector field $\widehat{F}_0$ which are all homogeneous polynomials, namely

$$I_1 = x_1, \qquad I_k = \sum_{j=0}^{k} \frac{(-1)^j}{j!} x_1^{k-1-j}\, x_2^j\, x_{k+1-j}, \quad 2 \le k \le n-1. \tag{3.21}$$

PROOF. The operator $\widehat{F}_0$ takes the simple form

$$\widehat{F}_0 = x_1\,\partial_{x_2} + \cdots + x_{n-1}\,\partial_{x_n}. \tag{3.22}$$

Equation (3.12) becomes

$$\frac{\mathrm{d}\tilde{x}_1(t)}{\mathrm{d}t} = 0, \qquad \frac{\mathrm{d}\tilde{x}_k(t)}{\mathrm{d}t} = \tilde{x}_{k-1}(t).$$

The integral curves are given by the formula

$$\tilde{x}_1(t) = x_1, \quad \tilde{x}_k(t) = \sum_{j=0}^{k-1} \frac{t^j}{j!}\, x_{k-j}. \tag{3.23}$$

In this case we obviously cannot choose our hyperplane as $x_1 = 1$ because x_1 is constant along the integral curves (3.23). A convenient choice of the hyperplane is

$$x_2 = 0$$

which implies

$$t = -\frac{x_2}{x_1}. \tag{3.24}$$

Substituting (3.24) into (3.23) we obtain the invariants

$$I_1 = x_1, \qquad \tilde{I}_k = \sum_{j=0}^{k-1} \frac{x_{k-j}}{j!} (-1)^j \left(\frac{x_2}{x_1}\right)^j, \quad k = 3, \ldots, n.$$

Multiplying $\tilde{I}_k$ by the invariant $(x_1)^{k-1}$ and shifting the label k in the sum we obtain the invariants in the form presented in the lemma. □

Lemma 3.2. *Consider the vector field $\widehat{F}_1$ defined by the matrix F as in* (3.19) *with $q = n$. The $n-1$ functionally independent invariants of the vector field $\widehat{F}_1$ are*

$$\begin{aligned} L_1 &= \alpha\,\frac{x_2}{x_1} - \ln(x_1), && \alpha \neq 0, \\ R_k &= \frac{I_k}{(x_1)^k}, && k = 2, \ldots, n-1, \end{aligned} \tag{3.25}$$

where I_k were defined in (3.21).

PROOF. The operator

$$\widehat{F}_1 = \alpha \sum_{j=1}^{n} x_j\,\partial_{x_j} + \sum_{j=1}^{n-1} x_j\,\partial_{x_{j+1}} \tag{3.26}$$

differs from $\widehat{F}_0$ of (3.22) considered in the previous lemma by a dilation operator

$$\widehat{D} = \alpha \sum_{j=1}^{n} x_j\,\partial_{x_j}$$

which acts on any homogeneous polynomial by multiplication by its degree together with an overall multiplication by α. Therefore, any ratio of two polynomials of the same degree such that both are invariant with respect to $\widehat{F}_0$ is an invariant of the vector field $\widehat{F}_1$. This gives us $n-2$ functionally independent invariants R_k (3.25). The invariant L_1 is obtained from the first two coordinates of the integral curves of $\widehat{F}_1$

$$\tilde{x}_k(t) = \sum_{j=1}^{k} \frac{x_{k-j+1}}{(j-1)!} t^{j-1} e^{\alpha t}$$

via the choice of hyperplane $x_1 = 1$, i.e., $t = -(1/\alpha)\ln x_1$ and consequently we have an invariant

$$\tilde{I}_1 = \frac{L_1}{\alpha} = \frac{x_2}{x_1} - \frac{1}{\alpha}\ln x_1. \qquad \square$$

Remark 3.3. Notice that proceeding as before, i.e., determining all functionally independent invariants as the coordinates of the intersection of the integral curve with the chosen hyperplane, we can immediately obtain a different but equivalent complete set of invariants

$$I_{k-1} = \sum_{j=1}^{k} \frac{x_{k-j+1}}{(j-1)!x_1}\left(-\frac{1}{\alpha}\ln x_1\right)^{j-1}.$$

This set is less convenient than the one chosen in the statement of the lemma due to the presence of powers of logarithms.

It is now clear that the invariants in the general case with F as in (3.18) will be of 4 types:

(1) Invariants of the type (3.25), associated with indecomposable nonnilpotent Jordan blocks.
(2) Invariants of the type (3.21), associated with indecomposable nilpotent Jordan blocks.
(3) Invariants involving block J_1 and the blocks J_{q_j}, $2 \le j \le M$.
(4) Invariants involving block J_1 and block J_{p_k}, $1 \le k \le K$.

Indeed to find all functionally independent invariants of the vector field (3.20) we can use the following procedure:

(1) Solve equations within each block. This yields invariants of the type (3.21) or (3.25).

(2) Connect all the blocks to block J_1 in the simplest possible manner. We choose $M-1$ "quasirational" invariants

$$Q_{s_{j-1}+1} = \frac{x_{s_{j-1}+1}}{(x_1)^{\alpha_j}}, \quad j = 2, \dots, M, \tag{3.27}$$

and K "logarithmic" invariants

$$L_{t_{k-1}+2} = \frac{x_{t_{k-1}+2}}{x_{t_{k-1}+1}} - \ln x_1, \quad k = 1, \dots, K, \tag{3.28}$$

(s_j and t_k are defined in (3.19)).

We call $Q_{s_{j-1}+1}$ quasirational, because α_j is not necessarily a rational number. Even so, ratios of the type $Q_{s_{j-1}+1}/Q_{s_{m-1}+1}$ may be rational.

The results can now be summed up as a theorem.

Theorem 3.4. *Let $\widehat{F} = \sum_{k=1}^{n}\sum_{j=1}^{n} \alpha_k^j x_j \partial/\partial x_k$ be a vector field such that the corresponding matrix F has the form as in (3.18) with $\alpha_1 = 1$. The $n-1$ functionally independent invariants of $\widehat{F}$ can be obtained as follows:*

(1) Each nonnilpotent block $J_{q_j}(\alpha_j)$ in F $(1 \le j \le M)$ provides $(q_j - 1)$ invariants, namely

$$L_{s_{j-1}+2} = \alpha_j \frac{x_{s_{j-1}+2}}{x_{s_{j-1}+1}} - \ln(x_{s_{j-1}+1}), \tag{3.29}$$

$$R_{s_{j-1}+l+1} = I_l(x_{s_{j-1}+1}, \ldots, x_{s_j}), \qquad 2 \le l \le q_j - 1. \tag{3.30}$$

where $I_l(x_{s_{j-1}+1}, \ldots, x_{s_j})$ is as in (3.21), after an appropriate relabelling of x_i.

(2) Each nonnilpotent block $J_{q_j}(\alpha_j)$, combined with $J_{q_1}(1)$ provides one more invariant, namely $Q_{s_{j-1}+1}$ of (3.27).

(3) Each nilpotent block $J_{p_k}(0)$ in F $(1 \le k \le K)$ provides $p_k - 1$ polynomial invariants

$$P_{t_{k-1}+1} = x_{t_{k-1}+1}, \qquad P_{t_{k-1}+l+1} = I_l(x_{t_{k-1}+1}, \ldots, x_{t_k}), \quad 2 \le l \le p_k - 1. \tag{3.31}$$

(4) Each nilpotent block $J_{p_k}(0)$, combined with block $J_{q_1}(1)$, provides one further invariant, namely $L_{t_{k-1}+2}$ of (3.28).

PROOF. The fact that expressions (3.29), (3.30), (3.31), (3.27) and (3.28) provide invariants was proven above. That all $n-1$ of them are functionally independent is seen by calculating the Jacobian $\partial(L_1, \ldots)/\partial(x_1, \ldots, x_n)$. Its rank is $n-1$. □

Remark 3.5. (1) If F is diagonal, then all invariants are of the type (3.27). They are rational functions if and only if all eigenvalues α_j are rational.

(2) In all other cases at least one of the invariants is logarithmic: one of the type (3.29) for each $q_j \ge 2$, one of the type (3.28) for each nilpotent block present.

(3) When dealing with real matrices F we may not be able to reduce F to its Jordan canonical form over the field $\mathbb{R}$. In this case we encounter in addition submatrices $J_q(A)$ where $A = \left(\begin{smallmatrix} a & b \\ -b & a \end{smallmatrix}\right) \in \mathbb{R}^{2\times 2}$ in the real canonical form of the matrix J. The computation is similar as in the complex case, only more subcases need to be considered. For canonical forms of real matrices see, e.g., [42, 75].

Example 3.6. Consider the case with $n = 13$ when the matrix F has the form (3.18) with $M = 4$, $K = 2$, $q_1 = q_2 = 1$, $q_3 = 2$, $q_4 = 4$, $p_1 = 2$, $p_2 = 3$, i.e.,

$$F = \begin{pmatrix}
\alpha_1 &&&&&&&&&&&& \\
& \alpha_2 &&&&&&&&&&& \\
&& \alpha_3 &&&&&&&&&& \\
&& 1 & \alpha_3 &&&&&&&&& \\
&&&& \alpha_4 &&&&&&&& \\
&&&& 1 & \alpha_4 &&&&&&& \\
&&&&& 1 & \alpha_4 &&&&&& \\
&&&&&& 1 & \alpha_4 &&&&& \\
&&&&&&&& 0 &&&& \\
&&&&&&&& 1 & 0 &&& \\
&&&&&&&&&& 0 && \\
&&&&&&&&&& 1 & 0 & \\
&&&&&&&&&&& 1 & 0
\end{pmatrix}.$$

Invariants of the type (3.29) come from blocks J_3 and J_4, type (3.30) from block J_4 only:

$$\begin{aligned} L_4 &= \alpha_3 \frac{x_4}{x_3} - \ln x_3, \\ L_6 &= \alpha_4 \frac{x_6}{x_5} - \ln x_5, \\ R_7 &= \frac{2x_5x_7 - (x_6)^2}{(x_5)^2}, \\ R_8 &= \frac{6x_5^2x_8 - 6x_5x_6x_7 + 2x_1^3}{(x_5)^3}. \end{aligned}$$

Invariants of the type (3.27) are

$$Q_2 = \frac{x_2}{x_1^{\alpha_2}}, \quad Q_3 = \frac{x_3}{x_1^{\alpha_3}}, \quad Q_5 = \frac{x_5}{x_1^{\alpha_4}}.$$

Invariants of the type (3.31) come from blocks J_5 and J_6 and are

$$P_9 = x_9, \quad P_{11} = x_{11}, \quad P_{13} = 2x_{11}x_{13} - x_{12}^2.$$

Invariants of the type (3.28) are

$$L_{10} = \frac{x_{10}}{x_9} - \ln x_1, \quad L_{12} = \frac{x_{12}}{x_{11}} - \ln x_1.$$

Notice that the subscripts of the invariants go from 2 and 13 so that say x_8 is first introduced in R_8. This shows that the twelve invariants are functionally independent.

3.2.3. Solution of the system of PDEs (3.2). Let us now assume that one of the equations of the system (3.2) has been solved by the method of characteristics as in Section 3.2.2, i.e., we already know the $n-1$ invariants I_k of the first vector field $\widehat{X}_1$. Next, we transform all remaining vector fields to a new set of coordinates

$$(x_1, \ldots, x_n) \to (I_1, \ldots, I_{n-1}, s) \tag{3.32}$$

where s is a coordinate along the integral curves of $\widehat{X}_1$. We obtain

$$\begin{aligned} \widehat{X}_1 &= \frac{\partial}{\partial s}, \\ \widehat{X}_k &= \sum_{c=1}^{n-1} \phi_k^c(I_1, \ldots, I_{n-1}, s)\frac{\partial}{\partial I_c} + \phi_k^s(I_1, \ldots, I_{n-1}, s)\frac{\partial}{\partial s}, \\ \phi_k^c &= \sum_{a,b=1}^{n} x_b c_{ka}^b \frac{\partial I_c}{\partial x_a}, \quad 2 \le k \le n. \end{aligned} \tag{3.33}$$

Any function J of $I_1, \ldots, I_{n-1}$ is an invariant of the vector field $\widehat{X}_1$. For J to be an invariant of the entire Lie algebra it must be a solution of the system of equations

$$\sum_{c=1}^{n-1} \phi_k^c(I_1, \ldots, I_{n-1}, s)\frac{\partial J}{\partial I_c} = 0, \quad 2 \le k \le n \tag{3.34}$$

for all values of the noninvariant parameter s. Since the vector fields $\widehat{X}_k$, $1 \le k \le n$ span a Lie algebra, that is an integrable distribution in the sense of the Frobenius theorem (see, e.g., [66]), the system (3.34) is compatible. It will have precisely n_I functionally independent solutions, as stated in (3.5). We can continue by solving another chosen equation of the system (3.34) using the method of characteristics. In this way we may be able to fully solve the system (3.2) equation by equation. However after the first step, i.e., the substitution of invariants of the first vector field $\widehat{X}_1$ into the system, the vector fields no longer have linear coefficients. Consequently, it may be difficult or indeed impossible to find the solution in closed form.

3.3. Calculation of generalized Casimir invariants by the method of moving frames

Another method of computation of the generalized Casimir invariants is based on the method of moving frames. The *method of moving frames* goes back to É. Cartan [23, 24], its modern formulation is due to M. Fels and P. Olver [48]. A related method was developed and applied in [94] for calculating Casimir invariants of inhomogeneous classical groups. V. Boyko et al. adapted the method of moving frames to the case of coadjoint representations. They presented an algebraic procedure for calculating (generalized) Casimir invariants and applied it to a large number of solvable Lie algebras [7–11].

Here we outline the method of moving frames from the practical viewpoint, i.e., an algorithm. For a more detailed description of the method with all the necessary proofs and technical assumptions see [48].

The method of moving frames can be roughly divided into the following steps.

(1) Integration of the coadjoint action of the Lie algebra $\mathfrak{g}$ on its dual $\mathfrak{g}^*$ as given by the vector fields (3.1) to the (local) action of the group G.

This is usually realized by choosing a convenient (local) parameterization of G in terms of one-parameter subgroups, e.g.,

$$g(\vec{\alpha}) = \exp(\alpha_N x_N) \cdot \dots \cdot \exp(\alpha_2 x_2) \cdot \exp(\alpha_1 x_1) \in G, \quad \vec{\alpha} = (\alpha_1, \dots, \alpha_N) \tag{3.35}$$

and correspondingly composing the flows $\Psi^{\alpha_k}_{\widehat{X}_k}$ of the vector fields $\widehat{X}_k$ defined in (3.1)

$$\frac{\mathrm{d}\Psi^{\alpha_k}_{\widehat{X}_k}(p)}{\mathrm{d}\alpha_k} = \widehat{X}_k\big(\Psi^{\alpha_k}_{\widehat{X}_k}(p)\big), \quad p \in \mathfrak{g}^*, \tag{3.36}$$

i.e., we have

$$\Psi(g(\vec{\alpha})) = \Psi^{\alpha_N}_{\widehat{X}_N} \circ \dots \circ \Psi^{\alpha_2}_{\widehat{X}_2} \circ \Psi^{\alpha_1}_{\widehat{X}_1}. \tag{3.37}$$

For a given point $p \in \mathfrak{g}^*$ with coordinates $x_k = x_k(p)$, $\vec{x} = (x_1, \dots, x_N)$ we denote the coordinates of the transformed point $\Psi(g(\vec{\alpha}))p$ by $\tilde{x}_k$

$$\tilde{x}_k \equiv \Psi_k(\vec{\alpha})(\vec{x}) = x_k\big(\Psi\big(g(\vec{\alpha})\big)p\big). \tag{3.38}$$

We consider $\tilde{x}_k$ to be a function of both the group parameters $\vec{\alpha}$ and the coordinates $\vec{x}$ of the original point p.

(2) Choice of a section cutting through the orbits of the action Ψ.

We need to choose in a smooth way a single point on each of the (generic) orbits of the action of the group G. Typically this is done as follows: we find a subset of r coordinates, say $(x_{\pi(i)})_{i=1}^r$, on which the group G acts transitively, at least locally in an open neighborhood of chosen values $(x^0_{\pi(i)})_{i=1}^r$. Here π denotes a suitable injection $\pi\colon \{1, \dots, r\} \to \{1, \dots, N\}$ and r is the rank of the matrix C in (3.4). Points whose coordinates satisfy

$$x_{\pi(i)} = x^0_{\pi(i)}, \quad 1 \le i \le r \tag{3.39}$$

form our section Σ, intersecting each generic orbit once.

(3) Construction of invariants.

For a given point $p \in \mathfrak{g}^*$ we find group elements transforming p into $\tilde{p} \in \Sigma$ by the action Ψ. We express as many of their parameters as possible (i.e., r of them) in terms of the original coordinates $\vec{x}$ and substitute them back into (3.38). This gives us $\tilde{x}_k$ as functions of $\vec{x}$ only. Out of them, $\tilde{x}_{\pi(i)}$, $i = 1, \ldots, r$ have the prescribed fixed values $x^0_{\pi(i)}$. The remaining $N - r$ functions $\tilde{x}_k$ are by construction invariant under the coadjoint action of G, i.e., define the invariants of the coadjoint representation.

Technically, it may not be necessary to evaluate all the functions $\tilde{x}_k$ so that a suitable choice of the basis in $\mathfrak{g}$ can substantially simplify the whole procedure. This happens when only a smaller subset of say r_0 group parameters α_k enters into the computation of $N - r + r_0$ functions $\tilde{x}_k$, $k = 1, \ldots, N - r + r_0$ (possibly after a rearrangement of x_k's). In this case the remaining parameters can be ignored throughout the computation. They are specified by the remaining equations

$$\tilde{x}_i = x^0_i, \quad N - r + r_0 + 1 \leq i \leq N \tag{3.40}$$

but do not enter into the expressions for $\tilde{x}_k$, $1 \leq k \leq N - r + r_0$ which define the invariants.

The method of moving frames exploits the fact that the flows (3.36) can be computed as in Section 3.2.2, provided one is able to find the respective eigenvalues and Jordan canonical form. That is due to the linear dependence on the coordinates in the coefficients of the vector fields (3.1). Consequently, the problem is reduced to a suitable choice of the section and the elimination of the group parameters, i.e., to a system of algebraic equations. Unfortunately, the resulting equations may be difficult or impossible to solve explicitly. In addition, the complexity of the computation strongly depends on the arbitrary choices involved in the selection of the section.

To sum up, both methods of solving the system (3.2) have their own advantages and disadvantages and it is hard to predict which of the two methods will lead to more efficient computation in any individual case. Obviously the two methods give equivalent results.

We mention that the invariants found using either of the methods may not be in the most convenient form. That can be remedied once we find them. For example, as we already mentioned, the generalized Casimir invariants of a nilpotent Lie algebra can be always chosen as polynomials. The method of moving frames may naturally give us nonpolynomial ones. Nevertheless, it is usually quite easy to construct polynomials out of them.

Example 3.7. Let us consider the Lie algebra $\mathfrak{s}_{4,7}$ of Section 17.3 with the nonvanishing Lie brackets

$$[e_2, e_3] = e_1, \quad [e_2, e_4] = e_3, \quad [e_3, e_4] = -e_2. \tag{3.41}$$

The vector fields (3.1) are

$$\begin{aligned} \widehat{E}_1 &= 0, & \widehat{E}_2 &= e_1 \partial_{e_3} + e_3 \partial_{e_4}, \\ \widehat{E}_3 &= -e_1 \partial_{e_2} - e_2 \partial_{e_4}, & \widehat{E}_4 &= -e_3 \partial_{e_2} + e_2 \partial_{e_3}. \end{aligned} \tag{3.42}$$

Consequently, the matrix C takes the form

$$C = \begin{pmatrix} 0 & 0 & 0 & 0 \\ 0 & 0 & e_1 & e_3 \\ 0 & -e_1 & 0 & -e_2 \\ 0 & -e_3 & e_2 & 0 \end{pmatrix}. \tag{3.43}$$

The generic rank of C is 2 and the number (3.5) of functionally independent Casimir invariants is

$$n_I = 4 - 2 = 2.$$

Since the first column of C consists of zeros, e_1 is a solution. We take $\widehat{E}_2$ as the first vector field to which we apply the method of characteristics. We have

$$\frac{de_3}{e_1} = \frac{de_4}{e_3}$$

and the invariants of $\widehat{E}_2$ are e_1, e_2 and $\xi = e_3^2 - 2e_1e_4$. Therefore, any Casimir invariant of the algebra $\mathfrak{s}_{4,7}$ must be of the from $J = J(e_1, e_2, \xi)$. When we apply $\widehat{E}_3$ to such J we get

$$\widehat{E}_3 J = e_1 \left(2e_2 \frac{\partial J}{\partial \xi} - \frac{\partial J}{\partial e_2} \right) \tag{3.44}$$

and we obtain two independent solution of $\widehat{E}_3 J = 0$ in the form $\eta = e_2^2 + e_3^2 - 2e_1e_4$ and e_1. Both e_1 and η are also annihilated by $\widehat{E}_4$. Altogether, we have found that the algebra $\mathfrak{s}_{4,7}$ has two Casimir invariants

$$I_1 = e_1, \quad I_2 = e_2^2 + e_3^2 - 2e_1e_4. \tag{3.45}$$

Let us redo the same calculation using the method of moving frames. The flows of the vector fields $\widehat{E}_1, \ldots, \widehat{E}_4$ are

$$\begin{aligned}
\Psi_{\widehat{E}_1}^{\alpha_1}(e_1, e_2, e_3, e_4) &= (e_1, e_2, e_3, e_4), \\
\Psi_{\widehat{E}_2}^{\alpha_2}(e_1, e_2, e_3, e_4) &= \left(e_1, e_2, \alpha_2 e_1 + e_3, \frac{\alpha_2^2}{2} e_1 + \alpha_2 e_3 + e_4 \right), \\
\Psi_{\widehat{E}_3}^{\alpha_3}(e_1, e_2, e_3, e_4) &= \left(e_1, -\alpha_3 e_1 + e_2, e_3, \frac{\alpha_3^2}{2} e_1 - \alpha_3 e_2 + e_4 \right), \\
\Psi_{\widehat{E}_4}^{\alpha_4}(e_1, e_2, e_3, e_4) &= (e_1, e_2 \cos\alpha_4 - e_3 \sin\alpha_4, e_2 \sin\alpha_4 + e_3 \cos\alpha_4, e_4).
\end{aligned}$$

We compose the flows as in (3.37) and obtain

$$\begin{aligned}
\Psi\big(g(\alpha_1, \alpha_2, \alpha_3, \alpha_4)\big) &= \Psi_{\widehat{E}_4}^{\alpha_4} \circ \Psi_{\widehat{E}_3}^{\alpha_3} \circ \Psi_{\widehat{E}_2}^{\alpha_2} \circ \Psi_{\widehat{E}_1}^{\alpha_1}, \\
\Psi\big(g(\alpha_1, \alpha_2, \alpha_3, \alpha_4)\big)&(e_1, e_2, e_3, e_4) \\
&= \Big(e_1, \cos\alpha_4(-\alpha_3 e_1 + e_2) - \sin\alpha_4(\alpha_2 e_1 + e_3), \\
&\qquad \sin\alpha_4(-\alpha_3 e_1 + e_2) + \cos\alpha_4(\alpha_2 e_1 + e_3), \\
&\qquad \frac{\alpha_2^2 + \alpha_3^2}{2} e_1 - \alpha_3 e_2 + \alpha_2 e_3 + e_4 \Big).
\end{aligned} \tag{3.46}$$

We choose a section Σ given by the equations

$$e_2 = 0, \quad e_3 = 1. \tag{3.47}$$

The intersection of our section Σ with the orbit $\Psi(g(\alpha_1, \alpha_2, \alpha_3, \alpha_4))(e_1, e_2, e_3, e_4)$ starting from the point (e_1, e_2, e_3, e_4) has the following values of α_2, α_3

$$\alpha_2 = \frac{\cos \alpha_4 - e_3}{e_1}, \qquad \alpha_3 = \frac{e_2 - \sin \alpha_4}{e_1} \tag{3.48}$$

(generically, i.e., when $e_1 \neq 0$). The coordinates of the intersection

$$\left(e_1, 0, 1, \frac{2e_1 e_4 - e_2^2 - e_3^2 + 1}{2e_1} \right). \tag{3.49}$$

are independent of the remaining two parameters α_1, α_4. That means that we have found using the method of moving frames that two functionally independent functions e_1 and $(2e_1e_4 - e_2^2 - e_3^2 + 1)/(2e_1)$ are generalized Casimir invariants. Equivalently, I_1 and I_2 of (3.45) are Casimir invariants of the algebra $\mathfrak{s}_{4,7}$.

Part 2

Recognition of a Lie Algebra Given by Its Structure Constants

CHAPTER 4

Identification of Lie Algebras through the Use of Invariants

In this chapter we shall discuss the possible ways of identification of a given Lie algebra in a known list of algebras. The methods explained here can be used to demonstrate whether two given algebras are isomorphic or not.

An obvious way of establishing the equivalence of two algebras is an explicit construction of a change of basis which transforms the structure constants between them. Such a task in general involves a solution of a nonlinear set of equations. Namely, we have two algebras of dimension n spanned by (e_i) and $(\tilde{e}_i)$, respectively, given by their Lie brackets

$$[e_i, e_j] = f_{ij}{}^k e_k, \quad [\tilde{e}_i, \tilde{e}_j] = g_{ij}{}^k \tilde{e}_k$$

and look for a regular transformation A such that

$$\bar{e}_i = e_k A_i^k, \quad [\bar{e}_i, \bar{e}_j] = g_{ij}{}^k \bar{e}_k$$

which reduces to the set of $n^2(n-1)/2$ quadratic equations for the components of A_i^k

$$A_i^k A_j^l f_{kl}{}^m = g_{ij}{}^k A_k^m, \quad \det A \neq 0. \tag{4.1}$$

The algebras are by definition isomorphic if and only if such a transformation exists. Unfortunately, the existence of a solution of (4.1) is in general rather difficult to establish.

Consequently one uses invariant, i.e., basis independent, characteristics of Lie algebras in order to distinguish classes of algebras which are mutually nonisomorphic. Examples of these are dimensions of unique ideals, properties of the Killing form etc. In the following we review some of them. For practical purposes we split our discussion into two parts: elementary invariants, most of which can be computed by hand once we write down the Lie brackets, and more sophisticated invariants whose efficient computation necessarily involves use of computer algebra systems in all but the lowest dimensions.

4.1. Elementary invariants

The first, very rough, invariant characteristic which one should establish is the type of the Lie algebra in question. Namely whether it is indecomposable or decomposable; semisimple, solvable or has a nontrivial Levi decomposition.

The solvability and semisimplicity is easily established from Cartan's criteria, Theorem 2.11. Namely, the algebra is semisimple if and only if its Killing form is nondegenerate; the algebra is solvable if and only if its Killing form restricted to the derived algebra vanishes. The decomposability can be established by the algorithm described in Chapter 5. The Levi decomposition into the radical and a semisimple subalgebra can be performed as described in Chapter 6. Both these algorithms are often realized on a computer since they involve the factorization of polynomials and the solution of a relatively large system of linear equations, respectively. The factors obtained by either of these algorithms should themselves be identified. For Levi decomposable algebras one also has to analyze how the radical and semisimple subalgebra are combined together since more than one nonequivalent possibility may exist (for a more detailed discussion see Chapter 14).

Another class of invariants which we have already encountered is the dimension of ideals in the characteristic series of the Lie algebra. We recall that we use the symbols DS, CS and US for sets of integers denoting the dimensions of subalgebras in the derived, lower central and upper central series, respectively.

These invariants are very easy to compute, essentially by inspection of the Lie brackets. The amount of information they contain depends strongly on the type of Lie algebra. They are useful for nilpotent and solvable algebras but do not provide any information for semisimple ones.

One can of course analyze the structure of characteristic series in more detail, e.g., identify the structure of each lower-dimensional ideal contained in them by calculating its respective characteristic series. This may provide more information in certain cases. However, often it doesn't help in the identification. Indeed the equality of dimensions sometimes implies that the algebraic structure of the ideals is the same. The difference between algebras then lies in different ways these ideals are arranged inside the entire algebra and cannot be identified in this way.

When the algebra in question is solvable but not nilpotent we shall identify its nilradical by the algorithm of Section 7. The (faithful) representation of the solvable factor algebra $\mathfrak{g}/C(\mathfrak{g})$ on the nilradical can provide further invariant characteristics, e.g., whether the representation is completely reducible or not, what are the dimensions of its invariant subspaces etc. This appears to be the most straightforward way of identifying the given algebra in our lists given in Part 4.

The invariant of particular importance for simple and semisimple algebras is the *rank* of the Lie algebra, i.e., the dimension of its Cartan subalgebra. Its determination together with the dimension of the algebra brings us very close to full identification of a simple complex Lie algebra in the Cartan's classification. The only remaining ambiguity to be resolved is the differentiation between B_l and C_l algebras, i.e., between $\mathfrak{so}(2l+1,\mathbb{C}))$ and $\mathfrak{sp}(2l,\mathbb{C})$. Exceptionally, for $l=6$ also E_6 shares the same values of rank and dimension with B_6 and C_6. The rank may provide nontrivial information also for other nonnilpotent algebras. For a nilpotent algebra the Cartan subalgebra coincides with the entire algebra and the term *rank* is often used for a different property, see below.

For semisimple Lie algebras another well-known invariant is the Dynkin diagram and other properties related to the structure of the simple roots [44]. For example, in order to distinguish B_l and C_l algebras given in an arbitrary basis one may find the Cartan subalgebra, construct the corresponding root system, identify the simple roots and consider their relative lengths. In the case of B_l one of the simple roots is shorter than the rest, C_l has one longer simple root. E_6 has all the simple roots of the same lengths.

Most of the invariants introduced up to now, in particular all the ones determined by the dimension of some distinguished subalgebra, may allow us to distinguish complex Lie algebras but are not suitable for identification of nonisomorphic real forms of the same complex algebra. They necessarily give the same answer for all of them. The most elementary invariant allowing such a discrimination is the signature of the Killing form, i.e., the number of its positive, negative and zero eigenvalues once it is brought to diagonal form. Because over the field $\mathbb{C}$ the Killing form is bilinear, not sesquilinear, its signature may change under a complex linear transformation but remains invariant once we restrict ourselves to real transformations. Therefore, it may allow the identification of a particular real form of a semisimple or solvable Lie algebra. For simple Lie algebras it is known that it allows complete identification of the real form. Of course, it is of no use in the case of nilpotent algebras since their Killing form automatically vanishes.

Finally, several invariant properties can be extracted also from the invariants of coadjoint representation, i.e., generalized Casimir invariants. An obvious invariant property is the number of functionally independent generalized Casimir invariants, (3.5). Related to it is the *Dixmier invariant*, i.e., the dimension of a maximal orbit of the coadjoint representation [53]. It is equal to the difference between the dimension of the algebra and the number of independent generalized Casimir invariants, i.e., to the generic rank of matrix C in (3.4).

Another possibility available for some solvable and Levi decomposable Lie algebras is the maximal number of proper, i.e., polynomial, Casimir invariants. For semisimple and nilpotent algebras it always coincides with the total number of independent generalized Casimir invariants.

Under certain assumptions about the chosen basis of Casimir invariants we can go even further. Provided we choose the lowest degree polynomials as the generators, their degrees become further invariant characteristics of the given Lie algebra $\mathfrak{g}$. Indeed a change of basis in $\mathfrak{g}$ induces a linear transformation of the functionals $e_1, \ldots, e_n$ which does not change the degrees of polynomials expressed in terms of them.

Similarly, the type of transcendent functions present in the invariants can be helpful; e.g., different real forms may be distinguishable by presence of exponential versus trigonometric functions in the generalized Casimir invariants. They cannot be transformed into each other by a change of basis over the field $\mathbb{R}$, nor by a different choice of functionally independent generalized Casimir invariants.

Example 4.1. Let us consider two real 6-dimensional solvable Lie algebras $\mathfrak{s}_{6,151}$ and $\mathfrak{s}_{6,152,\epsilon=1}$ (cf. Section 19.6) with the nonvanishing Lie brackets

$$\begin{aligned}
\mathfrak{s}_{6,151}: \quad & [e_3, e_4] = e_2, \quad [e_3, e_5] = e_1, \quad [e_4, e_5] = e_3, \quad [e_6, e_1] = e_1, \\
& [e_6, e_2] = -e_2, \quad [e_6, e_4] = e_2 - e_4, \quad [e_6, e_5] = e_5, \\
\mathfrak{s}_{6,152,\epsilon=1}: \quad & [e_3, e_4] = e_2, \quad [e_3, e_5] = e_1, \quad [e_4, e_5] = e_3, \quad [e_6, e_1] = e_2, \\
& [e_6, e_2] = -e_1, \quad [e_6, e_4] = -e_5, \quad [e_6, e_5] = e_2 + e_4.
\end{aligned}$$

These two algebras $\mathfrak{s}_{6,151}$, $\mathfrak{s}_{6,152,\epsilon=1}$ are real forms of a single complex Lie algebra $\mathfrak{s}_{\mathbb{C}} = \mathbb{C} \otimes \mathfrak{s}_{6,151} \simeq \mathbb{C} \otimes \mathfrak{s}_{6,152,\epsilon=1}$. The question is whether they are equivalent as real algebras or define two distinct real forms of $\mathfrak{s}_{\mathbb{C}}$.

Their independent generalized Casimir invariants can be written as

$$\begin{aligned}
\mathfrak{s}_{6,151}: \quad & e_1 e_2, \quad e_1^2 \exp\left(\frac{e_3^2 - 2e_1e_4 + 2e_2e_5}{e_1 e_2}\right), \\
\mathfrak{s}_{6,152,\epsilon=1}: \quad & e_1^2 + e_2^2, \quad (e_1^2 + e_2^2) \arctan\frac{e_2}{e_1} - e_1e_2 - 2e_1e_4 + 2e_2e_5 + e_3^2.
\end{aligned}$$

Since no real transformation can convert trigonometric functions into exponentials and vice versa we immediately see that the algebras $\mathfrak{s}_{6,151}$ and $\mathfrak{s}_{6,152,\epsilon=1}$ cannot be isomorphic, i.e., must define two different real forms of $\mathfrak{s}_{\mathbb{C}}$.

4.2. More sophisticated invariants

Let us review some of the more involved constructions of invariants of Lie algebras. The list cannot be exhaustive since new ways of characterizing Lie algebras suited for their particular classes are constantly being developed.

Going in the other direction than the characteristic series is the construction of Lie algebras which are typically bigger than the one we started with. The most natural possibility is to construct the algebra of derivations of the given Lie algebra and investigate its properties. Although it is of no use for semisimple Lie algebras whose all derivations are inner, for other classes of algebras we may obtain interesting information in this way. For example, a low-dimensional (up to 6) complex nilpotent Lie algebra can be fully identified once we determine its DS, CS and US and the dimension of its algebra of derivations. Their nonisomorphic real forms are then discerned by the signature of the Killing form of their algebras of derivations.

Such a procedure can be extended indefinitely, by constructing the algebra of derivations of the algebra of derivations etc. In principle one can obtain an infinite sequence of integers (i.e., dimensions) characterizing the given Lie algebra. Of course, its usefulness beyond certain limited number of terms is questionable in most applications; even its construction may become more and more difficult with the increasing dimensions of the terms.

Other invariants address the problem of the identification of possible isomorphisms inside a class of algebras involving a parameter. Such invariants can be constructed, e.g., using the notion of (α, β, γ)-derivations introduced in [85].

For given $\alpha, \beta, \gamma \in \mathbb{F}$ we call a linear operator $A\colon \mathfrak{g} \to \mathfrak{g}$ an (α, β, γ)-*derivation* when

$$\alpha A[x, y] = \beta[Ax, y] + \gamma[x, Ay] \tag{4.2}$$

for every $x, y \in \mathfrak{g}$. The vector spaces of such (α, β, γ)-derivations are denoted by $\mathcal{D}(\alpha, \beta, \gamma)$. As was shown in [85], in fact only the following values of (α, β, γ)

$$(\alpha, 0, 0), \quad (\alpha, 1, -1), \quad (\alpha, 1, 0), \quad (\alpha, 1, 1), \quad \alpha \in \mathbb{F}$$

should be used; any vector space $\mathcal{D}(\alpha, \beta, \gamma)$ with different values of parameters α, β, γ is equal to one in the above introduced range. The dimension of each of these vector spaces provides us with basis independent information about the given Lie algebra $\mathfrak{g}$. Their importance lies in the fact that they depend on the continuous parameter α; namely, in the language of [85] we have the "invariant function"

$$\psi_{\mathfrak{g}}(\alpha) = \dim \mathcal{D}(\alpha, 1, 1). \tag{4.3}$$

If we are given a family of Lie algebras depending on one or more continuous parameters and want to establish possible isomorphisms between members of the family, we can compute the invariant function $\psi_{\mathfrak{g}}(\alpha)$. The members of the family whose invariant functions differ are necessarily nonisomorphic. When the invariant functions of two algebras are the same, the algebras may be isomorphic. Other criteria or an explicit search for a basis transformation must be employed in order to complete the analysis in this case.

To illustrate the use of (α,β,γ)-derivations let us consider the problem of identifying 3-dimensional solvable Lie algebras up to isomorphisms (cf. Section 9.2, p. 102).

Example 4.2. Let us first consider the 3-dimensional Lie algebras parameterized by two parameters a, b with nonvanishing Lie brackets in the following form

$$\mathfrak{g}_{a,b}: \quad [e_1, e_3] = ae_1, \quad [e_2, e_3] = be_2 \tag{4.4}$$

where $(a, b) \neq (0, 0)$. We will show how $(\alpha, 1, 1)$-derivations allow us to split the algebras $\mathfrak{g}_{a,b}$ into mutually nonisomorphic subclasses.

Let us write an $(\alpha, 1, 1)$-derivation A as

$$A(e_j) = \sum_{k=1}^{3} A_j^k e_k. \tag{4.5}$$

Equation (4.2) with $\beta = \gamma = 1$ implies that

$$\begin{aligned} 0 &= -bA_1^3 e_2 + aA_2^3 e_1, \\ \alpha a \sum_k A_1^k e_k &= a(A_1^1 + A_3^3)e_1 + bA_1^2 e_2, \\ \alpha b \sum_k A_2^k e_k &= aA_2^1 e_1 + b(A_2^2 + A_3^3)e_2. \end{aligned} \tag{4.6}$$

Thus the vector of constants $\vec{A} = (A_1^1, A_1^2, A_1^3, A_2^1, A_2^2, A_2^3, A_3^1, A_3^2, A_3^3)$ defines an $(\alpha, 1, 1)$-derivation of algebra (4.4) if and only if it is annihilated by the following matrix

$$M = \begin{pmatrix} 0 & 0 & b & 0 & 0 & 0 & 0 & 0 & 0 \\ 0 & 0 & 0 & 0 & 0 & -a & 0 & 0 & 0 \\ a(\alpha-1) & 0 & 0 & 0 & 0 & 0 & 0 & 0 & -a \\ 0 & a\alpha - b & 0 & 0 & 0 & 0 & 0 & 0 & 0 \\ 0 & 0 & a\alpha & 0 & 0 & 0 & 0 & 0 & 0 \\ 0 & 0 & 0 & b\alpha - a & 0 & 0 & 0 & 0 & 0 \\ 0 & 0 & 0 & 0 & b(\alpha-1) & 0 & 0 & 0 & -b \\ 0 & 0 & 0 & 0 & 0 & \alpha b & 0 & 0 & 0 \end{pmatrix}. \tag{4.7}$$

In order to evaluate the function $\psi_{\mathfrak{g}_{a,b}}(\alpha)$ we have to compute the rank of the matrix M as a function of the parameters α, a, b,

$$\psi_{\mathfrak{g}_{a,b}}(\alpha) = 9 - \operatorname{rank} M. \tag{4.8}$$

We find that

$$\begin{aligned} \psi_{\mathfrak{g}_{a,b}}(\alpha) &= 3, && (\alpha a - b)(\alpha b - a) \neq 0, && ab \neq 0, && \alpha \neq 0, 1, \\ &= 4, && (\alpha a - b)(\alpha b - a) = 0, && ab \neq 0, && a \neq -b, \quad \alpha \neq 0, 1, \\ & && \text{or} \quad ab = 0, && \alpha \neq 0, 1, \\ &= 5, && a = -b, && a \neq 0, && \alpha = -1, \\ \psi_{\mathfrak{g}_{a,b}}(0) &= 3, && ab \neq 0, \\ &= 6, && ab = 0, \\ \psi_{\mathfrak{g}_{a,b}}(1) &= 4, && a \neq b, \\ &= 6, && a = b. \end{aligned} \tag{4.9}$$

We divide the algebras $\mathfrak{g}_{a,b}$ of the form (4.4) into classes C_p

$$C_p = \{\mathfrak{g}_{a,pa}, \mathfrak{g}_{pa,a} \mid a \in \mathbb{F}, a \neq 0\}, \quad p \in \mathbb{F}. \tag{4.10}$$

and conclude that

$$C_p = C_{1/p} \quad (p \neq 0), \qquad C_p \cap C_q = 0 \quad \text{whenever } p \neq q,\ p \neq 1/q, \tag{4.11}$$

i.e., up to the identification $p \simeq 1/p$ algebras belonging to different classes (4.10) are not isomorphic. By an explicit construction one may demonstrate that each of the different classes (4.10) contains precisely one Lie algebra up to isomorphism. The algebras in the class C_0 are decomposable.

Next, one similarly considers the remaining complex 3-dimensional solvable and nilpotent non-Abelian algebras which can be cast in the form

$$\mathfrak{g}_a: \quad [e_1, e_3] = ae_1, \quad [e_2, e_3] = e_1 + ae_2 \tag{4.12}$$

by a change of basis. We find that

$$\begin{aligned} \psi_{\mathfrak{g}_a}(\alpha) &= 3, \qquad a \neq 0, \quad \alpha \neq 1, \\ &= 6, \qquad a = 0, \\ \psi_{\mathfrak{g}_a}(1) &= 4, \qquad a \neq 0, \end{aligned} \tag{4.13}$$

i.e., we have two distinct classes $\widetilde{C} = \{\mathfrak{g}_0\}$ and $\widehat{C} = \{\mathfrak{g}_a \mid a \neq 0\}$. An explicit computation shows that all algebras in class $\widehat{C}$ are mutually isomorphic, i.e., the class $\widehat{C}$ can be represented by just one algebra, e.g., $\mathfrak{g}_1$.

Altogether we have demonstrated that all solvable and nilpotent algebras listed in Chapter 16 are not isomorphic over the field of complex numbers (taking into account the range of parameter a of the algebras $\mathfrak{s}_{3,1}(a)$, i.e., $0 < |a| \leq 1$, if $|a| = 1$ then $\arg(a) \leq \pi$). Their values of the invariant function $\psi_{\mathfrak{g}}(\alpha)$ are as follows:

$$\begin{aligned} \mathfrak{n}_{3,1}: \quad & \psi_{\mathfrak{n}_{3,1}}(\alpha) = 6, \\ \mathfrak{s}_{3,1}(a): \quad & \psi_{\mathfrak{s}_{3,1}(a)}(\alpha) = 3, \quad a \neq 0, -1,\ \alpha \neq 1, a, 1/a, \\ & \qquad = 4, \quad a \neq 0, \pm 1,\ \alpha = 1, a, 1/a, \\ & \qquad = 6, \quad a = 1, \qquad \alpha = 1, \\ & \qquad = 3, \quad a = -1, \quad \alpha \neq \pm 1, \\ & \qquad = 4, \quad a = -1, \quad \alpha = 1, \\ & \qquad = 5, \quad a = -1, \quad \alpha = -1, \\ \mathfrak{s}_{3,2}: \quad & \psi_{\mathfrak{s}_{3,2}}(\alpha) = 3, \quad \alpha \neq 1, \\ & \qquad = 4, \quad \alpha = 1. \end{aligned}$$

The fact that 3-dimensional complex Lie algebras are completely characterized by the function $\psi_{\mathfrak{g}}(\alpha)$ was first observed in [85].

To establish the classification of real forms one may employ the structure of the Casimir invariants or the signature of the Killing form of the algebra of derivations, as mentioned above.

In this low-dimensional example it is easy to arrive at the same classification without using (α, β, γ)-derivations, i.e., by considering all changes of bases leaving the nilradical invariant. However, in more complicated examples the values of the invariant function $\psi_{\mathfrak{g}}(\alpha)$ may provide the simplest proof that two algebras sharing the same dimensions of their characteristic series are not isomorphic.

Through a generalization of the (α, β, γ)-derivations we arrive at the notion of a twisted cocycle [56] which may be useful when classes of algebras involving more parameters are encountered. These cocycles arise as a generalization, or deformation, of cocycles encountered in the Chevalley cohomology theory of Lie algebras, cf. Section 2.4. We shall restrict ourselves to the cocycles defined through the adjoint representation. Let $k \in \mathbb{N}$ and let κ be a $(k+1) \times (k+1)$ symmetric matrix. An alternating (i.e., totally antisymmetric) map $c\colon \mathfrak{g}^k \to \mathfrak{g}$ is called a *κ-twisted cocycle* if and only if

$$
\begin{aligned}
(4.14)\quad 0 = &\sum_{i=1}^{k+1} (-1)^{i+1} \kappa_{ii} [x_i, c(x_1, \ldots, \hat{x}_i, \ldots, x_{k+1})] \\
&+ \sum_{1 \le i < j \le k+1} (-1)^{i+j} \kappa_{ij} c([x_i, x_j], x_1, \ldots, \hat{x}_i, \ldots, \hat{x}_j, \ldots, x_{k+1}),
\end{aligned}
$$

for any $(k+1)$-tuple of vectors $x_1, \ldots, x_{k+1} \in \mathfrak{g}$ ($\hat{x}_i$ indicates that the term is omitted). The space of such maps for a given matrix κ is denoted by $Z^k(\mathfrak{g}, \mathfrak{g}, \kappa)$. When $\kappa = \mathbf{1}$ we get back to the usual space of cocycles

$$
Z^k(\mathfrak{g}, \mathfrak{g}, \mathrm{ad}) = Z^k(\mathfrak{g}, \mathfrak{g}, \mathbf{1}). \tag{4.15}
$$

For the lowest value of $k = 1$ we find that the space $Z^1(\mathfrak{g}, \mathfrak{g}, \kappa)$ coincides with the space of (α, β, γ)-derivations upon setting

$$
\kappa_{11} = \gamma, \qquad \kappa_{12} = \alpha, \qquad \kappa_{22} = \beta. \tag{4.16}
$$

The dimensions of $Z^k(\mathfrak{g}, \mathfrak{g}, \kappa)$ do not depend on the choice of basis in $\mathfrak{g}$ and are therefore invariant characteristics of the Lie algebra $\mathfrak{g}$.

Convenient additional invariant functions defined in this way in [56] are

$$
\begin{gathered}
\phi_{\mathfrak{g}}(\alpha) = \dim Z^2(\mathfrak{g}, \mathfrak{g}, \kappa_1), \quad \phi^0_{\mathfrak{g}}(\alpha) = \dim Z^2(\mathfrak{g}, \mathfrak{g}, \kappa_2), \\
\text{where } \kappa_1 = \begin{pmatrix} \alpha & 1 & 1 \\ 1 & \alpha & 1 \\ 1 & 1 & \alpha \end{pmatrix}, \quad \kappa_2 = \begin{pmatrix} \alpha & 1 & 1 \\ 1 & 1 & 0 \\ 1 & 0 & 1 \end{pmatrix}.
\end{gathered} \tag{4.17}
$$

The functions $\psi_{\mathfrak{g}}(\alpha)$ and $\phi_{\mathfrak{g}}(\alpha)$ themselves allow a full identification of any Lie algebra in dimension 4, i.e., the functions of two 4-dimensional algebras coincide exactly when the algebras are isomorphic. (For proofs and all the values of these functions for 4-dimensional algebras see [56].)

For computational reasons one usually restricts to low values of k because equation (4.14) implies $(\dim \mathfrak{g})^{k+2}$ linear homogeneous equations for $(\dim \mathfrak{g}) \cdot \binom{\dim \mathfrak{g}}{k}$ components of the map c. Not all of the equations are in general independent. The number of independent solutions of these equations give us the value of the invariant function for the particular value of the parameter α.

We see that the computation of invariant functions suffers from a practical difficulty. Namely, their determination requires the computation of the number of linearly independent solutions of the set of homogeneous linear equations coming from (4.2) or (4.14). This depends on parameters α, β, γ (or κ_{ij}) and whatever parameters are present in the structure constants. In other words, we need a reliable and efficient procedure in some computer algebra system for the computation of the rank of a relatively large parameter-dependent matrix. Such a procedure is not readily available in most systems and needs to be implemented for this particular purpose.

Other invariants used in the literature for nilpotent algebras $\mathfrak{n}$ include the following (see [52]):

• The *characteristic sequence* defined in terms of the properties of $\mathrm{ad}(x)$ in the following way: let $c(x)$ denote the sequence of the dimensions of Jordan blocks of $\mathrm{ad}(x)$ in descending order. Introduce the lexicographic ordering of such sequences. The characteristic sequence in the language of [52] is

$$c(\mathfrak{g}) = \max_{x \in \mathfrak{n} \setminus \mathfrak{n}^2} \big(c(x)\big).$$

• The dimension of the maximal tori of derivations, i.e., maximal Abelian subalgebras of the algebra of derivations consisting of semisimple elements. This invariant is usually called the *rank* of the nilpotent algebra $\mathfrak{n}$ (not to be mistaken for the dimension of the Cartan subalgebra).

CHAPTER 5

Decomposition into a Direct Sum

The first step in identifying a Lie algebra $\mathfrak{g}$ is to decide whether $\mathfrak{g}$ is *decomposable*, i.e., whether, by an appropriate change of basis, we can decompose this algebra into a direct sum of two or more nonzero indecomposable Lie algebras. If $\mathfrak{g}$ is decomposable then we wish to find explicitly a basis transformation that will realize the decomposition. Below we shall give a simple criterion of decomposability and an algorithm for achieving a decomposition.

Although we restrict ourselves in this book to the fields of complex and real numbers, the procedure as described here can be used over an arbitrary field of characteristic zero. (The same is also true for procedures described in Chapters 6 and 7.) The proofs for the general case can be found in [102].

5.1. General theory and criteria

First we must introduce some concepts and general results that underlie the criterion and the algorithm.

We are given a Lie algebra $\mathfrak{g}$ of dimension n with basis $(x_1, \dots, x_n)$ and Lie brackets

$$[x_i, x_k] = \sum_{l=1}^{n} c_{ik}{}^{l} x_l, \quad 1 \le i,\ k \le n. \tag{5.1}$$

Let us recall that $C(\mathfrak{g})$ is the center of $\mathfrak{g}$, $D(\mathfrak{g})$ is the derived algebra and ad is the adjoint representation of $\mathfrak{g}$.

The *commuting algebra* (or the centralizer of S in R) $C_R(S)$ of a set S of square $f \times f$ matrices over the field $\mathbb{F}$ is the set of all elements of the full ring $R = \mathbb{F}^{f \times f}$ of square matrices of degree f, commuting elementwise with all elements of S:

$$C_R(S) = \{x \in R \mid [x, S] = 0\}. \tag{5.2}$$

An *idempotent* E is a nonzero element of the ring R satisfying

$$E^2 = E. \tag{5.3}$$

The unit element $E = \mathbf{1}$ is called a *trivial idempotent.* This is the only idempotent of rank f and hence the only nonsingular idempotent. Any idempotent is conjugated to a diagonal matrix with its diagonal matrix elements equal to 1 or 0.

We shall call two idempotents E_1 and E_2 *orthogonal* if $E_1 E_2 = E_2 E_1 = 0$.

The matrix set S is *decomposable* if there exists a nonsingular matrix $Y \in R$ such that

$$\begin{aligned} &YaY^{-1} = D_1 a \oplus D_2 a = \begin{pmatrix} D_1 a & 0 \\ 0 & D_2 a \end{pmatrix}, \quad \forall a \in S, \\ &D_i a \in \mathbb{F}^{f_i \times f_i}, \quad f_i \in \mathbb{N}, \quad i = 1, 2, \qquad f_1 + f_2 = f \end{aligned} \tag{5.4}$$

(i.e., if all elements $a \in S$ can be simultaneously transformed into block diagonal form with one block being the square matrix $D_1 a$ of degree f_1, the other the square matrix $D_2 a$ of degree f_2). The symbols D_1, D_2 denote the projection mappings of S onto the component sets of degree f_1, f_2 respectively.

Theorem 5.1. *Let S be a set of $f \times f$ matrices over the field $\mathbb{F}$. Then S is indecomposable if and only if the identity matrix $\mathbf{1}_f$ of degree f over $\mathbb{F}$ is the only idempotent of the commuting algebra $C_R(S)$ of S in the full algebra of matrices R.*

Proof. Let S be decomposable. By definition, the matrix Y of (5.4) exists and hence two nontrivial idempotents exist in $C_R(S)$, namely

$$E_1 = Y^{-1}(\mathbf{1}_{f_1} \oplus \mathbf{0}_{f_2})Y, \quad E_2 = Y^{-1}(\mathbf{0}_{f_1} \oplus \mathbf{1}_{f_2})Y, \quad E_1 + E_2 = \mathbf{1}_f. \tag{5.5}$$

Conversely, let $E \in C_R(S)$ be an idempotent of rank f_1 where $0 < f_1 < f$. Then there exists a nonsingular matrix $Y \in R$ for which we have

$$YEY^{-1} = \mathbf{1}_{f_1} \oplus \mathbf{0}_{f-f_1}. \tag{5.6}$$

The matrix Y will also realize the decomposition (5.4). □

The *Jacobson radical* [59, 139] $J(S)$ of an associative algebra $S \subset R$ is the maximal nilpotent ideal of S. Since the field $\mathbb{F}$ is of characteristic zero, we have

$$J(S) = \{x \in S \mid \mathrm{Tr}(xy) = 0, \ \forall y \in S\} \tag{5.7}$$

and $J(S)$ consists entirely of nilpotent matrices. Here $\mathrm{Tr}(Z)$ denotes the trace of the matrix Z.

A *division ring* is a ring with unity in which every nonzero element has a multiplicative inverse.

We recall

Theorem 5.2 (Wedderburn structure theorem). *Any finite-dimensional associative algebra A over the field $\mathbb{F}$ such that it contains a unit and its Jacobson radical vanishes, is isomorphic to a direct sum*

$$A = (R_1)^{f_1 \times f_1} \oplus (R_2)^{f_2 \times f_2} \oplus \cdots \oplus (R_K)^{f_K \times f_K} \tag{5.8}$$

of full matrix algebras $(R_j)^{f_j \times f_j}$ over some division rings R_j. The division rings R_j are in turn algebras over $\mathbb{F}$. The decomposition is unique up to the order of the terms.

For a proof see, e.g., [134, Section 10.10, p. 158 – 163]. Notice that associative algebras which satisfy hypotheses of the Wedderburn structure theorem are called *semisimple*.

Let us consider the commuting algebra $C_R(S)$, its Jacobson radical $J\big(C_R(S)\big)$, and the subalgebra $C_0(S) \subset C_R(S)$, defined as

$$C_0(S) = \{x \in C_R(S) \mid \mathrm{Tr}(x) = 0\}. \tag{5.9}$$

Their explicit construction is a straightforward task of linear algebra. We have

$$C_R(S) = C_0(S) \oplus \mathrm{span}\{\mathbf{1}_f\}, \tag{5.10}$$

where the symbol $\oplus$ indicates decomposition of the module on the left into the direct sum of the modules on the right-hand side, and

$$J\big(C_R(S)\big) \subseteq C_0(S). \tag{5.11}$$

We introduce the notation

$$Q = C_R(S)/J\big(C_R(S)\big), \quad Q_0 = C_0(S)/J\big(C_R(S)\big) \tag{5.12}$$

for the corresponding factor algebras.

Lemma 5.3 ([138]). *Idempotents can be lifted from $C_R(S)/J\big(C_R(S)\big)$ to $C_R(S)$, i.e., any idempotent $E_0+J\big(C_R(S)\big)$ in the factor algebra $C_R(S)/J\big(C_R(S)\big)$ gives rise to an idempotent E in $C_R(S)$.*

Proof. Let us assume that a matrix $E_0 \in A \equiv C_R(S)$ is given such that

$$N_0 = E_0^2 - E_0 \in J(A).$$

Thus, an integer $K \in \mathbb{N}$ exists such that $(N_0)^K = 0$ and for every $X \in A$ the products XN_0 and N_0X belong to $J(A)$. We shall construct a matrix $\widetilde{E}_0$ such $\widetilde{E}_0 - E_0 \in J(A)$, i.e., $\widetilde{E}_0$ represents the same idempotent $E_0 + J(A) \in A/J(A)$, and $(\widetilde{N}_0)^{K-1} = (\widetilde{E}_0^2 - \widetilde{E}_0)^{K-1} = 0$.

We choose

$$\widetilde{E}_0 = 3E_0^2 - 2E_0^3 = E_0 + (1 - 2E_0)N_0 \tag{5.13}$$

which by construction satisfies $\widetilde{E}_0 + J(A) = E_0 + J(A) \in A/J(A)$. We have

$$\begin{aligned}\widetilde{N}_0 &= \widetilde{E}_0^2 - \widetilde{E}_0\\ &= (E_0 + (1-2E_0)N_0)(E_0 + (1-2E_0)N_0) - E_0 - (1-2E_0)N_0\\ &= -3N_0^2 + 4N_0^3 = N_0^2(-3 + 4N_0).\end{aligned}$$

There are two possibilities: either $K > 1$, i.e., $2(K-1) \geq K$, and we have

$$\big(\widetilde{N}_0\big)^{K-1} = 0;$$

or $K = 1$ and we have $N_0^2 = 0$, i.e., $\widetilde{N}_0 = 0$. Therefore, in both cases we can conclude that

$$\big(\widetilde{N}_0\big)^{K-1} = 0. \tag{5.14}$$

Repeating the same procedure sufficiently many times (at most K times) we find an idempotent E in $C_R(S)$ such that $E + J\big(C_R(S)\big) = E_0 + J\big(C_R(S)\big)$. □

Theorem 5.4. *The set S of square matrices of degree f over a field $\mathbb{F}$ is indecomposable if and only if the factor algebra $Q = C_R(S)/J\big(C_R(S)\big)$ is a division ring, where $C_R(S)$ is the commuting algebra of S in R and $J\big(C_R(S)\big)$ is its Jacobson radical.*

Proof. It follows from Theorem 5.1 and Lemma 5.3 that S is indecomposable precisely if the semisimple factor algebra $C_R(S)/J\big(C_R(S)\big)$ contains only one, i.e., trivial, idempotent.

It follows from the Weddeburn structure Theorem 5.2 that $C_R(S)/J\big(C_R(S)\big)$ contains only the trivial idempotent precisely when $C_R(S)/J\big(C_R(S)\big)$ is a division ring. □

Remark 5.5. If $\mathbb{F} = \mathbb{C}$, the proof of Theorem 5.4 can be formulated also in the following, more straightforward way: It follows from the Schur lemma 2.5, 2.6 that the matrix set S is indecomposable if and only if

$$J\big(C_R(S)\big) = C_0(S) \tag{5.15}$$

(every traceless matrix commuting with all elements of S is nilpotent). Thus the matrix set S is indecomposable if and only if $Q = \mathbb{C}$, i.e., is a division ring.

Let us recall that a theorem due to Frobenius (cf. [39, p. 10–12; 58, p. 452–453]) implies that the algebra Q over $\mathbb{R}$ is a division ring if and only if one of the following possibilities occurs:

(5.16a)
$$\dim Q_0 = 0;$$

(5.16b)
$$\dim Q_0 = 1, \qquad Q_0 = \operatorname{span}\{Y\},\ Y^2 = -k^2\mathbf{1}\ (k \in \mathbb{R}, k \neq 0);$$

(5.16c)
$$\dim Q_0 = 3, \qquad Q_0 = \operatorname{span}\{y_1, y_2, y_3\} = \text{algebra of traceless quaternions.}$$

As we shall see below in Theorem 5.8, the case (5.16c) can never happen for the set of operators $S = \operatorname{ad}(\mathfrak{g})$ which is of our interest here.

Let us now return to the task at hand, namely the decomposability of a Lie algebra $\mathfrak{g}$.

Theorem 5.6. *A Lie algebra $\mathfrak{g}$ is decomposable into the direct sum of two ideals*

$$\mathfrak{g} = \mathfrak{g}_1 + \mathfrak{g}_2, \quad [\mathfrak{g}_a, \mathfrak{g}_a] \subseteq \mathfrak{g}_a\ (a = 1, 2), \quad [\mathfrak{g}_1, \mathfrak{g}_2] = 0, \tag{5.17}$$

if and only if its adjoint representation, as a set of matrices, is decomposable.

Proof. Let $\mathfrak{g}$ be decomposable, as in (5.17). Then bases of $\mathfrak{g}$ exist that realize the decomposition, i.e., satisfy

$$\begin{aligned} &[x_i, x_k] = f_{ik}{}^m x_m, \quad [y_a, y_b] = g_{ab}{}^c y_c, \quad [x_i, y_a] = 0, \\ &x_i \in \mathfrak{g}_1, \quad i = 1, \dots, n, \qquad y_a \in \mathfrak{g}_2, \quad a = n+1, \dots, n+m. \end{aligned} \tag{5.18}$$

(Summation over repeated indices applies throughout.)

In the adjoint representation we then have

$$\begin{aligned} \operatorname{ad}(x_h)_i^k &= \begin{cases} f_{hi}{}^k, & 1 \le h, i,\ k \le n, \\ 0, & \text{otherwise}, \end{cases} \\ \operatorname{ad}(y_c)_a^b &= \begin{cases} g_{ca}{}^b, & n+1 \le c,\ a,\ b \le n+m, \\ 0, & \text{otherwise}, \end{cases} \end{aligned} \tag{5.19}$$

Thus, the adjoint representation, in the appropriate basis, consists of block diagonal matrices. In other bases it will be decomposable (rather than already decomposed). Elements in the center $C(\mathfrak{g})$ are represented by zero matrices (which are trivially block diagonal).

Conversely, let ad be decomposable as a set of matrices. By conjugacy (change of basis in $\mathfrak{g}$) we can transform $\operatorname{ad}(z)$, $z \in \mathfrak{g}$ into a block diagonal form, with invariant subspaces $\tilde{\mathfrak{g}}_1$ and $\tilde{\mathfrak{g}}_2$. Let $x \in \tilde{\mathfrak{g}}_1$, $y \in \tilde{\mathfrak{g}}_2$. We have both $\operatorname{ad}(x)y \in \tilde{\mathfrak{g}}_2$ and $\operatorname{ad}(x)y = -\operatorname{ad}(y)x \in \tilde{\mathfrak{g}}_1$. Thus $\operatorname{ad}(x)y = \operatorname{ad}(y)x = 0$ and $\operatorname{ad}(x_h)$, $\operatorname{ad}(y_c)$ take the form (5.19) for some constants $f_{hi}{}^k, g_{ca}{}^b$. The commutation relations of (5.19) are those of (5.18); hence $\mathfrak{g}$ is decomposable. □

An important special case of a decomposable Lie algebra is one that admits a *central decomposition*, i.e., one that has a central component in the decomposition:

$$\mathfrak{g} = \mathfrak{g}_1 \oplus \mathfrak{g}_2, \quad [\mathfrak{g}_1, \mathfrak{g}_1] = 0, \quad [\mathfrak{g}_2, \mathfrak{g}_2] \subseteq \mathfrak{g}_2, \quad [\mathfrak{g}_1, \mathfrak{g}_2] = 0. \tag{5.20}$$

In other words, one of the components in (5.17), namely $\mathfrak{g}_1$, is Abelian.

Theorem 5.7. *A central decomposition of $\mathfrak{g}$, if it exists, can be obtained by choosing a complement $\mathfrak{g}_1$ of the intersection $C(\mathfrak{g}) \cap D(\mathfrak{g})$ in $C(\mathfrak{g})$, thus decomposing*

$$C(\mathfrak{g}) = \mathfrak{g}_1 \oplus C(\mathfrak{g}) \cap D(\mathfrak{g}). \tag{5.21}$$

The algebra $\mathfrak{g}_2$ is then chosen so that it satisfies

$$D(\mathfrak{g}) \subseteq \mathfrak{g}_2, \quad \mathfrak{g}_2 \cap C(\mathfrak{g}) \subseteq D(\mathfrak{g}). \tag{5.22}$$

Proof. Let us write a central decomposition of $\mathfrak{g}$ as

$$\mathfrak{g} = \tilde{\mathfrak{g}}_1 \oplus \tilde{\mathfrak{g}}_2, \quad [\tilde{\mathfrak{g}}_1, \mathfrak{g}] = 0, \quad [\tilde{\mathfrak{g}}_2, \tilde{\mathfrak{g}}_2] \subseteq \tilde{\mathfrak{g}}_2. \tag{5.23}$$

We have $\tilde{\mathfrak{g}}_1 \subseteq C(\mathfrak{g})$. Since $D(\mathfrak{g}) \subseteq \tilde{\mathfrak{g}}_2$ and $\tilde{\mathfrak{g}}_1 \cap \tilde{\mathfrak{g}}_2 = 0$ we find that $\tilde{\mathfrak{g}}_1 \cap D(\mathfrak{g}) = 0$. Therefore $C(\mathfrak{g}) \neq C(\mathfrak{g}) \cap D(\mathfrak{g})$ and any chosen complement $\mathfrak{g}_1$ of $C(\mathfrak{g}) \cap D(\mathfrak{g})$ in $C(\mathfrak{g})$ is a nonzero Abelian ideal commuting with $\mathfrak{g}$. Consequently any complement $\mathfrak{g}_2$ of $\mathfrak{g}_1$ in $\mathfrak{g}$ containing $D(\mathfrak{g})$ is also an ideal in $\mathfrak{g}$. We thus explicitly realize the decomposition $\mathfrak{g} = \mathfrak{g}_1 \oplus \mathfrak{g}_2$ and moreover $\mathfrak{g}_1 \sim C(\mathfrak{g})/C(\mathfrak{g}) \cap D(\mathfrak{g})$, i.e., this particular central decomposition is a maximal one. □

We mention that neither $\mathfrak{g}_1$ nor $\mathfrak{g}_2$ is unique, but both are unique up to a *central automorphism* of $\mathfrak{g}$, i.e., an automorphism α of $\mathfrak{g}$ over $\mathbb{F}$ for which

$$(\alpha - \mathbf{1}_{\mathfrak{g}})(\mathfrak{g}) \subseteq C(\mathfrak{g}). \tag{5.24}$$

Once the maximal central component of an algebra $\mathfrak{g}$ is removed, the remaining problem is the following one. Given a finite-dimensional Lie algebra $\mathfrak{g}$ satisfying

$$C(\mathfrak{g}) \subseteq D(\mathfrak{g}), \tag{5.25}$$

how does one decide whether $\mathfrak{g}$ is decomposable, and if it is, how does one obtain an algebraic decomposition? The task is facilitated by the following theorem.

Theorem 5.8. *Consider a Lie algebra $\mathfrak{g}$ of finite dimension n over the field $\mathbb{F}$, satisfying condition* (5.25). *The factor algebra $Q = A/J(A)$ of the centralizer $A = C_R(\operatorname{ad}\mathfrak{g})$ of the adjoint representation of $\mathfrak{g}$ in R (the algebra of all matrices in $\mathbb{F}^{n\times n}$) over its Jacobson radical $J(A)$, is Abelian.*

Proof. With no loss of generality we can assume that the ground field $\mathbb{F}$ is the field of complex numbers $\mathbb{C}$. Indeed, if $\mathfrak{g}$ is a real Lie algebra we can consider its complexification $\mathfrak{g}_{\mathbb{C}}$. The centers and derived algebras of the real algebra $\mathfrak{g}$ and its complexification $\mathfrak{g}_{\mathbb{C}}$ satisfy

$$D(\mathfrak{g}_{\mathbb{C}}) = \mathbb{C} \otimes D(\mathfrak{g}), \quad C(\mathfrak{g}_{\mathbb{C}}) = \mathbb{C} \otimes C(\mathfrak{g}),$$

i.e., also the complexification $\mathfrak{g}_{\mathbb{C}}$ satisfies the hypothesis (5.25).

Similarly, since $C_R(\operatorname{ad}\mathfrak{g}_{\mathbb{C}})$ is determined by linear conditions, we have $C_R(\operatorname{ad}\mathfrak{g}_{\mathbb{C}}) = \mathbb{C} \otimes C_R(\operatorname{ad}\mathfrak{g})$ and analogous relations hold also for the factor algebras Q, Q_0. Thus Q of the algebra $\mathfrak{g}$ is Abelian if and only if $Q_{\mathbb{C}}$ of the complexification $\mathfrak{g}_{\mathbb{C}}$ is Abelian.

From now on we assume that the algebra $\mathfrak{g}$ is over the field of complex numbers $\mathbb{C}$ and that the factor algebra Q of $C_{\mathbb{C}^{n\times n}}(\operatorname{ad}\mathfrak{g})$ over its Jacobson radical is noncommutative. It follows from Theorem 5.2 that Q contains a subalgebra Q_1, isomorphic to $\mathbb{C}^{2\times 2}$. Similarly as for the idempotents (cf. Lemma 5.3), matrix algebras can be lifted (see [138]), so that $C_{\mathbb{F}^{n\times n}}(\operatorname{ad}\mathfrak{g})$ itself contains a subalgebra Q_1, isomorphic to $\mathbb{C}^{2\times 2}$. In other words, Q_1 has four basis elements e_{ik} with the multiplication rule of the standard basis matrices:

$$e_{ik}e_{rs} = \delta_{kr}e_{is}, \quad i,k,r,s = 1,2. \tag{5.26}$$

We shall see that this assertion, following from the noncommutativity of Q, leads to a contradiction. Indeed, there holds the decomposition

$$\mathbf{1}_n = e_{11} + e_{22} + (\mathbf{1}_n - e_{11} - e_{22}) \tag{5.27}$$

of $\mathbf{1}_n$ into the sum of three orthogonal idempotents in A (where $\mathbf{1}_n - e_{11} - e_{22} = 0$ for $n = 2$.). It follows that the Lie algebra $\mathfrak{g}$ can be decomposed into the direct sum of three ideals:

$$\mathfrak{g} = \mathfrak{g}_1 \oplus \mathfrak{g}_2 \oplus \mathfrak{g}_3, \quad \mathfrak{g}_1 = e_{11}\mathfrak{g}, \quad \mathfrak{g}_2 = e_{22}\mathfrak{g}, \quad \mathfrak{g}_3 = (\mathbf{1}_n - e_{11} - e_{22})\mathfrak{g}, \tag{5.28}$$

of which at least the first two are not empty. Moreover, $\mathfrak{g}_1$ and $\mathfrak{g}_2$ are isomorphic and the mappings

$$\begin{gathered} \tilde{e}_{12}\colon \mathfrak{g}_2 \to \mathfrak{g}_1, \quad \tilde{e}_{21}\colon \mathfrak{g}_1 \to \mathfrak{g}_2, \\ \tilde{e}_{12}(x) = e_{12}x, \quad \tilde{e}_{21}(y) = e_{21}y, \quad (x \in \mathfrak{g}_2, y \in \mathfrak{g}_1) \end{gathered} \tag{5.29}$$

realize the isomorphism. In view of the assumptions made, $\mathfrak{g}_1$ and $\mathfrak{g}_2$ are non-Abelian. Hence $\mathfrak{g}_1$ contains two elements, x and y, for which $[x,y] \neq 0$. On the other hand, we have $[\mathfrak{g}_1, \mathfrak{g}_2] = 0$,

$$0 = [x, e_{21}y] = \operatorname{ad}(x)e_{21}(y) = e_{21}\operatorname{ad}(x)y = e_{21}[x,y].$$

Thus e_{21} applied to $[x,y]$ from the left is 0. But the restriction of e_{21} to $\mathfrak{g}_1$ is the isomorphism $\tilde{e}_{21}$ of $\mathfrak{g}_1$ on $\mathfrak{g}_2$, which implies that $[x,y] = 0$, a contradiction. Thus we have shown that Q is Abelian. □

Let us now assume that a basis of $A = C_R(\operatorname{ad}\mathfrak{g})$ is given: $(a_1 = \mathbf{1}_n, a_2, \dots, a_s)$, $\operatorname{Tr} a_j = 0$, $j \geq 2$. It is a task of linear algebra to transform to a new basis $\{x_1, \dots, x_\nu, b_1, \dots, b_\mu\}$, $\nu + \mu = s$, where $\{x_1, \dots, x_\nu\}$ form a basis of the Jacobson radical $J(A)$. The complementary basis elements $b_1, \dots, b_\mu$ give rise to a basis of the factor algebra $Q = A/J(A)$:

$$\bar{b}_i = b_i + J(A). \tag{5.30}$$

The basis of $J(A)$ is obtained in the form

$$x_i = \sum_{k=2}^{s} \xi_{ik}a_k, \quad i = 1, \dots, \nu, \tag{5.31}$$

where the s-tuples $\{\xi_{i1}, \dots, \xi_{is}\}$ form a solution basis of the system of linear homogeneous equations

$$\sum_{k=2}^{s} \xi_k \operatorname{Tr}(a_i a_k) = 0, \quad 2 \leq i \leq s, \tag{5.32}$$

for the unknowns $\xi_1, \dots, \xi_s$.

We choose the complementary basis elements $b_1, \ldots, b_\mu$ in such a way that

$$b_1 = \mathbf{1}_n, \quad \operatorname{Tr} b_i = 0, \quad 2 \le i \le \mu. \tag{5.33}$$

We can establish a trace function $\overline{\operatorname{Tr}}$ on the factor algebra $Q = A/J(A)$ by setting $\overline{\operatorname{Tr}}\, \bar{b}_i \equiv \operatorname{Tr} b_i$, $1 \le i \le \mu$ (because $J(A)$ consists of nilpotent matrices). We thus arrive at the traceless part of Q, namely

$$Q_0 = \{\bar{b}_2, \ldots, \bar{b}_\mu\}. \tag{5.34}$$

When the field $\mathbb{F}$ is the field of real numbers we distinguish between *absolutely indecomposable* Lie algebras and indecomposable, but *not absolutely indecomposable* ones. A real Lie algebra $\mathfrak{g}$ is defined to be *absolutely indecomposable* if and only if it is indecomposable and stays indecomposable after complexification. For complex Lie algebras the notions of indecomposability and absolute indecomposability are equivalent.

A convenient criterion of absolute indecomposability is given by the following theorem.

Theorem 5.9. *The finite-dimensional Lie algebra $\mathfrak{g}$ is absolutely indecomposable if and only if the traceless subalgebra A_0 of the centralizer algebra $A = C_R(\operatorname{ad} \mathfrak{g})$ of the adjoint representation of $\mathfrak{g}$ is closed under multiplication.*

Proof. It follows from Theorems 5.1 and 5.6 that the absolute indecomposability of $\mathfrak{g}$ is tantamount to the nonexistence of nontrivial idempotents in the complex algebra $\mathbb{C} \otimes A$. Since idempotents can be lifted modulo Jacobson radical (Lemma 5.3), the absolute indecomposability of $\mathfrak{g}$ is also equivalent to the nonexistence of nontrivial idempotents in the algebra $\mathbb{C} \otimes Q$, where $Q = A/J(A)$ is the factor algebra of A over its Jacobson radical $J(A)$.

From now on, without loss of generality, we assume that the ground field is the field of complex numbers. In view of Theorems 5.4 and 5.8 we have

$$Q = \bigoplus_{i=1}^{\mu} E_i, \quad \mu \in \mathbb{N}, \tag{5.35}$$

where $E_1, \ldots, E_\mu$ are isomorphic to the field $\mathbb{C}$.

The absolute indecomposability of $\mathfrak{g}$ is equivalent to the condition $\mu = 1$ in (5.33). That condition means that the traceless part Q_0 of Q vanishes. Consequently we have $A_0 = J(A)$, i.e., A_0 is closed under multiplication as claimed.

Conversely, let us assume that A_0 is closed under multiplication. Thus also Q_0 is closed. We wish to show that $\mu = 1$. Indeed, we have

$$\overline{\operatorname{Tr}}\ \mathbf{1}_{E_i} = n_i \tag{5.36}$$

for some natural numbers n_i and therefore

$$n_2 \mathbf{1}_{E_1} - n_1 \mathbf{1}_{E_2} \in Q_0. \tag{5.37}$$

By assumption we have

$$(n_2 \mathbf{1}_{E_1} - n_1 \mathbf{1}_{E_2})^2 = n_2^2 \mathbf{1}_{E_1} + n_1^2 \mathbf{1}_{E_2} \in Q_0, \tag{5.38}$$

but the evaluation of its trace leads to

$$\overline{\operatorname{Tr}}\, (n_2 \mathbf{1}_{E_1} - n_1 \mathbf{1}_{E_2})^2 = n_2^2 n_1 + n_1^2 n_2 > 0, \tag{5.39}$$

a contradiction with the definition of Q_0. Thus $\mu = 1$ and $\mathfrak{g}$ is absolutely indecomposable. □

The importance of Theorem 5.9 lies in the fact that it is not necessary to compute the radical $J(A)$ of the centralizer algebra A in order to apply it. However, computation of $J(A)$ is necessary for investigating indecomposability (as opposed to absolute indecomposability) of $\mathfrak{g}$ and to perform a decomposition when one exists.

In view of the fact that Q is an algebraic sum of finitely many fields and that idempotents can be lifted (Lemma 5.3), it follows that $\mathfrak{g}$ is *indecomposable* if and only if Q is a field. In case $\mathbb{F} = \mathbb{C}$ this condition is equivalent to $\mu = 1$, as stated above.

In case $\mathbb{F} = \mathbb{R}$, the condition is equivalent to either $\mu = 1$ or

$$\mu = 2, \quad \bar{b}_2^2 = \lambda \bar{b}_1, \quad \text{and} \quad \lambda < 0. \tag{5.40}$$

This is the only possibility for indecomposability which is not absolute.

For either field $\mathbb{F}$, the condition for indecomposability can be expressed in terms of a primitive element b of Q, i.e., an element of Q with minimal polynomial of degree μ. The *minimal polynomial* is the monic polynomial $m_{\bar{b}}$ of the lowest degree in t over $\mathbb{F}$ for which $m_{\bar{b}}(\bar{b}) = 0$. That such an element exists and can be formed with near 1 probability by random choice from the $2^\mu - 1$ linear combinations

$$\sum_{i=1}^{\mu} \varepsilon_i \bar{b}_i \neq 0 \quad (\varepsilon_i = 0 \text{ or } 1,\ 1 \leq i \leq \mu) \tag{5.41}$$

can be shown by the method established in [119].

With the help of a primitive element $\bar{b}$ we find $\mathfrak{g}$ to be indecomposable if and only if the minimal polynomial $m_{\bar{b}}(t) \in \mathbb{F}[t]$ of $\bar{b}$ is irreducible over $\mathbb{F}$. On the other hand, if $\mathfrak{g}$ is decomposable, then we know that there is a nontrivial factorization $m_{\bar{b}} = \bar{f}_1 \bar{f}_2$ of $m_{\bar{b}}$ in $\mathbb{F}[t]$ such that $\bar{f}_1$, $\bar{f}_2$ are nonconstant monic polynomials in t over $\mathbb{F}$.

It follows from the semisimplicity of the commutative algebra Q that the polynomials $\bar{f}_1$, $\bar{f}_2$ are mutually prime. We use this fact to construct a nontrivial idempotent of A in the following way.

The minimal polynomial m_b of a representative b of the residue class $\bar{b}$ modulo the nilpotent ideal $J(A)$ is of the form

$$m_b = f_1 f_2 \tag{5.42}$$

where $f_i = \gcd(m_b, \bar{f}_i^\nu)$, $i = 1, 2$, so that $\gcd(f_1, f_2) = \gcd(\bar{f}_1, \bar{f}_2) = 1$ (ν is the dimension of the Jacobson radical $J(A)$.)

Hence by the Euclidean division algorithm we find two polynomials P_1, P_2 in t over $\mathbb{F}$ satisfying

$$P_1 f_1 + P_2 f_2 = 1. \tag{5.43}$$

It follows that the matrices

$$e_i = P_i(b) f_i(b) \quad (i = 1, 2) \tag{5.44}$$

are two orthogonal idempotents of A. In other words

$$e_1 = P_1(b) f_1(b) \tag{5.45}$$

is a nontrivial idempotent of A, providing us with a nontrivial decomposition of $\mathfrak{g}$.

In case we want to avoid the use of a primitive element we have the following alternative.

Theorem 5.10. *Let $m_{\bar{b}_i}$ be the minimal polynomial of the basis element $\bar{b}_i$ of Q_0 over $\mathbb{F} = \mathbb{C}, \mathbb{R}$ $(1 < i \le \mu)$. If all $m_{\bar{b}_i}$ are irreducible over $\mathbb{F}$, then $\mathfrak{g}$ is indecomposable over $\mathbb{F}$. Otherwise, for some i, there holds a factorization*

$$m_{\bar{b}_i} = \bar{f}_1 \bar{f}_2 \tag{5.46}$$

of $m_{\bar{b}_i}$ into the product of two nonconstant monic polynomials $\bar{f}_1, \bar{f}_2$ in the variable t over $\mathbb{F}$. It follows that

$$\gcd(\bar{f}_1, \bar{f}_2) = 1. \tag{5.47}$$

Moreover, there holds the factorization

$$m_{b_i} = f_1 f_2, \quad f_j = \gcd(m_{b_i}, \bar{f}_j^{\nu}), \quad j = 1, 2, \tag{5.48}$$

of the minimal polynomial m_{b_i} over $\mathbb{F}$ into the product of two mutually prime nonconstant monic polynomials f_1, f_2 in t over $\mathbb{F}$. By the Euclidean division algorithm two polynomials P_1, P_2 in t over $\mathbb{F}$ are found satisfying

$$P_1 f_1 + P_2 f_2 = 1. \tag{5.49}$$

Now the matrix

$$P_1(b_i) f_1(b_i) \tag{5.50}$$

is a nontrivial idempotent of A providing a nontrivial decomposition of $\mathfrak{g}$.

Proof. The second part was already explained above for $\bar{b}$ in place of $\bar{b}_i$. In regard to the first part we remark that the irreducibility of all minimal polynomials $m_{\bar{b}_i}$ $(1 < i \le \mu)$ implies that

$$\bar{b}_i = \sum_{j=1}^{\rho} \bar{b}_{ij} \quad (\bar{b}_{ij} \in E_j, 1 \le j \le \rho), \tag{5.51}$$

where

$$m_{\bar{b}_i}(\bar{b}_{ij}) = 0 \quad (1 \le j \le \rho), \tag{5.52}$$

(no summation over i) and

$$Q = \oplus_{j=1}^{\rho} E_j, \quad \rho \le \mu \tag{5.53}$$

is the decomposition of Q into a direct sum of division rings, as implied by Theorems 5.2 and 5.8.

Hence the regular trace of $\bar{b}_j$ from E_j to $\operatorname{span}\{\mathbf{1}_{E_j}\}$ equals $\lambda \mathbf{1}_{E_j}$ with λ in $\mathbb{F}$ independent of j. As a matter of fact λ equals the product of the negative second highest coefficient of $m_{\bar{b}_i}$ and the quotient of the degree of E_j over $\mathbb{F}$ and the degree of $m_{\bar{b}_i}$. Hence

$$\overline{\operatorname{Tr}}\, \bar{b}_{ij} = n_j \lambda, \quad 0 = \overline{\operatorname{Tr}}\, \bar{b}_i = \sum_{j=1}^{\rho} n_j \lambda = \lambda \sum_{j=1}^{\rho} n_j, \quad \sum_{j=1}^{\rho} n_j \in \mathbb{N}, \tag{5.54}$$

so we have $\lambda = 0$. Thus it follows that every component $\bar{b}_{ij}$ has zero trace.

Since $\bar{b}_2, \ldots, \bar{b}_\mu$ provide a basis of the traceless subspace Q_0, it follows that for every element of Q_0 the trace of every component is zero. But if $\rho > 1$, then there is the element $n_2 \mathbf{1}_{E_1} - n_1 \mathbf{1}_{E_2}$ of Q_0 which has two components with nonvanishing trace, a contradiction.

It follows that $\rho = 1$, and $\mathfrak{g}$ is indecomposable. □

5.2. Algorithm

The results of the previous section can be summarized as an algorithm for establishing whether a Lie algebra $\mathfrak{g}$ of finite dimension n is decomposable or not. Moreover, if $\mathfrak{g}$ is decomposable, the algorithm will perform the decomposition.

Step 1. Remove the maximal central component of $\mathfrak{g}$, if one exists. This is done using Theorem 5.7 and more specifically equations (5.21) and (5.22). From now on assume $C(\mathfrak{g}) \subseteq D(\mathfrak{g})$.

Step 2. Determine the $n \times n$ matrices of the adjoint representation of $\mathfrak{g}$ and find the centralizer $A = C_R(\operatorname{ad} \mathfrak{g})$ of the adjoint representation in $R = \mathbb{F}^{n \times n}$. Choose a basis for A in the form $\{a_1 = \mathbf{1}_n, a_2, \dots, a_s\}$ with $\operatorname{Tr} a_i = 0$, $2 \le i \le s$. Denote by A_0 the traceless subset of A: $A_0 = \{a_2, \dots, a_s\}$.

Step 3. Determine whether $\mathfrak{g}$ is absolutely indecomposable by calculating the traces of the products $a_i a_k$, $2 \le i, k \le s$. The algebra $\mathfrak{g}$ is absolutely indecomposable if and only if

$$\operatorname{Tr} a_i a_k = 0, \quad 2 \le i, k \le s. \tag{5.55}$$

If $\mathbb{F} = \mathbb{C}$ and (5.55) holds, then $\mathfrak{g}$ is indecomposable. If (5.55) does not hold, or if $\mathbb{F} = \mathbb{R}$, proceed further.

Step 4. Determine the Jacobson radical $J(A)$ using the definition (5.7) (for $S = A$). Choose a basis $x_1, \dots, x_\nu$ for $J(A)$ and a complementary basis $b_1 = \mathbf{1}_n, \dots, b_\mu$ as in (5.33). If $\mathbb{F} = \mathbb{R}$ and $\mu = 2$, then verify whether the relation

$$b_2^2 = k\mathbf{1}_n \bmod J(A), \quad k < 0, \tag{5.56}$$

holds. If (5.56) does hold, then $\mathfrak{g}$ is indecomposable but not absolutely indecomposable. In all other cases the algebra $\mathfrak{g}$ is decomposable and we proceed to decompose it.

Step 5. Run through the basis elements $b_2, \dots, b_\mu$ until one is found with a reducible minimal polynomial as in (5.48). Call this element b_r. We now have a nonnilpotent traceless matrix b_r in $C_R(A)$. Using the invariant factors of b_r, or the rational roots theorem, and if necessary a more powerful factorization procedure, factor the minimal polynomial m_b into two mutually prime monic nonconstant polynomials f_1 and f_2, as in (5.48). We define the polynomials P_1, P_2 as in equation (5.49). The matrix

$$M = P_1(b_r) f_1(b_r) \tag{5.57}$$

is a nontrivial idempotent in $A/J(A)$, i.e.,

$$M^2 = M \bmod J(A). \tag{5.58}$$

Step 6. Perform a change of basis that diagonalizes M and realizes the decomposition of $\mathfrak{g}$. This is done using a matrix G obtained by placing the row space of M in its first r rows and the row space of $M - \mathbf{1}$ in the last $n - r$ rows. Thus columns $1, \dots, r$ and $r + 1, \dots, n$ of G^{-1} are the bases of the eigenspaces of M corresponding to the eigenvalues 1 and 0, respectively. The new basis of $\mathfrak{g}$ is then given as

$$e'_i = G_i^j e_j. \tag{5.59}$$

Step 7. We have decomposed $\mathfrak{g}$ into the direct sum of two algebras, $\mathfrak{g}_1$ and $\mathfrak{g}_2$. Repeat the algorithm, starting at Step 2, for each component and continue until we arrive at a decomposition into indecomposable components.

The algorithm described above is efficient if the dimension ν of $J(A)$ is not very large. If ν is large then a faster algorithm exists. We refer the reader to [102] for its formulation.

5.3. Examples

Let us demonstrate the whole procedure on several examples.

Example 5.11. First, we consider the algebras $\mathfrak{so}(4)$ and $\mathfrak{so}(1,3)$. It is well known that $\mathfrak{so}(4)$ is decomposable and $\mathfrak{so}(1,3)$ is indecomposable but not absolutely indecomposable. Let us derive these conclusions using the algorithm described above and also find an explicit decomposition.

The algebras $\mathfrak{so}(4)$ and $\mathfrak{so}(1,3)$ have the Lie brackets

$$\begin{aligned} &[x_1,x_2]=x_3, \quad &&[x_1,x_3]=-x_2, \quad &&[x_2,x_3]=x_1, \quad &&[x_1,x_5]=x_6,\\ &[x_1,x_6]=-x_5, &&[x_2,x_4]=-x_6, &&[x_2,x_6]=x_4, &&[x_3,x_4]=x_5,\\ &[x_3,x_5]=-x_4, &&[x_4,x_5]=\epsilon x_3, &&[x_4,x_6]=\epsilon x_2, &&[x_5,x_6]=\epsilon x_1 \end{aligned} \tag{5.60}$$

where $\epsilon=1$ for $\mathfrak{so}(4)$ and $\epsilon=-1$ for $\mathfrak{so}(1,3)$.

Step 1. The center of $\mathfrak{so}(4)$ and $\mathfrak{so}(1,3)$ vanishes, i.e., there is no central component.

Step 2. The centralizer $C_R(\operatorname{ad}\mathfrak{g})$ of the adjoint representation is

$$C_R(\operatorname{ad}\mathfrak{g})=\left\{\begin{pmatrix} u\mathbf{1}_3 & \epsilon v\mathbf{1}_3\\ v\mathbf{1}_3 & u\mathbf{1}_3\end{pmatrix}\,\middle|\, u,v\in\mathbb{F}\right\}. \tag{5.61}$$

We set

$$a_1=\mathbf{1}_6,\quad a_2=\begin{pmatrix} 0 & \epsilon\mathbf{1}_3\\ \mathbf{1}_3 & 0\end{pmatrix}. \tag{5.62}$$

We have $A_0=\operatorname{span}\{a_2\}$.

Step 3. We have $\operatorname{Tr} a_2^2=6\epsilon$. Therefore neither $\mathfrak{so}(4)$ nor $\mathfrak{so}(1,3)$ is absolutely indecomposable.

Step 4. The Jacobson radical $J\big(C_R(\operatorname{ad}\mathfrak{g})\big)$ vanishes and we can set $b_1=a_1$, $b_2=a_2$. We have

$$b_2^2=\epsilon b_1. \tag{5.63}$$

Consequently, $\mathfrak{so}(4)$ is decomposable both as a real and as a complex Lie algebra, whereas the real algebra $\mathfrak{so}(1,3)$ is indecomposable but not absolutely indecomposable.

We proceed to decompose $\mathfrak{so}(4)$ and the complexification $\mathfrak{so}_{\mathbb{C}}(1,3)$ of $\mathfrak{so}(1,3)$.

Step 5. The matrix b_2 has the reducible minimal polynomial

$$m_{b_2}(t)=t^2-\epsilon=\begin{cases}(t-1)(t+1) & \text{when } \epsilon=1,\\ (t-\mathrm{i})(t+\mathrm{i}) & \text{when } \epsilon=-1.\end{cases} \tag{5.64}$$

We consider the two cases separately.

When $\epsilon = 1$ we set $f_1(t) = t-1$ and $f_2(t) = t+1$. Solving (5.49) for polynomials P_1 and P_2 we find a particular solution $P_1(t) = \frac{1}{2}t$ and $P_2(t) = 1 - \frac{1}{2}t$. Thus we have

$$M_{\epsilon=1} = P_1(b_2)f_1(b_2) = \frac{1}{2}\begin{pmatrix} 1 & 0 & 0 & -1 & 0 & 0 \\ 0 & 1 & 0 & 0 & -1 & 0 \\ 0 & 0 & 1 & 0 & 0 & -1 \\ -1 & 0 & 0 & 1 & 0 & 0 \\ 0 & -1 & 0 & 0 & 1 & 0 \\ 0 & 0 & -1 & 0 & 0 & 1 \end{pmatrix}. \tag{5.65}$$

When $\epsilon = -1$ we set $f_1(t) = t - \mathrm{i}$ and $f_2(t) = t + \mathrm{i}$. Solving (5.49) we find $P_1(t) = \frac{1}{2}\mathrm{i}$ and $P_2(t) = -\frac{1}{2}\mathrm{i}$. Thus we have

$$M_{\epsilon=-1} = \frac{1}{2}\begin{pmatrix} 1 & 0 & 0 & -\mathrm{i} & 0 & 0 \\ 0 & 1 & 0 & 0 & -\mathrm{i} & 0 \\ 0 & 0 & 1 & 0 & 0 & -\mathrm{i} \\ \mathrm{i} & 0 & 0 & 1 & 0 & 0 \\ 0 & \mathrm{i} & 0 & 0 & 1 & 0 \\ 0 & 0 & \mathrm{i} & 0 & 0 & 1 \end{pmatrix} \tag{5.66}$$

Step 6. We find changes of bases diagonalizing the matrices (5.65), (5.66), respectively. We can choose the following new bases

$$\begin{aligned} \epsilon = 1: \quad & x'_1 = \tfrac{1}{2}(x_6 - x_3), \quad x'_2 = \tfrac{1}{2}(x_5 - x_2), \quad x'_3 = \tfrac{1}{2}(x_4 - x_1), \\ & x'_4 = \tfrac{1}{2}(x_2 + x_5), \quad x'_5 = \tfrac{1}{2}(x_1 + x_4), \quad x'_6 = \tfrac{1}{2}(x_3 + x_6), \\ \epsilon = -1: \quad & x'_1 = \tfrac{1}{2}(x_3 + \mathrm{i}x_6), \quad x'_2 = \tfrac{1}{2}(x_2 + \mathrm{i}x_5), \quad x'_3 = \tfrac{1}{2}(x_1 + \mathrm{i}x_4), \\ & x'_4 = \tfrac{1}{2}(\mathrm{i}x_6 - x_3), \quad x'_5 = \tfrac{1}{2}(\mathrm{i}x_5 - x_2), \quad x'_6 = \tfrac{1}{2}(\mathrm{i}x_4 - x_1). \end{aligned} \tag{5.67}$$

In these new bases the algebras $\mathfrak{so}(4)$ and $\mathfrak{so}_{\mathbb{C}}(1,3)$, respectively, split explicitly into direct sums of simple subalgebras spanned by x'_1, x'_2, x'_3 and x'_4, x'_5, x'_6. We have

$$\begin{aligned} &[x'_1, x'_2] = x'_3, \quad [x'_1, x'_3] = -x'_2, \quad [x'_2, x'_3] = x'_1, \\ &[x'_4, x'_5] = -x'_6, \quad [x'_4, x'_6] = x'_5, \quad [x'_5, x'_6] = -x'_4 \end{aligned} \tag{5.68}$$

when $\epsilon = 1$, and

$$\begin{aligned} &[x'_1, x'_2] = -x'_3, \quad [x'_1, x'_3] = x'_2, \quad [x'_2, x'_3] = -x'_1, \\ &[x'_4, x'_5] = x'_6, \quad [x'_4, x'_6] = -x'_5, \quad [x'_5, x'_6] = x'_4 \end{aligned} \tag{5.69}$$

for $\epsilon = -1$.

Notice that these new bases are not unique, any bases of the row spaces of the matrices M and $M - \mathbf{1}$ can be used. Consequently, the Lie brackets of the summands can be obtained in another form when the same procedure is repeated.

Step 7. The subalgebras arising in the decompositions (5.68) and (5.69) are simple, i.e., further indecomposable.

Thus, we have decomposed the algebra $\mathfrak{so}(4)$ into a direct sum of two subalgebras, both of which turn out to be isomorphic to $\mathfrak{so}(3)$ upon further inspection (e.g., calculating the signatures of their Killing forms). We have

$$\mathfrak{so}(4) = \mathfrak{so}(3) \oplus \mathfrak{so}(3). \tag{5.70}$$

Similarly, we have found that the algebra $\mathfrak{so}(1,3)$ is indecomposable but not absolutely indecomposable. Its complexification $\mathfrak{so}_{\mathbb{C}}(1,3)$ decomposes into a direct sum of two three-dimensional simple Lie algebras. Any such algebra is isomorphic to $\mathfrak{sl}(2,\mathbb{C})$, i.e., we have

$$\mathfrak{so}_{\mathbb{C}}(1,3) = \mathfrak{sl}(2,\mathbb{C}) \oplus \mathfrak{sl}(2,\mathbb{C}). \tag{5.71}$$

Example 5.12. As a second example let us consider a solvable 6-dimensional Lie algebra $\mathfrak{g}$ with the Lie brackets

$$[x_2, x_3] = x_2 - x_1 + x_6, \quad [x_2, x_4] = x_2 - x_1 + x_6, \quad [x_4, x_5] = x_1 - x_6. \tag{5.72}$$

We have the following dimensions of the ideals in the characteristic series

$$\mathrm{US} = [2,4], \quad \mathrm{CS} = [6,2,1], \quad \mathrm{DS} = [6,2,0]. \tag{5.73}$$

The derived algebra is spanned by $x_1 - x_6$ and $x_2 - x_1 + x_6$ whereas the two-dimensional center is

$$C(\mathfrak{g}) = \mathrm{span}\{x_1, x_6\}. \tag{5.74}$$

Thus the algebra $\mathfrak{g}$ has a one-dimensional central component, which can be chosen as $\mathrm{span}\{x_6\}$. We have a decomposition

$$\mathfrak{g} = \mathfrak{g}_1 \oplus \mathfrak{g}_2, \qquad \mathfrak{g}_1 = \mathrm{span}\{x_6\}, \quad \mathfrak{g}_2 = \mathrm{span}\{\tilde{x}_1, \tilde{x}_2, \tilde{x}_3, \tilde{x}_4, \tilde{x}_5\} \tag{5.75}$$

where we have defined

$$\tilde{x}_1 = x_1 - x_6, \quad \tilde{x}_2 = x_2, \quad \tilde{x}_3 = x_3, \quad \tilde{x}_4 = x_4, \quad \tilde{x}_5 = x_5. \tag{5.76}$$

Notice that the characteristic series of the subalgebra $\mathfrak{g}_2$ have dimensions

$$\mathrm{US} = [1,3], \quad \mathrm{CS} = [5,2,1], \quad \mathrm{DS} = [5,2,0]. \tag{5.77}$$

No such algebra appears in the list of indecomposable 5-dimensional Lie algebras in Chapter 18; thus it must be decomposable. We proceed to demonstrate this using our algorithm of Section 5.2 and to find the actual decomposition.

The algebra $\mathfrak{g}_2$ has no central component. The centralizer $C_R(\operatorname{ad}\mathfrak{g}_2)$ of the adjoint representation is

$$A = C_R(\operatorname{ad}\mathfrak{g}_2) = \left\{ \left. \begin{pmatrix} u & u-v & w & x & y \\ 0 & v & 0 & 0 & 0 \\ 0 & 0 & v & v-u & 0 \\ 0 & 0 & 0 & u & 0 \\ 0 & 0 & 0 & 0 & u \end{pmatrix} \right| u, v, w, x, y \in \mathbb{F} \right\}. \tag{5.78}$$

The traceless part A_0 of A consists of matrices as in (5.78) with $3u + 2v = 0$. The condition (5.55) does not hold; thus, $\mathfrak{g}_2$ is not indecomposable over the field $\mathbb{C}$. The Jacobson radical is

$$J(A) = \left\{ \left. \begin{pmatrix} 0 & 0 & w & x & y \\ 0 & 0 & 0 & 0 & 0 \\ 0 & 0 & 0 & 0 & 0 \\ 0 & 0 & 0 & 0 & 0 \\ 0 & 0 & 0 & 0 & 0 \end{pmatrix} \right| w, x, y \in \mathbb{F} \right\}. \tag{5.79}$$

We choose the basis of the complement of $J(a)$ in A in the form

$$b_1 = \mathbf{1}_5, \quad b_2 = \begin{pmatrix} 2 & 5 & 0 & 0 & 1 \\ 0 & -3 & 0 & 0 & 0 \\ 0 & 0 & -3 & -5 & 0 \\ 0 & 0 & 0 & 2 & 0 \\ 0 & 0 & 0 & 0 & 2 \end{pmatrix}. \tag{5.80}$$

We have $\operatorname{Tr} b_2^2 = 30$, i.e., the traceless part A_0 is not an (associative) subalgebra of A and the algebra $\mathfrak{g}_2$ is decomposable, as we already concluded using results of Chapter 18.

The minimal polynomial

$$m_{b_2}(t) = t^3 - t^2 - 8t + 12 \tag{5.81}$$

of the matrix b_2 is reducible, $m_{b_2}(t) = f_1(t) f_2(t)$ with

$$f_1(t) = t + 3, \quad f_2(t) = (t-2)^2. \tag{5.82}$$

We find the polynomials

$$P_1(t) = \frac{1}{25}(7 - t), \quad P_2(t) = \frac{1}{25} \tag{5.83}$$

such that (5.49) holds. The corresponding idempotent in $C_R(\operatorname{ad} \mathfrak{g}_2)$ is

$$M = P_1(b_2) f_1(b_2) = \begin{pmatrix} 1 & 1 & 0 & 0 & 0 \\ 0 & 0 & 0 & 0 & 0 \\ 0 & 0 & 0 & -1 & 0 \\ 0 & 0 & 0 & 1 & 0 \\ 0 & 0 & 0 & 0 & 1 \end{pmatrix}. \tag{5.84}$$

A change of basis diagonalizing the matrix M is

$$x_1' = \tilde{x}_1, \quad x_2' = \tilde{x}_4 - \tilde{x}_3, \quad x_3' = \tilde{x}_5, \quad x_4' = \tilde{x}_2 - \tilde{x}_1, \quad x_5' = \tilde{x}_3. \tag{5.85}$$

The vector spaces $\operatorname{span}\{x_1', x_2', x_3'\}$ and $\operatorname{span}\{x_4', x_5'\}$ are ideals in $\mathfrak{g}_2$. The Lie brackets are

$$[x_2', x_3'] = x_1', \quad [x_4', x_5'] = x_4'. \tag{5.86}$$

Thus the algebra $\mathfrak{g}_2$ decomposes into a direct sum of the Heisenberg algebra $\mathfrak{h}(1) = \mathfrak{n}_{3,1}$ and the solvable two-dimensional algebra $\mathfrak{s}_{2,1}$.

Altogether we have found that a change of basis

$$\begin{aligned} x_1' &= x_1 - x_6, & x_2' &= x_4 - x_3, & x_3' &= x_5, \\ x_4' &= x_2 - x_1 + x_6, & x_5' &= x_3, & x_6' &= x_6 \end{aligned} \tag{5.87}$$

performs an explicit decomposition of the six-dimensional solvable algebra $\mathfrak{g}$ into a direct sum of three algebras

$$\mathfrak{g} = \mathfrak{h}(1) \oplus \mathfrak{s}_{2,1} \oplus \mathfrak{a}_1, \tag{5.88}$$

where

$$\mathfrak{h}(1) = \operatorname{span}\{x_1 - x_6, x_4 - x_3, x_5\}, \quad \mathfrak{s}_{2,1} = \operatorname{span}\{x_3, x_2 - x_1 + x_6\}, \quad \mathfrak{a}_1 = \operatorname{span}\{x_6\}.$$

Notice that the polynomial $m_{b_2}(t)$ depends on the choice of the matrix b_2. Had we chosen instead of b_2 a matrix b_2' with vanishing $(1,5)$-entry, the minimal polynomial would be $m_{b_2'}(t) = t^2 + t - 6 = (t+3)(t-2)$ and the rest of the computation, e.g., the factors f_1, f_2, would be different. Nevertheless, the decomposition into subalgebras $\operatorname{span}\{\tilde{x}_1, \tilde{x}_4 - \tilde{x}_3, \tilde{x}_5\}$ and $\operatorname{span}\{\tilde{x}_2 - \tilde{x}_1, \tilde{x}_3\}$ would be the same.

CHAPTER 6

Levi Decomposition. Identification of the Radical and Levi Factor

The Levi theorem 2.12 in its full generality states that for an arbitrary finite-dimensional Lie algebra $\mathfrak{g}$ over a field $\mathbb{F}$ of characteristic zero it is possible to construct a semidirect sum decomposition

$$\mathfrak{g} = \mathfrak{p} \niplus \mathfrak{r}. \tag{6.1}$$

Here $\mathfrak{r} \equiv \mathrm{R}(\mathfrak{g})$ is the *radical* of $\mathfrak{g}$, i.e., its maximal solvable ideal, and $\mathfrak{p}$ is a semisimple Lie algebra. The radical is unique. The complement $\mathfrak{p}$, i.e., the *Levi factor*, is isomorphic to the factor algebra $\mathfrak{g}/\mathfrak{r}$ and is unique up to automorphisms of $\mathfrak{g}$.

In other words, given a basis of $\mathfrak{g}$, say $(x_1, \ldots, x_n)$, it is always possible to find a new basis

$$\{s_1, s_2, \ldots, s_\sigma, r_1, r_2, \ldots, r_\rho\}, \quad \sigma + \rho = n, \tag{6.2}$$

such that $\mathfrak{r} = \operatorname{span}\{r_1, \ldots, r_\rho\}$, $\mathfrak{p} = \operatorname{span}\{s_1, \ldots, s_\sigma\}$, and the commutation relations are such that

$$[\mathfrak{p}, \mathfrak{p}] = \mathfrak{p}, \quad [\mathfrak{r}, \mathfrak{r}] \subsetneq \mathfrak{r}, \quad [\mathfrak{p}, \mathfrak{r}] \subseteq \mathfrak{r}. \tag{6.3}$$

The question that we address here is: how does one find a convenient change of basis that realizes the Levi decomposition? Notice that a Levi decomposition can be performed for both decomposable and indecomposable Lie algebras. From the point of view of identifying a Lie algebra $\mathfrak{g}$, it is usually preferable to first perform a direct decomposition into indecomposable components and then construct a Levi decomposition for each component.

Since the factor algebra $\mathfrak{g}/\mathfrak{r} \cong \mathfrak{p}$ is semisimple and hence *perfect*, i.e., its derived algebra $D(\mathfrak{p})$ satisfies $D(\mathfrak{p}) = \mathfrak{p}$, we have

$$D(\mathfrak{g}) + \mathfrak{r} = \mathfrak{g}. \tag{6.4}$$

From (6.4) we obtain the isomorphism

$$D(\mathfrak{g})/[D(\mathfrak{g}) \cap \mathfrak{r}] \cong \mathfrak{g}/\mathfrak{r} \tag{6.5}$$

and hence $\dim(D(\mathfrak{g})/[D(\mathfrak{g}) \cap \mathfrak{r}]) = n - \rho = \sigma$.

The problem of obtaining a Levi decomposition is one of linear algebra. We shall address it in two ways.

6.1. Original algorithm

First, we present an algorithm as given in [102]. Its essence is a reduction of the general case to the case with an Abelian radical. We proceed in four steps.

Step 1. Find the radical $\mathfrak{r} = \mathrm{R}(\mathfrak{g})$. This is a simple task of linear algebra, since we can use the property [59, 139]

$$\mathrm{R}(\mathfrak{g}) = \{x \in \mathfrak{g} \mid K(x,y) = 0,\ \forall y \in D(\mathfrak{g})\}, \tag{6.6}$$

where $K(x,y)$ is the Killing form (2.31)

$$K(x,y) = \mathrm{Tr}\big(\mathrm{ad}(x)\,\mathrm{ad}(y)\big).$$

If $\mathfrak{g} = \mathrm{R}(\mathfrak{g})$, then $\mathfrak{g}$ is solvable and $\mathfrak{p} = 0$ in (6.1). If $\mathrm{R}(\mathfrak{g}) = 0$, then $\mathfrak{g} = \mathfrak{p}$ is semisimple. In both cases the Levi decomposition is trivial.

Step 2. If $0 \neq \mathfrak{p} \neq \mathfrak{g}$, we calculate the derived series (2.5) of $\mathfrak{g}$:

$$\mathfrak{g}^{(1)} = [\mathfrak{g},\mathfrak{g}], \dots, \mathfrak{g}^{(k+1)} = \big[\mathfrak{g}^{(k)}, \mathfrak{g}^{(k)}\big], \dots$$

till we arrive, after a finite number of steps, at a perfect Lie algebra

$$\mathfrak{g}^{(k+1)} = \mathfrak{g}^{(k)}, \quad \mathfrak{g}^{(k)} \neq \mathfrak{g}^{(k-1)}. \tag{6.7}$$

If we know the Levi decomposition of $\mathfrak{g}^{(k)}$, i.e.,

$$\mathfrak{g}^{(k)} = \mathfrak{p} \niplus R\big(\mathfrak{g}^{(k)}\big) \tag{6.8}$$

then we obtain the Levi decomposition of $\mathfrak{g}$ by extending the basis $(r_1, \dots, r_\tau)$ of $R\big(\mathfrak{g}^{(k)}\big)$ to a basis of $\mathrm{R}(\mathfrak{g})$: $(r_1, \dots, r_\tau, r_{\tau+1}, \dots, r_\rho)$.

Step 3. From now on we assume that $\mathfrak{g}$ is perfect which also implies that the radical of $\mathfrak{g}$ is nilpotent (since $D(\mathfrak{g}) \subseteq \mathfrak{p} \niplus NR(\mathfrak{g})$). We consider two cases separately, namely that of an Abelian and a non-Abelian radical, respectively. If $\mathrm{R}(\mathfrak{g})$ is Abelian, the problem is a simple task of linear algebra, solved below in Step 4. Let us first reduce the case of a non-Abelian radical to that of an Abelian one. We have

$$\mathfrak{g} = D(\mathfrak{g}), \quad D\big(\mathrm{R}(\mathfrak{g})\big) \neq 0. \tag{6.9}$$

We note that $D\big(\mathrm{R}(\mathfrak{g})\big)$ is an ideal in $\mathfrak{g}$.

Let us choose a basis for $\mathfrak{g}$ in the form

$$(e_1, \dots, e_\mu, r_1, \dots, r_\nu, x_1, \dots, x_\sigma), \qquad \mu + \nu = \rho \tag{6.10}$$

where $(e_1, \dots, e_\mu)$ is a basis for $D\big(\mathrm{R}(\mathfrak{g})\big)$, $(r_1, \dots, r_\nu)$ for a complement of $D\big(\mathrm{R}(\mathfrak{g})\big)$ in $\mathrm{R}(\mathfrak{g})$, and $(x_1, \dots, x_\sigma)$ for a complement of $\mathrm{R}(\mathfrak{g})$ in $\mathfrak{g}$. Next, we construct the factor algebra $\bar{\mathfrak{g}} = \mathfrak{g}/D\big(\mathrm{R}(\mathfrak{g})\big)$ satisfying $\bar{\mathfrak{g}} = D(\bar{\mathfrak{g}})$ and $D\big(\mathrm{R}(\bar{\mathfrak{g}})\big) = 0$. We have $\dim \bar{\mathfrak{g}} = n - \mu < n$ and the commutation relations for $\bar{\mathfrak{g}}$ are obtained by setting $e_1 = \dots = e_\mu = 0$ in the commutation relations for $\mathfrak{g}$. Using the method described below (for Abelian radicals), we obtain the Levi decomposition

$$\bar{\mathfrak{g}} = \bar{\mathfrak{p}} \niplus \mathrm{R}(\bar{\mathfrak{g}}) \tag{6.11}$$

of $\bar{\mathfrak{g}}$. From the residue classes in $\bar{\mathfrak{p}}$ we construct their representative elements in $\mathfrak{g}$ and denote by $\mathfrak{p}_1$ the vector space spanned by them. We obtain a proper subalgebra of $\mathfrak{g}$, namely

$$\mathfrak{g}_1 = D\big(\mathrm{R}(\mathfrak{g})\big) + \mathfrak{p}_1, \tag{6.12}$$

satisfying $\dim \mathfrak{g}_1 = n - \nu < n$. If $D\big(\mathrm{R}(\mathfrak{g})\big)$ is Abelian, we find a Levi decomposition of $\mathfrak{g}_1$; if not, we repeat Steps 2 and 3 until, after a finite number of steps, we arrive at an algebra $\mathfrak{g}_k$ with an Abelian radical. For it we obtain

$$\mathfrak{g}_k = \mathfrak{p} \niplus \mathrm{R}(\mathfrak{g}_k). \tag{6.13}$$

The Levi decomposition of $\mathfrak{g}$ is then (6.1) with $\mathfrak{p}$ as in (6.13).

Step 4. Finally, let us consider the case of an Abelian radical. We have $\mu = 0$ in (6.10), and the commutation relations for $\mathfrak{g}$ are

$$[x_i, x_k] = c_{ik}{}^l x_l + f_{ik}{}^q r_q, \quad [r_p, r_q] = 0, \quad [x_i, r_p] = h_{ip}{}^q r_q, \qquad 1 \le i, k, l \le \sigma, \quad 1 \le p, q \le \rho. \tag{6.14}$$

Summation over repeated indices over the appropriate ranges is understood throughout.

We now replace the basis elements x_i by

$$s_i = x_i + b_i^p r_p \tag{6.15}$$

and require that the s_i form a Lie algebra:

$$[s_i, s_k] = c_{ik}{}^l s_l. \tag{6.16}$$

Equations (6.14) – (6.16) imply that the unknown coefficients b_i^p must satisfy a system of inhomogeneous linear equations

$$c_{ik}{}^l b_l^q - h_{ip}{}^q b_k^p + h_{kp}{}^q b_i^p = f_{ik}{}^q, \quad 1 \le p, q \le \rho, \quad 1 \le i, k \le \sigma. \tag{6.17}$$

The system (6.17) involves $\rho\sigma$ unknowns b_i^p and $\frac{1}{2}\sigma(\sigma - 1)\rho/2$ equations. It follows from the Levi theorem that the system is compatible and that a solution b_i^p can always be found. It is actually possible to solve (6.17) explicitly and analytically, making use of the second order Casimir operator of the factor algebra $\mathfrak{p} = \mathfrak{g}/\mathfrak{r}$ [139]. From the computational point of view it is usually preferable to solve (6.17) directly in each specific case.

6.2. Modified algorithm

A streamlined alternative formulation of the algorithm combines Steps 3 and 4 into one. Its basic idea is a construction of a proper subalgebra of the Lie algebra $\mathfrak{g}$ which contains $\mathfrak{p}$, without constructing the factor algebra $\mathfrak{g}/D\big(\mathrm{R}(\mathfrak{g})\big)$. Repeating the procedure finitely many times one obtains a subalgebra of $\mathfrak{g}$ which has vanishing radical, i.e., coincides with $\mathfrak{p}$. From the computational point of view it is equivalent to the original algorithm.

We observe that $D(\mathfrak{r}) \equiv \mathfrak{r}^2$ is a characteristic ideal of $\mathfrak{g}$ and consequently the subalgebra $\tilde{\mathfrak{g}} = \mathfrak{p} \dotplus D(\mathfrak{r})$ of $\mathfrak{g}$ is a Levi decomposable algebra with the Levi factor $\mathfrak{p}$ and radical $\mathfrak{r}^2$. We shall proceed to construct its basis.

Let us suppose that we have a basis of $\mathfrak{g}$ of the form (6.10). Thus we have the following particular Lie brackets

$$[x_i, x_j] = c_{ij}{}^k x_k + d_{ij}{}^p r_p + f_{ij}{}^l e_l, \tag{6.18}$$

$$[x_i, r_p] = g_{ip}{}^q r_q + h_{ip}{}^m e_m. \tag{6.19}$$

Summation over repeated indices $k = 1, \dots, \sigma$, $l, m = 1, \dots, \mu$ and $p, q = 1, \dots, \nu$ applies throughout.

A basis of $\tilde{\mathfrak{g}} = \mathfrak{p} \dotplus D(\mathfrak{r})$ can be without loss of generality chosen in the form

$$\{e_1, \dots, e_\mu, \hat{x}_1, \dots, \hat{x}_\sigma\} \tag{6.20}$$

where $\hat{x}_k \in \operatorname{span}\{x_k, r_1, \dots, r_\nu\}$, i.e.,

$$\hat{x}_j = x_j + b_j^p r_p. \tag{6.21}$$

The span of the set (6.20) is by construction complementary to $\text{span}\{r_1, \dots, r_\nu\}$. Therefore it forms a basis of $\mathfrak{p} \dotplus D(\mathfrak{r})$ if and only if it closes under the Lie bracket. That is always true for commutators of the type $[\hat{x}_i, e_l]$ since $D(\mathfrak{r})$ is an ideal in $\mathfrak{g}$. It remains to satisfy

$$[\hat{x}_i, \hat{x}_j] = c_{ij}{}^k \hat{x}_k + \widehat{f_{ij}}^l e_l \tag{6.22}$$

for some constants $\widehat{f_{ij}}^l$. Notice that the structure constants $c_{jk}{}^k$ are the same in (6.18) and (6.22) because they are the structure constants of the semisimple factor algebra $\mathfrak{g}/\mathfrak{r}$ in the same basis $x_j \bmod \mathfrak{r} = \hat{x}_j \bmod \mathfrak{r}$, $j = 1, \dots, \sigma$.

Substituting (6.21) into (6.22) and dropping any term proportional to e_l we obtain the following set of equations, to be satisfied for all $1 \leq i < j \leq \sigma$

$$d_{ij}{}^q r_q + b_j^p g_{ip}{}^q r_q - b_i^p g_{jp}{}^q r_q = c_{ij}{}^k b_k^q r_q, \tag{6.23}$$

i.e., a set of $\frac{1}{2}\sigma(\sigma - 1)\nu$ linear inhomogeneous equations

$$g_{jp}{}^q b_i^p - g_{ip}{}^q b_j^p + c_{ij}{}^k b_k^q = d_{ij}{}^q \tag{6.24}$$

for $\sigma\nu$ unknowns b_i^p. Due to the Levi theorem the set of equations (6.24) always has a solution. Thus, once we find any particular solution of (6.24) we have a basis of $\tilde{\mathfrak{g}} = \mathfrak{p} \dotplus D(\mathfrak{r})$. We repeat the procedure until we arrive at $k \in \mathbb{N}$ such that

$$\mathfrak{r}^{(k)} = 0.$$

Notice that it is advantageous to suppose that $\mathfrak{g}$ is perfect but the streamlined version of the procedure does not depend on it.

Remark 6.1. An analogous procedure was given in [35] where the lower central series of the nilpotent radical $\mathfrak{r}$ of a perfect Lie algebra $\mathfrak{g}$ was used instead of the derived series employed here. Consequently, the procedure has to be repeated more times but the arising systems of equations (6.24) are smaller and thus easier to solve. However, a comparable computational simplification can be achieved in our algorithm by simply using a basis of the nilpotent radical $\mathfrak{r}$ which respects the lower central series, as in (9.19) in Section 9.3 below.

6.3. Examples

In order to illuminate the algorithms outlined above we conclude this chapter by two examples.

Example 6.2. Let us consider the finite dimensional part of the algebra of infinitesimal point symmetries of the heat equation [86]. It is spanned by the following vector fields in $\mathbb{R}^3$ with coordinates t, x, u

$$\begin{aligned} &Y_1 = 4t^2\partial_t + 4xt\partial_x - (2t + x^2)u\partial_u, && Y_2 = 4t\partial_t + 2x\partial_x, \\ &Y_3 = \partial_t, && Y_4 = -2t\partial_x + xu\partial_u, \\ &Y_5 = u\partial_u, && Y_6 = \partial_x. \end{aligned} \tag{6.25}$$

Evaluation of the commutators gives the following Lie brackets of an abstract Lie algebra (with y_i replacing the vector fields Y_i)

$$\begin{aligned} &[y_1, y_2] = -4y_1, && [y_1, y_3] = -2y_2 + 2y_5, && [y_1, y_6] = 2y_4, \\ &[y_2, y_3] = -4y_3, && [y_2, y_4] = 2y_4, && [y_2, y_6] = -2y_6, \\ &[y_3, y_4] = -2y_6, && [y_4, y_6] = -y_5. && \end{aligned} \tag{6.26}$$

In Steps 1 and 2 we find that the Lie algebra (6.26) is perfect and its radical is spanned by y_4, y_5, y_6. The radical (equal to the nilradical) is the Heisenberg algebra $\mathfrak{h}(1) \equiv \mathfrak{n}_{3,1}$. The derived algebra of the radical is spanned by y_5. The basis of the form (6.10) can be chosen as

$$e_1 = y_5, \quad r_1 = y_4, \quad r_2 = y_6, \quad x_1 = y_1, \quad x_2 = y_2, \quad x_3 = y_3 \tag{6.27}$$

with the Lie brackets

$$\begin{aligned} &[r_1, r_2] = -e_1, \quad [r_1, x_2] = -2r_1, \quad [r_1, x_3] = 2r_2, \quad &&[r_2, x_1] = -2r_1, \\ &[r_2, x_2] = 2r_2, \quad [x_1, x_2] = -4x_1, \quad [x_1, x_3] = 2e_1 - 2x_2, \quad &&[x_2, x_3] = -4x_3. \end{aligned} \tag{6.28}$$

Step 3 of the procedure now requires the construction of a 5-dimensional Lie algebra $\mathfrak{g}/\operatorname{span}\{e_1\}$ with an Abelian radical. Its Lie brackets are

$$\begin{aligned} &[\tilde{r}_1, \tilde{x}_2] = -2\tilde{r}_1, \quad [\tilde{r}_1, \tilde{x}_3] = 2\tilde{r}_2, \quad [\tilde{r}_2, \tilde{x}_1] = -2\tilde{r}_1, \quad [\tilde{r}_2, \tilde{x}_2] = 2\tilde{r}_2, \\ &[\tilde{x}_1, \tilde{x}_2] = -4\tilde{x}_1, \quad [\tilde{x}_1, \tilde{x}_3] = -2\tilde{x}_2, \quad [\tilde{x}_2, \tilde{x}_3] = -4\tilde{x}_3 \end{aligned} \tag{6.29}$$

where tildes were used to distinguish residue classes in $\mathfrak{g}/\operatorname{span}\{e_1\}$ from the corresponding elements in $\mathfrak{g}$. Its (Abelian) radical is spanned by $\tilde{r}_1$ and $\tilde{r}_2$.

In Step 4 the Levi factor of $\mathfrak{g}/\operatorname{span}\{e_1\}$ is obtained by an explicit solution of (6.17) or by inspection; it can be chosen to be the span of $\tilde{x}_1, \tilde{x}_2, \tilde{x}_3$.

Thus, we have $\mathfrak{p} \dotplus D(R(\mathfrak{g})) = \operatorname{span}\{e_1, x_1, x_2, x_3\}$. The same conclusion is immediate in our modified algorithm since e_1, x_1, x_2, x_3 span a subalgebra of $\mathfrak{g}$ by an inspection of (6.28).

Next, we construct the Levi decomposition of $\mathfrak{p} \dotplus D(R(\mathfrak{g})) = \operatorname{span}\{e_1, x_1, x_2, x_3\}$ with the nonvanishing Lie brackets

$$[x_1, x_2] = -4x_1, \quad [x_1, x_3] = 2e_1 - 2x_2, \quad [x_2, x_3] = -4x_3. \tag{6.30}$$

This step coincides in both versions of the algorithm. The solution of the set of linear equations (6.17) or, equivalently, (6.24) is

$$b_1^1 = 0, \quad b_2^1 = -1, \quad b_3^1 = 0$$

and leads to the desired basis for the Levi factor in the form

$$x_1 = y_1, \quad x_2 - e_1 = y_2 - y_5, \quad x_3 = y_3. \tag{6.31}$$

Notice that in this case the Levi factor of the subalgebra $\operatorname{span}\{e_1, x_1, x_2, x_3\}$ coincides with its derived algebra and is therefore unique. However, as a Levi factor of the whole algebra $\mathfrak{g}$ it is not unique since a different choice of the preimage $\mathfrak{p}_1$ of $\bar{\mathfrak{p}}$ was possible in Step 3. As stated in Theorem 2.13, the Levi factor $\mathfrak{p}$ is determined only up to inner automorphisms of $\mathfrak{g}$.

We conclude that the Levi decomposition of the "heat algebra" $\mathfrak{g}$ (6.25) is

$$\mathfrak{g} \simeq \mathfrak{sl}(2, \mathbb{R}) \ni \mathfrak{h}(1) \tag{6.32}$$

where the Levi factor $\mathfrak{sl}(2, \mathbb{R})$ has the basis

$$\begin{aligned} Y_1 &= 4t^2\partial_t + 4xt\partial_x - (2t + x^2)u\partial_u, \\ Y_2 - Y_5 &= 4t\partial_t + 2x\partial_x - u\partial_u, \quad Y_3 = \partial_t \end{aligned} \tag{6.33}$$

and the radical is the Heisenberg algebra spanned by Y_4, Y_5, Y_6.

As we have just seen explicitly, the difference between the two versions of the algorithm is conceptual rather than computational. Therefore, in our second example we shall employ only the second, streamlined version of it.

Example 6.3. Let us consider an 8-dimensional Lie algebra $\mathfrak{g}$ with the Lie brackets

$$
\begin{aligned}
&[y_1, y_2] = 2y_2, && [y_1, y_3] = -2y_3 + 2y_6, && [y_1, y_4] = 2y_4,\\
&[y_1, y_5] = 2y_5, && [y_1, y_7] = 2y_7, && [y_2, y_3] = y_1 + y_5 + y_8,\\
&[y_2, y_6] = y_5, && [y_2, y_8] = 2y_4, && [y_3, y_5] = -y_4 + y_6,\\
&[y_3, y_8] = y_6, && [y_4, y_8] = 2y_4, && [y_5, y_6] = y_4,\\
&[y_5, y_8] = y_5, && [y_6, y_8] = y_6, && [y_7, y_8] = 2y_7.
\end{aligned} \tag{6.34}
$$

Its radical is spanned by y_4, y_5, y_6, y_7, y_8 and the nilradical by y_4, y_5, y_6, y_7. In fact, the radical $\{y_4, \dots, y_8\}$ is isomorphic to the algebra $\mathfrak{s}_{5,22,a=1,b=2}$ of Section 18.3 and the nilradical is decomposable into $\mathfrak{n}_{3,1} \oplus \mathfrak{n}_{1,1}$ where $\mathfrak{n}_{3,1} \equiv \mathfrak{h}(1)$ is the Heisenberg algebra.

We start by Step 2. The derived series is

$$
\begin{aligned}
\mathfrak{g}^{(1)} &= D(\mathfrak{g}) = \mathrm{span}\{y_1 + y_8, y_2, \dots, y_7\},\\
\mathfrak{g}^{(2)} &= \mathfrak{g}^{(3)} = \mathrm{span}\{y_1 + y_8, y_2, y_3, y_4, y_5, y_6\}.
\end{aligned} \tag{6.35}
$$

Thus, the Levi factor of $\mathfrak{g}$ is found once the Levi factor of $\mathfrak{g}^{(2)}$ is constructed using the algorithm of Section 6.2. The nilpotent non-Abelian radical of $\mathfrak{g}^{(2)}$ is spanned by y_4, y_5, y_6 with a single nonvanishing Lie bracket

$$[y_5, y_6] = y_4.$$

We choose the basis as in (6.10)

$$e_1 = y_4, \quad r_1 = y_5, \quad r_2 = y_6, \quad x_1 = y_1 + y_8, \quad x_2 = y_2, \quad x_3 = y_3. \tag{6.36}$$

The Lie brackets of $\tilde{\mathfrak{g}} = \mathfrak{g}^{(2)}$ in this basis become

$$
\begin{aligned}
&[r_1, r_2] = e_1, && [r_1, x_1] = -r_1, && [r_1, x_3] = e_1 - r_2,\\
&[r_2, x_1] = r_2, && [r_2, x_2] = -r_1, && [x_1, x_2] = -2e_1 + 2x_2,\\
&[x_1, x_3] = r_2 - 2x_3, && [x_2, x_3] = r_1 + x_1.
\end{aligned} \tag{6.37}
$$

In order to find $\mathfrak{p} \dotplus D\big(R(\tilde{\mathfrak{g}})\big)$ we have to perform the change of basis (6.21). The conditions (6.24) reduce to the equations

$$b_2^2 = 0, \quad b_3^1 = 0, \quad b_2^1 + b_1^2 = 0, \quad b_3^2 - b_1^1 = 1.$$

Thus, a particular solution of (6.24) is

$$b_1^1 = 1, \qquad b_1^2 = 0, \qquad b_2^p = b_3^p = 0, \quad p = 1, 2,$$

which corresponds to the change of basis

$$\hat{x}_1 = x_1 + r_1 = y_1 + y_5 + y_8, \quad \hat{x}_2 = x_2 = y_2, \quad \hat{x}_3 = x_3 = y_3. \tag{6.38}$$

The vectors $e_1, \hat{x}_1, \hat{x}_2, \hat{x}_3$ form a basis of $\mathfrak{p} \dotplus D\big(R(\tilde{\mathfrak{g}})\big)$. In this basis we have the Lie brackets

$$[\hat{x}_1, \hat{x}_2] = -2e_1 + 2\hat{x}_2, \qquad [\hat{x}_1, \hat{x}_3] = e_1 - 2\hat{x}_3, \qquad [\hat{x}_2, \hat{x}_3] = \hat{x}_1. \tag{6.39}$$

The radical of $\mathfrak{p} \dotplus D\big(R(\tilde{\mathfrak{g}})\big)$ is spanned by e_1 which coincides with the center of $\mathfrak{p} \dotplus D\big(R(\tilde{\mathfrak{g}})\big)$. Thus, the Lie algebra (6.39) is decomposable into a direct sum of a simple algebra $\mathfrak{sl}(2, \mathbb{F})$ and a central component spanned by e_1, and the algorithm of the previous chapter can be used. Alternatively, we use the change of basis (6.21)

$$\bar{x}_1 = \hat{x}_1 + \hat{b}_1^1 e_1, \quad \bar{x}_2 = \hat{x}_2 + \hat{b}_2^1 e_1, \quad \bar{x}_3 = \hat{x}_3 + \hat{b}_3^1 e_1$$

once again, arriving at the conditions (6.24) expressed as

$$\hat{b}_1^1 = 0, \quad \hat{b}_2^1 = -1, \quad 2\hat{b}_3^1 = -1.$$

Thus, we have constructed a basis $\{\bar{x}_1, \bar{x}_2, \bar{x}_3\}$ of the Levi factor $\mathfrak{p}$ of $\mathfrak{g}$ in the form

$$\bar{x}_1 = \hat{x}_1 = y_1 + y_5 + y_8, \quad \bar{x}_2 = \hat{x}_2 - e_1 = y_2 - y_4, \quad \bar{x}_3 = \hat{x}_3 - \tfrac{1}{2}e_1 = y_3 - \tfrac{1}{2}y_4.$$

To sum up, the Levi factor of the algebra (6.34) is

$$\mathfrak{p} = \operatorname{span}\left\{y_1 + y_5 + y_8, y_2 - y_4, y_3 - \tfrac{1}{2}y_4\right\} \tag{6.40}$$

and is isomorphic to $\mathfrak{sl}(2, \mathbb{F})$.

Let us recall that the Levi factor (6.40) is generically far from unique. In our example, another choice for it is spanned by

$$\tilde{x}_1 = y_1 + y_8, \quad \tilde{x}_2 = y_2 - y_4, \quad \tilde{x}_3 = y_3 - y_6 \tag{6.41}$$

and, in fact, the choice (6.41) is more convenient because the Lie brackets of the algebra (6.34) are more compact when written in terms of $\tilde{x}_1, \tilde{x}_2, \tilde{x}_3$. Different choices of Levi factors arise through different choices of the particular solutions of the systems of linear equations involved.

CHAPTER 7

The Nilradical of a Lie Algebra

The aim of this section is to present an algorithm by means of which it is possible to explicitly calculate the maximal nilpotent ideal of a finite-dimensional Lie algebra $\mathfrak{g}$. This algorithm was first presented in [102]. It is completely rational: it requires no irrational operations such as eigenvalue calculations.

7.1. General theory

Let us first present several results that underlie the algorithm.

Lemma 7.1. *For any epimorphism (i.e., surjective homomorphism) $\varepsilon\colon \mathfrak{g} \to \mathfrak{g}'$ we have*

$$\varepsilon\big(\mathrm{NR}(\mathfrak{g})\big) \subseteq \mathrm{NR}(\mathfrak{g}'). \tag{7.1}$$

PROOF. For any nilpotent ideal $\mathfrak{i}$ of $\mathfrak{g}$ we have

$$[\mathfrak{g}, \mathfrak{i}] \subseteq \mathfrak{i}, \qquad \mathfrak{i}^n = 0, \quad n \in \mathbb{N}.$$

Upon application of ε we have

$$[\mathfrak{g}', \varepsilon(\mathfrak{i})] = [\varepsilon(\mathfrak{g}), \varepsilon(\mathfrak{i})] \subseteq \varepsilon(\mathfrak{i}), \quad \varepsilon(\mathfrak{i}^n) = 0.$$

Hence $\varepsilon(\mathfrak{i})$ is a nilpotent ideal of $\mathfrak{g}'$ and $\varepsilon\big(\mathrm{NR}(\mathfrak{g})\big) \subseteq \mathrm{NR}(\mathfrak{g}')$. □

For our purposes an important corollary of Lemma 7.1 is that for any ideal $\mathfrak{i}$ of $\mathfrak{g}$ we have

$$\mathrm{NR}(\mathfrak{g})/\mathfrak{i} \subseteq \mathrm{NR}(\mathfrak{g}/\mathfrak{i}). \tag{7.2}$$

Note that the equality sign alone in (7.1) and (7.2) would be inappropriate.

Example 7.2. Let us consider the algebra

$$\mathfrak{g} = \mathrm{span}\{e_1, e_2\}, \qquad [e_1, e_2] = e_1 \tag{7.3}$$

and take $\mathfrak{g}' = \mathrm{span}\{f\}$ (with $f \neq 0$). We consider an epimorphism $\varepsilon\colon \mathfrak{g} \to \mathfrak{g}'$, $\varepsilon(e_1) = 0$, $\varepsilon(e_2) = f$. Thus

$$\varepsilon\big(\mathrm{NR}(\mathfrak{g})\big) = \varepsilon(\mathrm{span}\{e_1\}) = 0 \subsetneq \mathrm{NR}(\mathfrak{g}') = \mathfrak{g}'.$$

Our example also shows that a finite-dimensional Lie algebra $\mathfrak{g}$ with a nilpotent ideal $\mathfrak{b}$ and a nilpotent factor algebra $\mathfrak{g}/\mathfrak{b}$ is not necessarily nilpotent — it suffices to consider $\mathfrak{g}$ as in (7.3) and $\mathfrak{b} = \ker \varepsilon = \mathrm{span}\{e_1\}$.

A *composition series* for a finite-dimensional solvable Lie algebra $\mathfrak{g}$ [59] is a chain of ideals

$$\mathfrak{g} = \mathfrak{g}_0 \supsetneq \mathfrak{g}_1 \supsetneq \mathfrak{g}_2 \supsetneq \cdots \supsetneq \mathfrak{g}_s = 0 \tag{7.4}$$

such that $\mathfrak{g}_i$ is maximal among the ideals of $\mathfrak{g}$ properly contained in $\mathfrak{g}_{i-1}$ $(1 \le i \le s)$. A composition series of a solvable Lie algebra always exists as a refinement of the derived series.

Lemma 7.3. *The nilradical of a finite-dimensional solvable Lie algebra* $\mathfrak{g}$ *is characterized as the set of all elements* $x \in \mathfrak{g}$ *for which*

$$[x, \mathfrak{g}_{i-1}] \subseteq \mathfrak{g}_i, \quad 1 \le i \le s \tag{7.5}$$

in any chosen composition series $(\mathfrak{g}_i)$.

PROOF. We shall use the notation

$$[x_1, x_2, \ldots, x_j] = \big[x_1, [x_2, \ldots, x_j]\big], \tag{7.6}$$

defining multiple commutators recursively. Let us denote by $\mathfrak{x}$ the set of all elements $x \in \mathfrak{g}$ satisfying (7.5). Equation (7.5) implies that $\mathfrak{x}$ is an ideal in $\mathfrak{g}$.

Let us first assume $x \in \mathrm{NR}(\mathfrak{g})$. We have

$$\mathfrak{g}_{i-1} \supseteq \mathfrak{a}_i \equiv [\mathrm{NR}(\mathfrak{g}), \mathfrak{g}_{i-1}] + \mathfrak{g}_i \supseteq \mathfrak{g}_i.$$

Let us suppose that $\mathfrak{a}_i \neq \mathfrak{g}_i$. Then by construction $\mathfrak{a}_i = \mathfrak{g}_{i-1}$. Hence

$$\begin{aligned}[\mathrm{NR}(\mathfrak{g}), \mathrm{NR}(\mathfrak{g}), \mathfrak{g}_{i-1}] + \mathfrak{g}_i &= [\mathrm{NR}(\mathfrak{g}), \mathfrak{a}_i] + \mathfrak{g}_i \\ &= [\mathrm{NR}(\mathfrak{g}), \mathfrak{g}_{i-1}] + \mathfrak{g}_i = \mathfrak{a}_i = \mathfrak{g}_{i-1}.\end{aligned}$$

By iteration we have $[\mathrm{NR}(\mathfrak{g}), \ldots, \mathrm{NR}(\mathfrak{g}), \mathfrak{g}_{i-1}] + \mathfrak{g}_i = \mathfrak{g}_{i-1}$. On the other hand, there exists a natural number n for which $\mathrm{NR}(\mathfrak{g})^n = 0$. Consequently,

$$[\underbrace{\mathrm{NR}(\mathfrak{g}), \ldots, \mathrm{NR}(\mathfrak{g})}_{n \text{ times}}, \mathfrak{g}_{i-1}] \subseteq [\mathrm{NR}(\mathfrak{g}), \ldots, \mathrm{NR}(\mathfrak{g})] = 0,$$

so that we obtain $\mathfrak{g}_i = \mathfrak{g}_{i-1}$, a contradiction. Hence we find $[\mathrm{NR}(\mathfrak{g}), \mathfrak{g}_{i-1}] \subseteq \mathfrak{g}_i$ for $1 \le i \le s$ and (7.5) holds for $x \in \mathrm{NR}(\mathfrak{g})$. Therefore we have $\mathrm{NR}(\mathfrak{g}) \subseteq \mathfrak{x}$.

Conversely, let $x_1, \ldots, x_s$ belong to the ideal $\mathfrak{x}$ (the x_j are not necessarily distinct). From (7.5) we have

$$\mathrm{ad}(x_1)\mathfrak{g} \subseteq \mathfrak{g}_1, \mathrm{ad}(x_2)\,\mathrm{ad}(x_1)\mathfrak{g} \subseteq \mathfrak{g}_2, \ldots, \mathrm{ad}(x_s)\,\mathrm{ad}(x_{s-1}) \cdots \mathrm{ad}(x_1)\mathfrak{g} = \mathfrak{g}_s = 0.$$

Consequently the sth element $\mathfrak{x}^s$ of the lower central series of the ideal $\mathfrak{x}$ is 0. Thus, by definition, the ideal $\mathfrak{x}$ is nilpotent, $\mathfrak{x} \subseteq \mathrm{NR}(\mathfrak{g})$. Since $\mathrm{NR}(\mathfrak{g}) \subseteq \mathfrak{x}$ we obtain $\mathrm{NR}(\mathfrak{g}) = \mathfrak{x}$. □

We already know (cf. Example 7.2) that a finite-dimensional Lie algebra $\mathfrak{g}$ with a nilpotent ideal $\mathfrak{b}$ and a nilpotent factor algebra $\mathfrak{g}/\mathfrak{b}$ is not necessarily nilpotent. However, the following lemma provides a useful nilpotency criterion.

Lemma 7.4. *Let* $\mathfrak{g}$ *be solvable,* $\mathfrak{i} \subset \mathfrak{g}$ *its nilpotent ideal and* $\mathrm{NR}(\mathfrak{g}/\mathfrak{i}^2)$ *be the nilradical of the factor algebra* $\mathfrak{g}/\mathfrak{i}^2$. *Then the nilradical of* $\mathfrak{g}$ *can be found using*

$$\mathrm{NR}(\mathfrak{g})/\mathfrak{i}^2 = \mathrm{NR}(\mathfrak{g}/\mathfrak{i}^2). \tag{7.7}$$

PROOF. It is clear that $\mathfrak{r} = \{x \in \mathfrak{g} | x + \mathfrak{i}^2 \in \mathrm{NR}(\mathfrak{g}/\mathfrak{i}^2)\}$ is an ideal since Lie brackets in $\mathfrak{g}$ and $\mathfrak{g}/\mathfrak{i}^2$ differ only by elements of $\mathfrak{i}^2$ and $\mathfrak{i}^2 \subset \mathfrak{r}$, and that if $\mathfrak{r}$ is nilpotent then it is maximal in $\mathfrak{g}$ by (7.2). Therefore we only have to prove that it is nilpotent. In order to do that we have to show that there exists a positive integer $N \in \mathbb{N}$ such that

$$(\mathrm{ad}(x)|_{\mathfrak{r}})^N = 0, \quad \forall x \in \mathfrak{r}.$$

By assumption we know that $K \in \mathbb{N}$ exists such that

$$(\mathrm{ad}(x)|_{\mathfrak{r}})^K : \mathfrak{r} \to \mathfrak{i}^2.$$

We use the identity

$$(\mathrm{ad}(x))^J [a, b] = \sum_{l=0}^{J} \binom{J}{l} \left[(\mathrm{ad}(x))^l a, (\mathrm{ad}(x))^{J-l} b \right] \tag{7.8}$$

and proceed by induction, showing in each step that there always exists an integer K such that

$$(\mathrm{ad}(x)|_{\mathfrak{r}})^K : \mathfrak{i}^j \to \mathfrak{i}^{j+1}, \quad j \leq n \in \mathbb{N}.$$

Let $M \in \mathbb{N}$ be such that $(\mathrm{ad}(x)|_{\mathfrak{r}})^M : \mathfrak{i}^j \to \mathfrak{i}^{j+1}$, $j \leq n-1$. The ideal $\mathfrak{i}^n = [\mathfrak{i}, \mathfrak{i}^{n-1}]$ is spanned by elements of the form $[a, b]$, $a \in \mathfrak{i}$, $b \in \mathfrak{i}^{n-1}$ and we shall consider only these. Then for any l we have $(\mathrm{ad}(x))^l a \in \mathfrak{i}$ and for $l \geq M$ it lies in $\mathfrak{i}^2$. Similarly $(\mathrm{ad}(x))^{J-l} b \in \mathfrak{i}^{n-1}$ and lies in $\mathfrak{i}^n$ for $J - l \geq M$. Finally for $J = 2M$ we have using the identity (7.8)

$$(\mathrm{ad}(x)|_{\mathfrak{r}})^{2M} \mathfrak{i}^n \subseteq [\mathfrak{i}, \mathfrak{i}^n] + [\mathfrak{i}^2, \mathfrak{i}^{n-1}] = \mathfrak{i}^{n+1}.$$

Since the lower central series of $\mathfrak{i}$ terminates after say Q iterations and we have shown that there is an integer K such that $(\mathrm{ad}(x))^K$ maps $\mathfrak{i}^j$ into $\mathfrak{i}^{j+1}$ for any $j \in \mathbb{N}$, taking $N = QK$ we have

$$(\mathrm{ad}(x)|_{\mathfrak{r}})^N : \mathfrak{r} \to 0, \quad \forall x \in \mathfrak{r},$$

i.e., $\mathfrak{r}$ is a nilpotent ideal in $\mathfrak{g}$ and consequently $\mathfrak{r}$ is the nilradical of $\mathfrak{g}$. □

Lemma 7.4 reduces the problem of finding the nilradical of $\mathfrak{g}$ to that of finding the nilradical of $\mathfrak{g}/\mathfrak{i}^2$, a lower-dimensional problem. Once $\mathrm{NR}(\mathfrak{g}/\mathfrak{i}^2)$ is found, the nilradical of $\mathfrak{g}$ is spanned by a basis of $\mathfrak{i}^2$ together with elements representing a basis of $\mathrm{NR}(\mathfrak{g}/\mathfrak{i}^2)$. By a slight abuse of notation we may write

$$\mathrm{NR}(\mathfrak{g}) = \mathrm{NR}(\mathfrak{g}/\mathfrak{i}^2) \dotplus \mathfrak{i}^2. \tag{7.9}$$

Example 7.5. Consider a 4-dimensional algebra $\mathfrak{g}$ spanned by e_1, e_2, e_3, e_4 with the nonvanishing Lie brackets

$$[e_1, e_4] = e_1, \quad [e_2, e_3] = e_1, \quad [e_2, e_4] = e_2. \tag{7.10}$$

and the nilpotent ideal $\mathfrak{i} = \mathrm{span}\{e_1, e_2, e_3\}$. We have $\mathfrak{i}^2 = \mathrm{span}\{e_1\}$. Upon factorization we obtain the Lie algebra $\mathfrak{g}/\mathfrak{i}^2$ spanned by $\bar{e}_2 \equiv e_2 + \mathfrak{i}^2, \bar{e}_3$ and $\bar{e}_4$ with the only nonvanishing bracket

$$[\bar{e}_2, \bar{e}_4] = \bar{e}_2. \tag{7.11}$$

The nilradical of $\mathfrak{g}/\mathfrak{i}^2$ is found to be $\mathrm{NR}(\mathfrak{g}/\mathfrak{i}^2) = \mathrm{span}\{\bar{e}_2, \bar{e}_3\}$. Using Lemma 7.4 we find that

$$\mathrm{NR}(\mathfrak{g}) = \mathrm{span}\{e_2, e_3\} \dotplus \mathfrak{i}^2 = \mathrm{span}\{e_1, e_2, e_3\} = \mathfrak{i}. \tag{7.12}$$

Thus by inspection of the nilradical and of the derived algebra $\mathfrak{g}^2 = \mathrm{span}\{e_1, e_2\}$ the algebra (7.10) is isomorphic to $\mathfrak{s}_{4,11}$ of Section 17.3.

A similar factorization is obtained also with respect to the hypercenter.

Lemma 7.6. *Let $\mathfrak{g}$ be a Lie algebra with the hypercenter $\mathfrak{z}_\infty(\mathfrak{g})$. Then the following relation holds*

$$\mathrm{NR}\big(\mathfrak{g}/\mathfrak{z}_\infty(\mathfrak{g})\big) = \mathrm{NR}(\mathfrak{g})/\mathfrak{z}_\infty(\mathfrak{g}). \tag{7.13}$$

PROOF. The inclusion $\supseteq$ is obvious, cf. (7.2). On the other hand, let $\mathfrak{i}$ be the ideal in $\mathfrak{g}$ such that $\mathfrak{z}_\infty(\mathfrak{g}) \subset \mathfrak{i}$, $\mathfrak{i}/\mathfrak{z}_\infty(\mathfrak{g}) = \mathrm{NR}(\mathfrak{g}/\mathfrak{z}_\infty(\mathfrak{g}))$. By definition $j, k \in \mathbb{N}$ exist such that

$$[\underbrace{\mathfrak{g}, \dots, \mathfrak{g}}_{j \text{ times}}, \mathfrak{z}_\infty(\mathfrak{g})] = 0$$

and $\mathfrak{i}^k \subset \mathfrak{z}_\infty(\mathfrak{g})$. Thus,

$$\mathfrak{i}^{j+k} \subset [\underbrace{\mathfrak{g}, \dots, \mathfrak{g}}_{j \text{ times}}, \mathfrak{i}^k] = 0,$$

i.e., $\mathfrak{i}$ is a nilpotent ideal in $\mathfrak{g}$, $\mathfrak{i} \subseteq \mathrm{NR}(\mathfrak{g})$. □

Example 7.7. Consider a 4-dimensional algebra $\mathfrak{g}$ spanned by e_1, e_2, e_3, e_4 with the nonvanishing Lie brackets

$$[e_2, e_4] = e_1, \quad [e_3, e_4] = e_3. \tag{7.14}$$

It has the 1-dimensional center $\mathfrak{z}_1 = \mathrm{span}\{e_1\}$ and the 2-dimensional second center $\mathfrak{z}_2 = \mathrm{span}\{e_1, e_2\}$ which coincides with the hypercenter. We have $\mathfrak{g}/\mathfrak{z}_\infty = \mathrm{span}\{\bar{e}_3, \bar{e}_4\}$ with

$$[\bar{e}_3, \bar{e}_4] = \bar{e}_3. \tag{7.15}$$

Thus the nilradical of $\mathfrak{g}/\mathfrak{z}_\infty$ is spanned by $\bar{e}_3$ and we find the nilradical of $\mathfrak{g}$ in the form

$$\mathrm{NR}(\mathfrak{g}) = \mathrm{span}\{e_3\} \dotplus \mathfrak{z}_\infty = \mathrm{span}\{e_1, e_2, e_3\}. \tag{7.16}$$

Consequently, the algebra (7.14) is isomorphic to $\mathfrak{s}_{4,1}$ of Section 17.2.

Theorem 7.8. *Let $\mathfrak{g}$ be a finite-dimensional Lie algebra, let $\mathfrak{b}$ be an ideal of $\mathfrak{g}$, and let $\mathfrak{m}$ be the ideal of $\mathfrak{g}$ containing $\mathfrak{b}$ for which*

$$\mathrm{NR}(\mathfrak{g}/\mathfrak{b}) = \mathfrak{m}/\mathfrak{b} \tag{7.17}$$

holds. We then have

$$\mathrm{NR}(\mathfrak{g}) = \mathrm{NR}(\mathfrak{m}). \tag{7.18}$$

PROOF. According to (7.1) we have

$$\mathrm{NR}(\mathfrak{g}) \subseteq \mathrm{NR}(\mathfrak{m}). \tag{7.19}$$

By Theorem 2.10 we know that $\mathrm{NR}(\mathfrak{m})$ is an ideal of $\mathfrak{g}$. By construction $\mathrm{NR}(\mathfrak{m})$ is nilpotent.

Hence

$$\mathrm{NR}(\mathfrak{m}) \subseteq \mathrm{NR}(\mathfrak{g}). \tag{7.20}$$

But (7.19) and (7.20) imply (7.18). □

Example 7.9. Consider the algebra $\mathfrak{s}_{5,39}$ of Section 18.5, i.e.,

$$\mathfrak{g} = \mathrm{span}\{e_1, e_2, e_3, e_4, e_5\}$$

with

$$[e_4, e_2] = e_2, \quad [e_5, e_3] = e_3, \quad [e_5, e_4] = e_1, \tag{7.21}$$

and the ideal $\mathfrak{b} = \mathrm{span}\{e_3\}$. We have $\mathfrak{g}/\mathfrak{b} = \mathrm{span}\{\bar{e}_1, \bar{e}_2, \bar{e}_4, \bar{e}_5\}$ with

$$[\bar{e}_4, \bar{e}_2] = \bar{e}_2, \quad [\bar{e}_4, \bar{e}_5] = \bar{e}_1. \tag{7.22}$$

The nilradical of $\mathfrak{g}/\mathfrak{b}$ is $\mathrm{NR}(\mathfrak{g}/\mathfrak{b}) = \mathrm{span}\{\bar{e}_1, \bar{e}_2, \bar{e}_5\}$. Thus, the ideal $\mathfrak{m}$ of Theorem 7.8 is

$$\mathfrak{m} = \mathrm{span}\{e_1, e_2, e_3, e_5\} \tag{7.23}$$

and we conclude that

$$\mathrm{NR}(\mathfrak{g}) = \mathrm{NR}(\mathfrak{m}) = \mathrm{span}\{e_1, e_2, e_3\}. \tag{7.24}$$

As a consequence of Lemma 7.4 we find that for solvable Lie algebras $\mathfrak{g}$ we have

$$\mathrm{NR}(\mathfrak{g})/(\mathfrak{g}^2)^2 = \mathrm{NR}(\mathfrak{g}/(\mathfrak{g}^2)^2). \tag{7.25}$$

Using Lemma 7.6 we next arrive at the algebra $\mathfrak{g}/(\mathfrak{g}^2)^2/\mathfrak{z}_\infty(\mathfrak{g}/(\mathfrak{g}^2)^2)$ with an Abelian derived algebra and vanishing (hyper)center. Such algebras have the following useful property.

Lemma 7.10. *Let $\mathfrak{g}$ be a finite-dimensional Lie algebra satisfying the conditions*

$$(\mathfrak{g}^2)^2 = 0, \tag{7.26}$$

$$C(\mathfrak{g}) = 0. \tag{7.27}$$

Then we have

$$\mathrm{cent}_{\mathfrak{g}}(\mathfrak{g}^2) = \{x \in \mathfrak{g} \mid [x, \mathfrak{g}^2] = 0\} = \mathfrak{g}^2. \tag{7.28}$$

Proof. The centralizer $\mathrm{cent}_{\mathfrak{g}}(\mathfrak{g}^2)$ of $\mathfrak{g}^2$ in $\mathfrak{g}$ is an ideal in $\mathfrak{g}$, i.e., a representation space for the representation Γ of $\mathfrak{g}$

$$\Gamma\colon \mathfrak{g} \to \mathrm{End}\big(\mathrm{cent}_{\mathfrak{g}}(\mathfrak{g}^2)\big),$$
$$\Gamma(x)(u) = [x, u], \quad \big(x \in \mathfrak{g}, u \in \mathrm{cent}_{\mathfrak{g}}(\mathfrak{g}^2)\big)$$

with $\mathfrak{g}^2$ in its kernel; hence $\Gamma(\mathfrak{g}) \cong \mathfrak{g}/\ker\Gamma$ is the homomorphic image of $\mathfrak{g}/\mathfrak{g}^2$, which means that $\Gamma(\mathfrak{g})$ is Abelian. It follows that

$$\mathrm{cent}_{\mathfrak{g}}(\mathfrak{g}^2) = \mathfrak{m}_0 + \widehat{\mathfrak{m}}_0,$$

where $\mathfrak{m}_0$ is the $\mathfrak{g}$-invariant subset of the elements of $\mathrm{cent}_{\mathfrak{g}}(\mathfrak{g}^2)$ which are annihilated by some power of $\Gamma(x)$ for each x of $\mathfrak{g}$, whereas the complementary subspace $\widehat{\mathfrak{m}}_0$ is the intersection of the $\mathfrak{g}$-invariant subspaces $\Gamma(\mathfrak{g})^i\big(\mathrm{cent}_{\mathfrak{g}}(\mathfrak{g}^2)\big)$ $(i \in \mathbb{N})$. It follows that $\widehat{\mathfrak{m}}_0 \subseteq \mathfrak{g}^2$. However, since $C(\mathfrak{g}) = 0$, it follows that $C(\mathfrak{g}) \cap \mathfrak{m}_0 = 0$ and consequently $\mathfrak{m}_0 = 0$. However, (7.26) implies that $\mathfrak{g}^2 \subseteq \mathrm{cent}_{\mathfrak{g}}(\mathfrak{g}^2)$. Hence $\mathrm{cent}_{\mathfrak{g}}(\mathfrak{g}^2) = \widehat{\mathfrak{m}}_0 = \mathfrak{g}^2$. □

7.2. Algorithm

The results of the previous section give rise to an algorithm for calculating the nilradical of a finite-dimensional Lie algebra $\mathfrak{g}$.

Step 1. Determine the radical $\mathrm{R}(\mathfrak{g})$ using the formulas (6.6), (2.31). In view of Lemma 2.4 we have

$$\mathrm{NR}(\mathfrak{g}) = \mathrm{NR}\big(\mathrm{R}(\mathfrak{g})\big). \tag{7.29}$$

From now on we replace $\mathrm{R}(\mathfrak{g})$ by $\mathfrak{g}$ and assume that $\mathfrak{g}$ is solvable. In view of Lemma 2.4 we can build up the nilradical gradually from smaller nilpotent ideals.

Step 2. Calculate the ideals $D(\mathfrak{g}) = \mathfrak{g}^2 = [\mathfrak{g}, \mathfrak{g}]$ and $D^2(\mathfrak{g}) = [D(\mathfrak{g}), D(\mathfrak{g})]$. In view of (7.25) we obtain $\mathrm{NR}(\mathfrak{g})$ from the nilradical of the factor algebra $\mathfrak{g}/D^2(\mathfrak{g})$ using

$$\mathrm{NR}(\mathfrak{g})/D^2(\mathfrak{g}) = \mathrm{NR}\big(\mathfrak{g}/D^2(\mathfrak{g})\big).$$

From now on we consider the algebra $\mathfrak{g}/D^2(\mathfrak{g})$, i.e., we assume that $\mathfrak{g}$ is solvable and its derived algebra $D(\mathfrak{g}) = [\mathfrak{g}, \mathfrak{g}]$ is Abelian:

$$[D(\mathfrak{g}), D(\mathfrak{g})] = 0. \tag{7.30}$$

Step 3. Calculate the hypercenter $\mathfrak{z}_\infty(\mathfrak{g})$ of $\mathfrak{g}$ (defined in (2.10)). In view of Lemma 7.6 we obtain $\mathrm{NR}(\mathfrak{g})$ from the nilradical of $\mathfrak{g}/\mathfrak{z}_\infty(\mathfrak{g})$ using

$$\mathrm{NR}(\mathfrak{g})/\mathfrak{z}_\infty(\mathfrak{g}) = \mathrm{NR}\big(\mathfrak{g}/\mathfrak{z}_\infty(\mathfrak{g})\big).$$

We have thus reduced the problem of finding the nilradical of an arbitrary finite-dimensional Lie algebra to that of finding $\mathrm{NR}(\mathfrak{g})$ for a solvable Lie algebra $\mathfrak{g}$ satisfying (7.30) and

$$C(\mathfrak{g}) = 0 \tag{7.31}$$

(since $\mathfrak{z}_\infty(\mathfrak{g}) = 0$ is equivalent to $C(\mathfrak{g}) = 0$).

Step 4. Introduce a basis for the derived algebra $D(\mathfrak{g})$ of the solvable algebra $\mathfrak{g}$, satisfying (7.30) and (7.31). Extend this basis to a basis of $\mathfrak{g}$:

$$D(\mathfrak{g}) = \mathrm{span}\{u_1, \dots, u_m\}, \quad \mathfrak{g} = \mathrm{span}\{u_1, \dots, u_m, x_1, \dots, x_{n-m}\}. \tag{7.32}$$

Step 5. Choose a specific element of $D(\mathfrak{g})$, say $u \equiv u_1$ and an element of the complement, say x_j. Define a chain of elements u_{kj} in $D(\mathfrak{g})$ as

$$u_{0j} = u, \quad u_{1j} = [x_j, u_{0j}], \dots, u_{pj} = [x_j, u_{p-1j}] \tag{7.33}$$

such that $\{u_{0j}, \dots, u_{p-1j}\}$ are linearly independent but u_{pj} is a linear combination of the preceding elements. Call the space spanned by these elements

$$X_j = \mathrm{span}\{u_{0j}, \dots, u_{p-1,j}\}. \tag{7.34}$$

Its dimension is p and X_j is by construction $\mathrm{ad}\, x_j$ invariant. The matrix representing $\mathrm{ad}\, x_j$ restricted to the space X_j has the form

$$\mathrm{ad}\, x_j|_{X_j} = \begin{pmatrix} 0 & 0 & \dots & 0 & a_{pj} \\ 1 & 0 & \dots & 0 & a_{p-1j} \\ & 1 & \ddots & \vdots & \vdots \\ & & \ddots & 0 & a_{2j} \\ & & & 1 & a_{1j} \end{pmatrix} \tag{7.35}$$

where a_{kj}, $1 \le k \le p$ are some numbers.

The characteristic polynomial of $\mathrm{ad}\, x_j|_{X_j}$ is

$$P_j(\lambda) = \lambda^p - a_{1j}\lambda^{p-1} - a_{2j}\lambda^{p-2} - \dots - a_{p-1j}\lambda - a_{pj}. \tag{7.36}$$

It follows from (7.35) that $P(\lambda)$ is also the minimal polynomial of $\mathrm{ad}\, x_j|_{X_j}$. Proceed to Step 6 if $\lambda = 0$ is an eigenvalue (i.e., if $a_{pj} = 0$), to Step 7 otherwise.

Step 6. Form the ideal

$$\mathfrak{b}_1 = [x_j, D(\mathfrak{g})]. \tag{7.37}$$

(The subspace $\mathfrak{b}_1$ is an ideal because $D(\mathfrak{g})$ is Abelian.) The fact that we have $a_{pj} = 0$ guarantees that $\mathfrak{b}_1$ is properly contained in $D(\mathfrak{g})$. Furthermore we have $\mathfrak{b}_1 \neq 0$ since $x_j \notin D(\mathfrak{g})$ and by Lemma 7.10 the centralizer of $D(\mathfrak{g})$ is $D(\mathfrak{g})$.

Having formed the ideal $\mathfrak{b}_1$ we use Theorem 7.8 and we have to determine the nilradical of $\mathfrak{g}/\mathfrak{b}_1$, a lower-dimensional task. We then form the ideal $\mathfrak{m}$ of $\mathfrak{g}$ defined by $\mathfrak{m}/\mathfrak{b}_1 = \mathrm{NR}(\mathfrak{g}/\mathfrak{b}_1)$. Finally we must find $\mathrm{NR}(\mathfrak{m})$ ($\dim \mathfrak{m} < \dim \mathfrak{g}$) and put

$$\mathrm{NR}(\mathfrak{g}) = \mathrm{NR}(\mathfrak{m}). \tag{7.38}$$

Step 7. We are now considering the case $a_{pj} \neq 0$. If the matrix $\operatorname{ad} x_j|_{X_j}$ has p distinct eigenvalues, i.e., if the polynomial $P_j(\lambda)$ is square-free, proceed to Step 8. If not, then find a nonconstant polynomial $g_j(\lambda)$ that divides $P_j(\lambda)$ and is square-free. Form the ideal

$$\mathfrak{b}_2 = g_j\big(\mathrm{ad}(x_j)\big)D(\mathfrak{g}). \tag{7.39}$$

Since $g_j(\lambda)$ is not a minimal polynomial of the matrix (7.35) we have $\mathfrak{b}_2 \neq 0$. Since $g_j(\operatorname{ad} x_j)$ is by construction a singular matrix $\mathfrak{b}_2$ is properly contained in $D(\mathfrak{g})$ $\big(\mathfrak{b}_2 \neq D(\mathfrak{g})\big)$. Since $D(\mathfrak{g})$ is Abelian, $\mathfrak{b}_2$ is an ideal in $\mathfrak{g}$. Proceed further as with the ideal $\mathfrak{b}_1$, i.e. put $\mathfrak{m}/\mathfrak{b}_2 = \mathrm{NR}(\mathfrak{g}/\mathfrak{b}_2)$, $\mathfrak{b}_2 \subseteq \mathfrak{m}$, and find $\mathrm{NR}(\mathfrak{m})$; we have $\mathrm{NR}(\mathfrak{g}) = \mathrm{NR}(\mathfrak{m})$.

Step 8. We finally arrive at the case when the polynomial $P_j(\lambda)$ is square-free and $a_{pj} \neq 0$. The label j is associated with the chosen element x_j. We run through all values of $1 \leq j \leq n-m$ and repeat Steps 5, 6 and 7. If $\mathrm{NR}(\mathfrak{g})$ is still not determined, i.e., we have $P_j(\lambda)$ square-free and $a_{pj} \neq 0$ for all j, we proceed as follows.

Take the element $u = u_{0j} \in D(\mathfrak{g})$ of Step 5 and find its centralizer in $\mathfrak{g}$

$$\mathfrak{m} = \mathrm{cent}_{\mathfrak{g}}(u) = \{y \in \mathfrak{g} \mid [y, u] = 0\}. \tag{7.40}$$

Since $D(\mathfrak{g})$ is Abelian we have

$$D(\mathfrak{g}) \subseteq \mathfrak{m}. \tag{7.41}$$

The crucial fact, justified below is

$$\mathrm{NR}(\mathfrak{g}) = \mathrm{NR}(\mathfrak{m}). \tag{7.42}$$

If $D(\mathfrak{g}) \neq \mathfrak{m}$, then $\mathrm{NR}(\mathfrak{m})$ is found by returning to Step 2 with $\mathfrak{m}$ used instead of $\mathfrak{g}$. If $D(\mathfrak{g}) = \mathfrak{m}$ we have

$$\mathrm{NR}(\mathfrak{g}) = D(\mathfrak{g}) \tag{7.43}$$

and the algorithm terminates.

Let us now justify (7.42).

Lemma 7.11. *Let $\mathfrak{g}$ be a Lie algebra satisfying all conditions of Step 8, namely (7.30), (7.31) and let the characteristic polynomial $P_j(\lambda)$ be square-free with $a_{pj} \neq 0$ for all elements x_j in the basis (7.32). Then $\mathrm{NR}(\mathfrak{g}) \subseteq \mathrm{cent}_{\mathfrak{g}}(u) = \mathfrak{m}$ and consequently we have $\mathrm{NR}(\mathfrak{g}) = \mathrm{NR}(\mathfrak{m})$.*

PROOF. The adjoint representation $\operatorname{ad} \mathfrak{g}|_{D(\mathfrak{g})}$ of $\mathfrak{g}$ restricted to $D(\mathfrak{g})$ is a commutative algebra (this follows from the Jacobi identity since $D(\mathfrak{g})$ is Abelian) of matrices

$$X \in \mathbb{C}^{m\times m}. \tag{7.44}$$

Any Abelian subalgebra of $\mathfrak{gl}(m,\mathbb{C})$ can be transformed into a normal form in which all matrices X simultaneously have the block diagonal form

$$X = \begin{pmatrix} X_1 & & & \\ & X_2 & & \\ & & \ddots & \\ & & & X_k \end{pmatrix}, \qquad \dim X_q = m_q, \quad 1 \le q \le k \tag{7.45}$$

where each block $X_q \in \mathfrak{gl}(m_q,\mathbb{C})$ has the form

$$X_q = a_q I + Y_q, \quad Y_q^{m_q} = 0. \tag{7.46}$$

The matrices Y_q are nilpotent matrices which form an Abelian subalgebra of $\mathfrak{sl}(m_q,\mathbb{C})$ [122]. We recall that an algebra $\mathfrak{g}$ is nilpotent if and only if a positive integer k exists such that

$$(\operatorname{ad} y)^k = 0, \quad \forall y \in \mathfrak{g} \tag{7.47}$$

(cf. the Engel theorem 2.7). Let y be an element of $\mathfrak{g}$ such that $(\operatorname{ad} y)^{k-1}(D(\mathfrak{g})) = 0$, then we have $(\operatorname{ad} y)^k\mathfrak{g} = 0$ (since $\operatorname{ad} y\colon \mathfrak{g} \to D(\mathfrak{g})$). Consequently, $\operatorname{ad} x|_{D(\mathfrak{g})}$ is nilpotent if and only if $\operatorname{ad} x|_{NR(\mathfrak{g})}$ is nilpotent, i.e., $\mathrm{NR}(\mathfrak{g})$ consists of elements $x \in \mathfrak{g}$ such that $\operatorname{ad} x|_{D(\mathfrak{g})}$ is a nilpotent operator.

Since $P_j(\lambda)$ is a square-free minimal polynomial such that $a_{pj} \neq 0$ we have

$$P_j(\lambda) = \prod_{k=1}^{p} (\lambda - K_{jk}) \tag{7.48}$$

with

$$\prod_{k=1}^{p} K_{jk} \neq 0, \quad K_{jk} \neq K_{jl} \quad \text{for } k \neq l. \tag{7.49}$$

We have

$$P_j(\operatorname{ad} x_j)u = \prod_{k=1}^{p} (\operatorname{ad} x_j - K_{jk})u = 0.$$

The matrix X in (7.45) acts on the vector space $V \sim D(\mathfrak{g}) \sim \mathbb{C}^m$. We decompose V into subspaces invariant under the matrices X of (7.45): $V = \bigoplus_{q=1}^{k} V_q$. We have

$$u = \sum_q u_q, \quad u_q \in V_q, \quad \operatorname{ad} x(u) = \sum_q \operatorname{ad} x(u_q), \tag{7.50}$$

$$0 = \sum_q P_j(\operatorname{ad} x_j)(u_q) = \sum_q \prod_{k=1}^{p} (\operatorname{ad} x_j - K_{jk})u_q. \tag{7.51}$$

Since all eigenvalues K_{jk} for j given are different, at most one term in each of the products in (7.51) annihilates u_q. We conclude that if $u_q \neq 0$ we have

$$\operatorname{ad} x_j u_q = \lambda_{jq} u_q \tag{7.52}$$

where $\lambda_{jq} = K_{jk}$ for some $1 \le k \le p$.

Suppose that $x \in \mathfrak{g}$ is in $\mathrm{NR}(\mathfrak{g})$ and expand x in the basis (7.32): $x = \sum_i \xi_i x_i + y$, $y \in D(\mathfrak{g})$. We have

$$(\mathrm{ad}\, x)^K u = \sum_q (\mathrm{ad}\, x)^K u_q = \sum_q \Big(\sum_j \xi^j \lambda_{jq} \Big)^K u_q = 0 \tag{7.53}$$

for K large enough (since we assumed $x \in \mathrm{NR}(\mathfrak{g})$). This implies that for a given q we must have

$$u_q = 0 \quad \text{or} \quad \sum_j \xi^k \lambda_{jq} = 0. \tag{7.54}$$

In turn this means

$$\mathrm{ad}\, x(u_q) = 0, \quad \forall q.$$

Consequently $\mathrm{ad}\, x(u) = 0$ and finally

$$x \in \mathrm{cent}_{\mathfrak{g}}(u),$$

i.e., $\mathrm{NR}(\mathfrak{g}) \subseteq \mathrm{cent}_{\mathfrak{g}}(u) = \mathfrak{m}$. If $\mathrm{cent}_{\mathfrak{g}}(u)$ is not nilpotent we must take its nilradical and we obtain (7.42). □

7.3. Examples

Let us present several examples to illustrate the nilradical algorithm. We shall also demonstrate how the methods described in this and previous Chapters can be used to identify the given Lie algebra.

For simplicity we shall denote various cosets arising in the procedure by the same letter used for the corresponding element in $\mathfrak{g}$, e.g., $e_k \equiv e_k + \mathfrak{b}_1$.

Example 7.12. Let $\mathfrak{g}$ be a 6-dimensional solvable Lie algebra with nonvanishing brackets

$$\begin{gathered} [e_1, e_6] = -e_1, \quad [e_2, e_6] = -e_2, \quad [e_3, e_5] = e_1, \\ [e_4, e_5] = e_2, \quad [e_5, e_6] = -e_5. \end{gathered} \tag{7.55}$$

Its derived algebra $D(\mathfrak{g}) = \mathrm{span}\{e_1, e_2, e_5\}$ is Abelian and the center $C(\mathfrak{g})$ vanishes. Thus we proceed immediately to Step 4 of the algorithm. We choose the basis

$$u_1 = e_1, \quad u_2 = e_2, \quad u_3 = e_5, \quad x_1 = e_3, \quad x_2 = e_4, \quad x_3 = e_6. \tag{7.56}$$

In Step 5 we consider $u = u_1$ and x_1. We find

$$[x_1, u_1] = [e_3, e_1] = 0.$$

Thus $P_1(\lambda) = \lambda$ and we proceed to Step 6. We have

$$\mathfrak{b}_1 = [x_1, D(\mathfrak{g})] = [e_3, D(\mathfrak{g})] = \mathrm{span}\{e_1\}.$$

The quotient algebra $\mathfrak{g}/\mathfrak{b}_1$ is spanned by $e_2 + \mathfrak{b}_1, \ldots, e_6 + \mathfrak{b}_1$. The algebra $\mathfrak{g}/\mathfrak{b}_1$ has nontrivial center $C(\mathfrak{g}/\mathfrak{b}_1) = \mathfrak{z}_\infty(\mathfrak{g}/\mathfrak{b}_1) = \mathrm{span}\{e_3\}$. Therefore, we construct another quotient $\tilde{\mathfrak{g}} = (\mathfrak{g}/\mathfrak{b}_1)/\mathrm{span}\{e_3\} = \mathfrak{g}/\mathrm{span}\{e_1, e_3\}$ and proceed to Step 5 with the basis

$$\tilde{u}_1 = e_2, \quad \tilde{u}_2 = e_5, \quad \tilde{x}_1 = e_4, \quad \tilde{x}_2 = e_6 \quad (+\,\mathrm{span}\{e_1, e_3\}). \tag{7.57}$$

We have again

$$[\tilde{x}_1, \tilde{u}_1] = [e_4, e_2] = 0 \pmod{\mathrm{span}\{e_1, e_3\}},$$

i.e., we proceed to Step 6, factor out $\tilde{\mathfrak{b}}_1 = [\tilde{x}_1, \operatorname{span}\{\tilde{u}_1, \tilde{u}_1\}] = \operatorname{span}\{e_2\}$ and the arising center $\operatorname{span}\{\tilde{x}_1\} \simeq \operatorname{span}\{e_4\}$ of the quotient algebra, arriving at the algebra

$$\begin{aligned} &\hat{\mathfrak{g}} = \operatorname{span}\{\hat{u}_1, \hat{x}_1\}, && [\hat{x}_1, \hat{u}_1] = \hat{u}_1, \\ &\hat{u}_1 = e_5 \quad (+\operatorname{span}\{e_1, \dots, e_4\}), && \hat{x}_1 = e_6 \quad (+\operatorname{span}\{e_1, \dots, e_4\}). \end{aligned} \tag{7.58}$$

In Step 5 we have $[\hat{x}_1, \hat{u}_1] = \hat{u}_1$ which implies $\widehat{P}_1(\lambda) = \lambda - 1$. The polynomial $\widehat{P}_1$ is square-free with a nonvanishing constant term. Thus we proceed to Step 8 and form the centralizer

$$\widehat{\mathfrak{m}} = \operatorname{cent}_{\hat{\mathfrak{g}}}(\hat{u}_1) = \operatorname{span}\{u_1\} = D(\hat{\mathfrak{g}}).$$

Equation (7.43) implies that we have found the nilradical $\operatorname{NR}(\hat{\mathfrak{g}}) = \operatorname{span}\{u_1\}$. Retracing back our quotients we find that the nilradical of $\tilde{\mathfrak{g}}$ coincides with the preimage $\widetilde{\mathfrak{m}}$ of $\operatorname{NR}(\hat{\mathfrak{g}})$ under the quotient

$$\operatorname{NR}(\tilde{\mathfrak{g}}) = \operatorname{span}\{\tilde{u}_1, \tilde{u}_2, \tilde{x}_1\} \simeq \operatorname{span}\{e_2, e_4, e_5\}$$

and similarly we find the nilradical of the original algebra $\mathfrak{g}$

$$\operatorname{NR}(\mathfrak{g}) = \operatorname{span}\{u_1, u_2, u_3, x_1, x_2\} = \operatorname{span}\{e_1, e_2, e_3, e_4, e_5\} \tag{7.59}$$

with the Lie brackets

$$[e_3, e_5] = e_1, \quad [e_4, e_5] = e_2. \tag{7.60}$$

Thus the algebra (7.55) must be isomorphic to one of the algebras listed in Section 19.5. By comparison of the dimensions of the derived algebra we conclude that it is the algebra $\mathfrak{s}_{6,149}$. As a matter of fact, it is expressed in the same basis.

In the example just presented we twice had to construct the ideal $\mathfrak{b}_1$ of (7.37). In our second example we will demonstrate the construction of the ideal $\mathfrak{b}_2$ of (7.39).

Example 7.13. Let $\mathfrak{g}$ be a 6-dimensional solvable Lie algebra with nonvanishing brackets

$$\begin{aligned} &[e_1, e_6] = -2e_1, && [e_2, e_4] = e_1, && [e_2, e_6] = -2e_2, \\ &[e_3, e_5] = e_1, && [e_3, e_6] = -e_3 - e_5, && [e_5, e_6] = -e_5. \end{aligned} \tag{7.61}$$

Its derived algebra $D(\mathfrak{g}) = \operatorname{span}\{e_1, e_2, e_3, e_5\}$ is non-Abelian and its derived algebra satisfies $D^2(\mathfrak{g}) = \operatorname{span}\{e_1\}$. Thus in Step 2 we factor out $D^2(\mathfrak{g}) = \operatorname{span}\{e_1\}$.

The quotient algebra $\mathfrak{g}/D^2(\mathfrak{g})$ has a nonvanishing center

$$C\big(\mathfrak{g}/D^2(\mathfrak{g})\big) = \mathfrak{z}_\infty\big(\mathfrak{g}/D^2(\mathfrak{g})\big) = \operatorname{span}\{e_4\},$$

i.e., Step 3 instructs us to further remove $\mathfrak{z}_\infty\big(\mathfrak{g}/D^2(\mathfrak{g})\big)$. Thus we obtain the algebra

$$\tilde{\mathfrak{g}} = \mathfrak{g}/\operatorname{span}\{e_1, e_4\} \simeq \operatorname{span}\{e_2, e_3, e_5, e_6\} \tag{7.62}$$

with the brackets

$$[e_2, e_6] = -2e_2, \quad [e_3, e_6] = -e_3 - e_5, \quad [e_5, e_6] = -e_5. \tag{7.63}$$

In Step 4 we choose the basis

$$u_1 = e_3, \quad u_2 = e_5, \quad u_3 = e_2, \quad x_1 = e_6. \tag{7.64}$$

In Step 5 we consider $u = u_1 = e_3$ and $x_1 = e_6$. We find

$$\begin{aligned} [x_1, u_1] &= [e_6, e_3] = e_3 + e_5 = u_1 + u_2 \equiv u_{11}, \\ [x_1, u_{11}] &= [e_6, e_3 + e_5] = e_3 + 2e_5 = 2u_{11} - u_1. \end{aligned}$$

Thus we have $\operatorname{ad} x_1|_{X_1}$ in (7.35) in the following form

$$\operatorname{ad} x_1|_{X_1} = \begin{pmatrix} 0 & -1 \\ 1 & 2 \end{pmatrix}. \tag{7.65}$$

Its characteristic polynomial

$$P_1(\lambda) = \lambda^2 - 2\lambda + 1 = (\lambda - 1)^2 \tag{7.66}$$

does not have a root $\lambda = 0$ and is not square-free. Thus in Step 7 we are instructed to form the ideal

$$\mathfrak{b}_2 = (\operatorname{ad} x_1 - \mathbf{1})D(\mathfrak{g}) = (\operatorname{ad} e_6 - \mathbf{1})D(\mathfrak{g}) = \operatorname{span}\{e_2, e_5\} \tag{7.67}$$

and factor it out of the algebra $\tilde{\mathfrak{g}}$. We obtain the algebra

$$\hat{\mathfrak{g}} = \tilde{\mathfrak{g}}/\operatorname{span}\{e_2, e_5\} \simeq \operatorname{span}\{e_3, e_6\}, \qquad [e_3, e_6] = -e_3.$$

The nilradical of the algebra $\hat{\mathfrak{g}}$ is found to be $\operatorname{NR}(\hat{\mathfrak{g}}) = \operatorname{span}\{e_3\}$ in Step 8, as in the previous example. Consequently, the nilradical of $\tilde{\mathfrak{g}}$ is the same as the nilradical of the preimage $\widetilde{\mathfrak{m}}$ of $\operatorname{NR}(\hat{\mathfrak{g}})$

$$\widetilde{\mathfrak{m}} = \operatorname{span}\{e_2, e_3, e_5\} \subset \tilde{\mathfrak{g}}.$$

The ideal $\widetilde{\mathfrak{m}}$ is nilpotent, i.e., $\operatorname{NR}(\tilde{\mathfrak{g}}) = \widetilde{\mathfrak{m}}$. Finally, we include the (hyper)center $C(\mathfrak{g}/D^2(\mathfrak{g}))$ and the second element of the derived series $\mathfrak{g}^{(2)} = D^2(\mathfrak{g}) = (\mathfrak{g}^2)^2$ to obtain the nilradical of $\mathfrak{g}$ in the form

$$\operatorname{NR}(\mathfrak{g}) = \operatorname{span}\{e_1, e_2, e_3, e_4, e_5\}. \tag{7.68}$$

with the Lie brackets

$$[e_2, e_4] = e_1, \quad [e_3, e_5] = e_1. \tag{7.69}$$

Thus $\operatorname{NR}(\mathfrak{g})$ is the Heisenberg algebra in two spatial dimensions $\mathfrak{n}_{5,4}$ and the algebra (7.55) must be isomorphic to one of the algebras listed in Section 19.8. By inspection of the list of algebras in Section 19.8 with the same structure of characteristic series we identify it as the algebra $\mathfrak{s}_{6,189}$.

Notice that which branch of the algorithm is followed is rather sensitive to the choice of basis in the algebra. Let us now consider an example where both $\mathfrak{b}_1$ and $\mathfrak{b}_2$ arise.

Example 7.14. Let $\mathfrak{g}$ be a 6-dimensional solvable Lie algebra spanned by $f_1, \dots, f_6$ with the following Lie brackets

$$\begin{array}{lll} [f_1, f_4] = -3f_1 - f_2, & [f_1, f_5] = -f_1 - f_2, & \\ [f_1, f_6] = -f_3, & [f_2, f_4] = f_1 - f_2, & \\ [f_2, f_5] = f_1 + f_2, & [f_2, f_6] = f_3, & [f_3, f_4] = -2f_3, \\ [f_4, f_5] = 2f_1 + 2f_2, & [f_4, f_6] = 2f_1 + 2f_3, & [f_5, f_6] = f_1 + f_2. \end{array} \tag{7.70}$$

The Abelian derived algebra is expressed as $D(\mathfrak{g}) = \operatorname{span}\{f_1, f_2, f_3\}$ and the center $C(\mathfrak{g})$ vanishes. Thus, we proceed to Step 4. The basis $(f_1, \dots, f_6)$ has the required form (7.32) with the identification

$$u_1 = f_1, \quad u_2 = f_2, \quad u_3 = f_3, \quad x_1 = f_4, \quad x_2 = f_5, \quad x_3 = f_6. \tag{7.71}$$

In Step 5 we consider $u_{01} = f_1$ and $x_1 = f_4$. We have

$$u_{01} = f_1, \quad u_{11} = [f_4, f_1] = 3f_1 + f_2, \quad u_{21} = [f_4, 3f_1 + f_2] = 4u_{11} - 4u_{01}, \tag{7.72}$$

i.e., $\operatorname{ad} x_1|_{X_1}$ in (7.35) takes the form

$$\operatorname{ad} x_1|_{X_1} = \begin{pmatrix} 0 & -4 \\ 1 & 4 \end{pmatrix}. \tag{7.73}$$

Its characteristic polynomial

$$P_1(\lambda) = \lambda^2 - 4\lambda + 4 = (\lambda - 2)^2 \tag{7.74}$$

does not have a root $\lambda = 0$ but is not square-free. We proceed to Step 7 and construct the ideal

$$\mathfrak{b}_2 = (\operatorname{ad} x_1 - 2 \cdot \mathbf{1})\mathcal{D}(\mathfrak{g}) = (\operatorname{ad} f_4 - 2 \cdot \mathbf{1})\mathcal{D}(\mathfrak{g}) = \operatorname{span}\{f_1 + f_2\} \tag{7.75}$$

and factor it out of the algebra $\mathfrak{g}$, i.e., we identify f_2 with $-f_1$. The factor algebra $\mathfrak{g}/\mathfrak{b}_2 \simeq \operatorname{span}\{f_1, f_3, f_4, f_5, f_6\}$ has the Lie brackets

$$[f_1, f_4] = -2f_1, \quad [f_1, f_6] = -f_3, \quad [f_3, f_4] = -2f_3, \quad [f_4, f_6] = 2f_1 + 2f_3. \tag{7.76}$$

It has the Abelian derived algebra $\mathcal{D}(\mathfrak{g}/\mathfrak{b}_2) = \operatorname{span}\{f_1, f_3\}$ and the 1-dimensional (hyper)center

$$C(\mathfrak{g}/\mathfrak{b}_2) = \operatorname{span}\{f_5\}. \tag{7.77}$$

In Step 3 we factor out $C(\mathfrak{g}/\mathfrak{b}_2)$, i.e., we consider the 4-dimensional algebra $\tilde{\mathfrak{g}} = \operatorname{span}\{f_1, f_3, f_4, f_6\}$ with the Lie brackets (7.76). Again, the basis is already ordered as in Step 4. We proceed to Step 5 with $\tilde{u} = f_1$ and $\tilde{x}_1 = f_4$. We find that

$$[\tilde{x}_1, \tilde{u}] = [f_4, f_1] = 2f_1 = 2\tilde{u}. \tag{7.78}$$

The characteristic polynomial is

$$\widetilde{P}_1(\lambda) = \lambda - 2, \tag{7.79}$$

i.e., square-free without vanishing root. We proceed to Step 8 and have to repeat the procedure for another choice of the vector $\tilde{x}$. We take the vector $\tilde{x}_2 = f_6$ and find

$$\tilde{u}_{12} = [\tilde{x}_2, \tilde{u}] = [f_6, f_1] = f_3, \quad \tilde{u}_{22} = [\tilde{x}_2, \tilde{u}_{12}] = [f_6, f_3] = 0. \tag{7.80}$$

The characteristic polynomial is

$$\widetilde{P}_2(\lambda) = \lambda^2, \tag{7.81}$$

with root $\lambda = 0$. We proceed to Step 6 and construct the ideal

$$\mathfrak{b}_1 = [\tilde{x}_2, \mathcal{D}(\mathfrak{g})] = [f_6, \operatorname{span}\{f_1, f_3\}] = \operatorname{span}\{f_3\}. \tag{7.82}$$

We factor $\mathfrak{b}_1$ out of the algebra $\tilde{\mathfrak{g}}$ and obtain the algebra $\hat{\mathfrak{g}} \simeq \operatorname{span}\{f_1, f_4, f_6\}$ with the Lie brackets

$$[f_1, f_4] = -2f_1, \quad [f_4, f_6] = 2f_1. \tag{7.83}$$

and return to Step 2. The algebra $\hat{\mathfrak{g}}$ has an Abelian derived algebra and the 1-dimensional (hyper)center

$$C(\hat{\mathfrak{g}}) = \operatorname{span}\{f_1 - f_6\}. \tag{7.84}$$

We factor $\mathfrak{z}(\hat{\mathfrak{g}})$ out of $\hat{\mathfrak{g}}$, i.e., identify f_6 with f_1, and we obtain the two-dimensional solvable Lie algebra

$$[f_1, f_4] = -2f_1 \tag{7.85}$$

which has the nilradical f_1, as can be found either by inspection or repeating Steps 4 to 8 once again.

Finally, we lift the ideal

$$\text{span}\{f_1\} = \text{NR}\big(\mathfrak{g}/\mathfrak{b}_2/C(\mathfrak{g}/\mathfrak{b}_2)/\mathfrak{b}_1/C(\hat{\mathfrak{g}})\big) \tag{7.86}$$

to an ideal in $\mathfrak{g}$,

$$\begin{aligned} \text{NR}(\mathfrak{g}) &= \text{span}\{f_1\}\dot{+}\text{span}\{f_1 - f_6\}\dot{+}\text{span}\{f_3\}\dot{+}\text{span}\{f_5\}\dot{+}\text{span}\{f_1 + f_2\} \\ &= \text{span}\{f_1, f_2, f_3, f_5, f_6\} \end{aligned} \tag{7.87}$$

since both ideals $\mathfrak{m}$ arising in Steps 6 and 7 are nilpotent. The nilradical $\text{NR}(\mathfrak{g})$ has the Lie brackets

$$\begin{gathered} [f_1, f_5] = -f_1 - f_2, \quad [f_1, f_6] = -f_3, \quad [f_2, f_5] = f_1 + f_2, \\ [f_2, f_6] = f_3, \quad [f_5, f_6] = f_1 + f_2. \end{gathered}$$

Computation of its characteristic series reveals that it is isomorphic to the nilpotent Lie algebra $\mathfrak{n}_{5,4}$. Thus, the algebra (7.70) must be isomorphic to one of the algebras listed in Section 19.5. The dimensions of its derived algebra and of its center imply that it is isomorphic to the algebra $\mathfrak{s}_{6,149}$, i.e., to the algebra (7.55) of Example 7.12. It is indeed so, the algebra (7.70) is obtained from the algebra $\mathfrak{s}_{6,149}$ by the change of basis

$$\begin{gathered} f_1 = e_1 + e_5, \quad f_2 = e_1 - e_5, \quad f_3 = e_2, \\ f_4 = 2(e_3 - e_5 + e_6), \quad f_5 = 2e_3, \quad f_6 = e_4 + e_5. \end{gathered} \tag{7.88}$$

The nilradical (7.87) of the algebra (7.70) is of course just the nilradical (7.59) expressed in another basis (7.88). However, as we have seen, the branching of the algorithm and the complexity of computation were different.

In our last example we will present an algebra for which neither ideals $\mathfrak{b}_1$ nor $\mathfrak{b}_2$ arise.

Example 7.15. Let $\mathfrak{g}$ be a 6-dimensional solvable Lie algebra with nonvanishing brackets

$$\begin{aligned} &[e_1, e_5] = -e_1, && [e_1, e_6] = e_1, && [e_2, e_5] = -2e_2, && [e_3, e_5] = -3e_3, \\ &[e_3, e_6] = -e_3, && [e_4, e_5] = -4e_4, && [e_4, e_6] = -2e_4. \end{aligned} \tag{7.89}$$

Its derived algebra $D(\mathfrak{g}) = \text{span}\{e_1, e_2, e_3, e_4\}$ is Abelian and $\mathfrak{g}$ has a vanishing center. Thus we proceed to Step 4. We choose the basis

$$u_1 = e_1, \quad u_2 = e_2, \quad u_3 = e_3, \quad u_4 = e_4, \quad x_1 = e_5, \quad x_2 = e_6. \tag{7.90}$$

In Step 5 we consider $u = u_1 = e_1$ and $x_1 = e_5$. We find

$$[x_1, u] = u.$$

Thus we have $X_1 = \text{span}\{e_1\}$ and $\text{ad}\, x_1|_{X_1} = \mathbf{1}$ in (7.35). The characteristic polynomial $P_1(\lambda) = \lambda - 1$ is square-free and without vanishing root. We proceed directly to Step 8 and are instructed to repeat Step 5 with $u = u_1 = e_1$ and $x_2 = e_6$. We find

$$[x_2, u] = -u, \quad X_2 = \text{span}\{u\} = \text{span}\{e_1\}, \quad \text{ad}\, x_2|_{X_2} = -\mathbf{1}, \quad P_2(\lambda) = \lambda + 1.$$

We proceed to Step 8. Since we have exhausted all x_i and in both cases found square-free characteristic polynomial of $\text{ad}\, x_j|_{X_j}$ without a root equal to 0, we construct

$$\mathfrak{m} = \text{cent}_{\mathfrak{g}}(e_1) = \text{span}\{e_1, e_2, e_3, e_4\} \tag{7.91}$$

which equals the derived algebra $D(\mathfrak{g})$. Thus the algorithm terminates and we have found the nilradical of $\mathfrak{g}$,

$$\mathrm{NR}(\mathfrak{g}) = \mathfrak{m} = \mathrm{span}\{e_1, e_2, e_3, e_4\}. \tag{7.92}$$

The nilradical is Abelian, thus the algebra (7.89) must be contained in the list presented in Section 19.11 among the algebras characterized by the values

$$\mathrm{US} = [0], \quad \mathrm{CS} = [6, 4], \quad \mathrm{DS} = [6, 4, 0]. \tag{7.93}$$

We compute the Casimir invariants using the procedure described in Section 3.2 and find two functionally independent rational invariants

$$I_1 = \frac{e_1 e_3}{e_2^2}, \quad I_2 = \frac{e_1^2 e_4}{e_2^3}. \tag{7.94}$$

The only algebra in Section 19.11 of the structure (7.93) which has two rational invariants is $\mathfrak{s}_{6,213}$ for special values of the parameters a, b, c, d. Thus the algebra (7.89) must be contained in the class $\mathfrak{s}_{6,213}$ and by an explicit calculation one finds that it is so for the values

$$a = -1, \quad b = -2, \quad c = 2, \quad d = 3. \tag{7.95}$$

The change of basis taking us from $\mathfrak{s}_{6,213}$ to the algebra (7.89) is

$$e_1' = e_4, \quad e_2' = e_3, \quad e_3' = e_1, \quad e_4' = e_2, \quad e_5' = e_5 + 2e_6, \quad e_6' = -e_5. \tag{7.96}$$

Notice that also in this example the sequence of steps depends on the chosen basis (7.90). If we had chosen instead $u_1 = e_2$ and $x_1 = e_6$, the ideal $\mathfrak{b}_1$ of Step 6 would arise in our computation and more steps would be necessary to perform.

7.4. Identification of the nilradical using the Killing form

Another method which is often used to find the nilradical of the given Lie algebra $\mathfrak{g}$ is based on the properties of the Killing form. Its essential advantage is its computational simplicity. Its major disadvantage is that it does not always lead to a conclusive answer.

Let $\mathfrak{g}$ be a solvable Lie algebra. From the general properties of derivations of solvable Lie algebras we know that

$$\mathrm{ad}\, x \colon \mathfrak{g} \to \mathrm{NR}(\mathfrak{g}) \tag{7.97}$$

for any $x \in \mathfrak{g}$ and

$$\mathrm{ad}\, y \colon \mathrm{NR}^k(\mathfrak{g}) \to \mathrm{NR}^{k+1}(\mathfrak{g}) \tag{7.98}$$

for any $k \geq 1$ and $y \in \mathrm{NR}(\mathfrak{g})$. Thus, the composition $\mathrm{ad}\, y \circ \mathrm{ad}\, x$ of these two operators is nilpotent and consequently

$$K(y, x) = \mathrm{Tr}(\mathrm{ad}\, y \circ \mathrm{ad}\, x) = 0. \tag{7.99}$$

To sum up, we have found that

$$\mathrm{NR}(\mathfrak{g}) \subseteq \mathfrak{g}^{\perp}, \tag{7.100}$$

where $\mathfrak{g}^{\perp}$ is the orthogonal complement of $\mathfrak{g}$ with respect to the Killing form,

$$\mathfrak{g}^{\perp} = \{x \in \mathfrak{g} \mid K(x, \mathfrak{g}) = 0\}. \tag{7.101}$$

Unfortunately, equality does not in general hold in (7.100), as the example of a complex 3-dimensional solvable algebra

$$\mathfrak{g} = \mathrm{span}\{e_1, e_2, e_3\}, \qquad [e_1, e_3] = e_1, \quad [e_2, e_3] = \mathrm{i} e_2 \tag{7.102}$$

with vanishing Killing form clearly demonstrates.

On the other hand, given the relative computational complexity of the algorithm described in Section 7.2 it appears to be reasonable to first compute the ideal $\mathfrak{g}^{\perp}$ and check whether it is nilpotent. If yes, we have identified the nilradical of $\mathfrak{g}$,

$$\text{NR}(\mathfrak{g}) = \mathfrak{g}^{\perp}. \tag{7.103}$$

If not, one should compute the codimension of $\mathfrak{g}^2 + \mathfrak{z}_\infty(\mathfrak{g})$ in $\mathfrak{g}^{\perp}$. If it equals one, the nilradical is

$$\text{NR}(\mathfrak{g}) = \mathfrak{g}^2 + \mathfrak{z}_\infty(\mathfrak{g}) \tag{7.104}$$

on dimensional grounds. Only if the codimension is larger than 1, the relations

$$\mathfrak{g}^2 + \mathfrak{z}_\infty(\mathfrak{g}) \subseteq \text{NR}(\mathfrak{g}) \subsetneq \mathfrak{g}^{\perp} \tag{7.105}$$

are insufficient to determine the nilradical of $\mathfrak{g}$ immediately and we have to perform the algorithm as described in Section 7.2.

We observe that in all examples discussed in Section 7.3 the nilradical coincides with $\mathfrak{g}^{\perp}$.

Part 3

Nilpotent, Solvable and Levi Decomposable Lie Algebras

CHAPTER 8

Nilpotent Lie Algebras

The classification of low-dimensional nilpotent Lie algebras as performed in [77] is based on the fact that such algebras necessarily possess a nonvanishing Abelian ideal. Examples of such ideals are the last nonvanishing ideal in the lower central series or the center.

Let us assume that we want to classify nilpotent Lie algebras $\mathfrak{n}$ of given dimension n with a k-dimensional maximal Abelian ideal $\mathfrak{a}$. The procedure starts with the classification of nonequivalent representations of $(n-k)$-dimensional nilpotent algebras on the maximal Abelian ideal $\mathfrak{a}$. We are only interested in representations by nilpotent matrices. The knowledge of these defines the Lie brackets of the form $[\mathfrak{n}, \mathfrak{a}]$. The remaining Lie brackets, i.e., between pairs of vectors in the set $\mathfrak{n} \backslash \mathfrak{a}$ are now specified modulo $\mathfrak{a}$ and different nonequivalent possibilities for them must be found.

Contrary to the case of the radical or the nilradical of a Lie algebra, the maximal Abelian ideal of a nilpotent Lie algebra is not necessarily unique. The reason for this is that the sum of Abelian ideals is not necessarily an Abelian ideal. Consequently, the maximal one cannot be constructed simply by taking the sum of all Abelian ideals. It may even happen that different maximal Abelian ideals in a given nilpotent algebra have different dimensions. Proper care must be taken in order to avoid redundancies in the classification obtained in this manner.

8.1. Maximal Abelian ideals and their extensions

Let us assume that $\mathfrak{n}$ is a nilpotent Lie algebra and $\mathfrak{a}$ is one of its maximal Abelian ideals (possibly unique). Let $(e_1, \dots, e_k)$ be a basis of $\mathfrak{a}$ and $(e_{k+1}, \dots, e_n)$ its extension to a basis of $\mathfrak{n}$. Let us construct the adjoint representation of $\mathfrak{n}$, restrict it to the ideal $\mathfrak{a}$ and find $\mathrm{ad}|_{\mathfrak{a}}(x)$, $x \in \mathfrak{n}$. For any $x \in \mathfrak{a}$ we have $\mathrm{ad}|_{\mathfrak{a}}(x) = 0$. On the other hand, all $\mathrm{ad}|_{\mathfrak{a}}(e_j)$, $k < j \leq n$ must be linearly independent nilpotent matrices; otherwise the ideal $\mathfrak{a}$ would not be maximal or the algebra $\mathfrak{n}$ would not be nilpotent. Therefore, the representation of $\mathfrak{n}$

$$x \in \mathfrak{n} \to \mathrm{ad}|_{\mathfrak{a}}(x)$$

gives rise to a faithful representation of the nilpotent factor algebra $\mathfrak{n}/\mathfrak{a}$ on $\mathfrak{a}$ by nilpotent matrices. Such a representation can have dimension at most $k(k-1)/2$ and we obtain the restriction

$$n - k \leq \frac{k(k-1)}{2}$$

which leads to the lower bound on the dimension k of the maximal Abelian ideal

$$\tag{8.1} \tfrac{1}{2}\big(\sqrt{1+8n} - 1\big) \leq k \leq n.$$

(The bound (8.1) is due to V. V. Morozov, [77].) This relation greatly facilitates the classification of nilpotent algebras in dimensions $n = 1, 2, 3, 4$. In these dimensions we have

n	k
1	1
2	2
3	2, 3
4	3, 4

and consequently either $\mathfrak{n}$ is Abelian ($\mathfrak{n} = \mathfrak{a}$) or $\mathfrak{n}/\mathfrak{a}$ is one-dimensional and the classification is completed by considering all possible canonical forms of nilpotent 2×2 or 3×3 matrices. In higher dimensions we may have $n - k \geq 2$ and we proceed as follows.

Assuming that n, k are given we consider step by step all nilpotent Lie algebras $\mathfrak{m}$ of dimension $n - k$. Since $n - k < n$ these are assumed to be already known. We find all their inequivalent representations on the k-dimensional vector space $\mathfrak{a}$ expressed entirely in terms of nilpotent matrices. For each of these representations we presume that $\mathfrak{m} = \mathfrak{n}/\mathfrak{a}$ and reconstruct from it the Lie brackets in $\mathfrak{n}$. As was noted above, some of these are defined only modulo $\mathfrak{a}$. Finally, we employ a convenient change of basis in $\mathfrak{n}$ that preserves the already fixed brackets $[e_i, e_j]$, $i \leq k < j$ in order to bring the nilpotent algebra $\mathfrak{n}$ to a chosen (and therefore unique) canonical form. Such a transformation allows a simplification of the brackets $[e_i, e_j]$, $k < i, j$.

We note that nilpotent algebras arising in this way may be decomposable and that algebras constructed starting from different maximal Abelian ideals may turn out to be isomorphic. These questions must be addressed once the computation is finished. The decomposable algebras must be deleted from the list. Isomorphic algebras must be identified and precisely one representative of each isomorphism class chosen and declared "canonical."

As an example showing all the aspects of the classification procedure let us consider the case $\dim \mathfrak{n} = 5$. The restriction (8.1) allows three possible values of $k = 3, 4, 5$. For the obtained nilpotent algebras we use the notation $\mathfrak{n}_{n,j}$, to be used consistently in the tables presented below. The letter $\mathfrak{n}$ stands for "nilpotent," the first subscript gives the dimension of the algebra, i.e., $n = 5$ here. The second subscript $j = 1, 2, 3, \ldots$ labels nonisomorphic indecomposable Lie algebras of the dimension n.

When $k = 5$ the nilpotent algebra itself is Abelian and consequently decomposable.

For $k = 4$ we have one element e_5 outside of the maximal Abelian ideal $\mathfrak{a} = \operatorname{span}\{e_1, e_2, e_3, e_4\}$. Therefore, the only information needed for the full knowledge of $\mathfrak{n}$ is the action of e_5 on the Abelian algebra $\mathfrak{a}$. This is specified in terms of the nilpotent linear operator A

$$A \colon \mathfrak{a} \to \mathfrak{a} \colon [x, e_5] = A(x). \tag{8.2}$$

Such an operator A can be chosen in its Jordan canonical form and will have zeros on the diagonal. When any of the irreducible blocks on the diagonal is 1×1 the corresponding algebra $\mathfrak{n}$ is decomposable. Indeed, it has a nontrivial central component. Consequently, there are only two nonequivalent possibilities

$$A = \begin{pmatrix} 0 & 1 & 0 & 0 \\ 0 & 0 & 1 & 0 \\ 0 & 0 & 0 & 1 \\ 0 & 0 & 0 & 0 \end{pmatrix}, \quad \tilde{A} = \left(\begin{array}{cc|cc} 0 & 1 & 0 & 0 \\ 0 & 0 & 0 & 0 \\ \hline 0 & 0 & 0 & 1 \\ 0 & 0 & 0 & 0 \end{array}\right). \tag{8.3}$$

The corresponding algebras are $\mathfrak{n}_{5,5}$ (cf. Section 18.1)

	e_1	e_2	e_3	e_4	e_5
e_1	0	0	0	0	0
e_2		0	0	0	e_1
e_3			0	0	e_2
e_4				0	e_3

and $\tilde{\mathfrak{n}}_{5,1}$

	e_1	e_2	e_3	e_4	e_5
e_1	0	0	0	0	0
e_2		0	0	0	e_1
e_3			0	0	0
e_4				0	e_3

.

The algebra $\tilde{\mathfrak{n}}_{5,1}$ is written in a form which naturally arose during the procedure but is not the most convenient one from the point of view of its algebraic structure, namely the characteristic series. Therefore we perform a change of basis in order to arrive at its form $\mathfrak{n}_{5,1}$ presented in Section 18.1

	e_1	e_2	e_3	e_4	e_5
e_1	0	0	0	0	0
e_2		0	0	0	0
e_3			0	0	e_1
e_4				0	e_2

.

Any 5-dimensional indecomposable nilpotent algebra with a 4-dimensional maximal Abelian nilradical is isomorphic to $\mathfrak{n}_{5,1}$ or $\mathfrak{n}_{5,5}$.

Finally, we consider the case $k = 3$. That means that we have two elements e_4, e_5 acting on the maximal Abelian ideal $\mathfrak{a} = \operatorname{span}\{e_1, e_2, e_3\}$. The two-dimensional factor algebra $\mathfrak{m} = \mathfrak{n}/\mathfrak{a}$ is Abelian (since no two-dimensional indecomposable nilpotent algebra exists). We write its representation on $\mathfrak{a}$ in terms of two operators

$$A_{1,2} = -\operatorname{ad}|_{\mathfrak{a}}(e_{4,5}), \qquad \text{i.e., } [x, e_4] = A_1(x),\ [x, e_5] = A_2(x). \tag{8.4}$$

Thus we need to classify all two-dimensional Abelian subalgebras of $\mathfrak{sl}(3, \mathbb{F})$ that are realized by nilpotent matrices. All subalgebras of $\mathfrak{sl}(3, \mathbb{F})$ are known [136]. Three nonequivalent representations of $\mathfrak{n}/\mathfrak{a}$ on $\mathfrak{a}$ in terms of nilpotent matrices exist, namely

$$A_1 = \begin{pmatrix} 0 & 0 & 1 \\ 0 & 0 & 0 \\ 0 & 0 & 0 \end{pmatrix}, \quad A_2 = \begin{pmatrix} 0 & 0 & 0 \\ 0 & 0 & 1 \\ 0 & 0 & 0 \end{pmatrix}, \tag{8.5}$$

$$\bar{A}_1 = \begin{pmatrix} 0 & 1 & 0 \\ 0 & 0 & 1 \\ 0 & 0 & 0 \end{pmatrix}, \quad \bar{A}_2 = \begin{pmatrix} 0 & 0 & 1 \\ 0 & 0 & 0 \\ 0 & 0 & 0 \end{pmatrix}, \tag{8.6}$$

$$\tilde{A}_1 = \begin{pmatrix} 0 & 1 & 0 \\ 0 & 0 & 0 \\ 0 & 0 & 0 \end{pmatrix}, \quad \tilde{A}_2 = \begin{pmatrix} 0 & 0 & 1 \\ 0 & 0 & 0 \\ 0 & 0 & 0 \end{pmatrix}, \tag{8.7}$$

distinguished by the maximal rank of a linear combination $a_1A_1 + a_2A_2$ and the dimension of the common nullspace of A_1 and A_2. For each of these representations we have to investigate the distinct possibilities for the Lie bracket $[e_4, e_5]$. In general we have

$$[e_4, e_5] = \alpha e_1 + \beta e_2 + \gamma e_3. \tag{8.8}$$

In the first case (8.5), we can perform a change of basis

$$e_4' = e_4 - \beta e_3, \quad e_5' = e_5 + \alpha e_3$$

to obtain $[e_4', e_5'] = \gamma e_3$. When $\gamma = 0$, the vector space span$\{e_1, e_2, e_4, e_5\}$ is a maximal Abelian ideal of dimension 4, the case already investigated above. When $\gamma \neq 0$ we rescale e_1, e_2, e_3 by γ to get $[e_4', e_5'] = e_3'$. Dropping the primes we find the algebra $\mathfrak{n}_{5,2}$

	e_1	e_2	e_3	e_4	e_5
e_1	0	0	0	0	0
e_2		0	0	0	0
e_3			0	e_2	e_1
e_4				0	e_3

.

Similarly, considering the case (8.6) we can simplify $[e_4, e_5]$ by a suitable addition of e_2, e_3 to e_4, e_5 into the form

$$[e_4, e_5] = \gamma e_3.$$

The two distinct subcases are $\gamma = 0$ and $\gamma \neq 0$. The first case gives rise to the algebra $\tilde{\mathfrak{n}}_{5,4}$

	e_1	e_2	e_3	e_4	e_5
e_1	0	0	0	0	0
e_2		0	0	0	e_1
e_3			0	e_1	e_2
e_4				0	0

which is again presented in Section 18.1 in a modified form $\mathfrak{n}_{5,4}$, better suited to the structure of the characteristic series and obtained by the change of basis $e_3' = -e_4$, $e_4' = e_3$

	e_1	e_2	e_3	e_4	e_5
e_1	0	0	0	0	0
e_2		0	0	0	e_1
e_3			0	e_1	0
e_4				0	e_2

.

The case $\gamma \neq 0$ is brought to the form $\mathfrak{n}_{5,6}$

	e_1	e_2	e_3	e_4	e_5
e_1	0	0	0	0	0
e_2		0	0	0	e_1
e_3			0	e_1	e_2
e_4				0	e_3

by a rescaling of $e_{1,2,3}$.

Finally, for the case (8.7) we add a multiple of e_3 to e_4 in order to obtain

$$[e_4, e_5] = \beta e_2 + \gamma e_3.$$

Now either $\beta = \gamma = 0$ and we obtain the algebra $\mathfrak{n}_{5,3}$

	e_1	e_2	e_3	e_4	e_5
e_1	0	0	0	0	0
e_2		0	0	e_1	0
e_3			0	0	e_1
e_4				0	0

or at least one of β, γ is nonvanishing. In that case we may assume without loss of generality that $\beta \neq 0$ (otherwise we interchange $e_2 \leftrightarrow e_3$ and $e_4 \leftrightarrow e_5$) and perform a change of basis

$$e_1' = \beta e_1, \quad e_2' = \beta e_2 + \gamma e_3, \quad e_3' = \beta e_3, \quad e_4' = e_4, \quad e_5' = e_5 - \frac{\gamma}{\beta} e_4.$$

This change of basis preserves the brackets defined by the matrices $\tilde{A}_{1,2}$, i.e., the Lie brackets between $e_{1,2,3}$ and $e_{4,5}$, and (8.8) reduces to

$$[e_4, e_5] = e_2.$$

The corresponding nilpotent algebra, however, turns out to be isomorphic to the algebra $\mathfrak{n}_{5,4}$ constructed above upon the interchange of e_4 and e_5 and the change of sign of e_2.

This completes the construction of all nonisomorphic indecomposable 5-dimensional nilpotent Lie algebras.

8.2. Classification of low-dimensional nilpotent Lie algebras

Complete lists of isomorphism classes of nilpotent Lie algebras have been published for dimensions $n \leq 7$ [51, 52, 77, 107, 110] and partial results exist for $n = 8$ and 9 [125–127]. The classification is performed along similar lines as that for $n = 5$ given in Section 8.1.

The method due to V. V. Morozov [77] consists of several steps.

(1) Choose a dimension $k \leq n - 1$ respecting the bound (8.1) and denote the basis for the corresponding maximal Abelian ideal $\mathfrak{a}$ by $(e_1, \ldots, e_k)$.

(2) The remaining basis elements $(e_{k+1}, \ldots, e_n)$ must form a nilpotent factor algebra $\mathfrak{n}/\mathfrak{a}$ represented by $k \times k$ matrices acting on $\mathfrak{a}$ as in (8.2) and (8.4). The factor algebra $\mathfrak{n}/\mathfrak{a}$ has dimension less than n, i.e., we can assume that it is already known and has a prescribed fixed structure. For each nilpotent $(n-k)$-dimensional nilpotent algebra $\mathfrak{n}/\mathfrak{a}$ we thus have to consider its representations, i.e., nonequivalent sets of linearly independent nilpotent matrices $A_1, \ldots, A_{n-k}$ with $A_j \in \mathbb{F}^{k,k}$. If $\mathfrak{n}/\mathfrak{a}$ is Abelian we have what is called an Abelian Nilpotent Subalgebra (ANS) of $\mathfrak{sl}(k, \mathbb{F})$ (i.e., Abelian Lie algebra realized by nilpotent matrices) [64, 122, 137].

(3) The classification of Abelian nilpotent subalgebras of $\mathfrak{sl}(k, \mathbb{F})$ is somewhat simplified by the existence of the so-called *Kravchuk normal form* of Maximal Abelian Nilpotent Subalgebras (MANS) of $\mathfrak{sl}(k, \mathbb{F})$, i.e., ANS that are their own centralizers in $\mathfrak{sl}(k, \mathbb{F})$. Each MANS is partly characterized by its *Kravchuk signature* $(\lambda\, \mu\, \nu)$ which is a triplet of nonnegative integers satisfying

$$\lambda \geq 1, \quad \nu \geq 1, \quad \lambda + \mu + \nu = k \tag{8.9}$$

where λ is the dimension of the common nullspace of the set of matrices A_j and ν is the codimension of their common image. Both λ and ν (and hence also μ) are basis independent (cf. Section 10, (10.11)).

In the case $n = 5$, $k = 3$ considered in Section 8.1, the factor algebra is indeed Abelian. The three cases (8.5), (8.6) and (8.7) correspond to Kravchuk signatures (2 0 1), (1 1 1) and (1 0 2), respectively.

If the factor algebra $\mathfrak{n}/\mathfrak{a}$ is nilpotent but not Abelian, it is also partly characterized by its Kravchuk signature and can be presented in the normal form [122]. For chosen values of $(\lambda\mu\nu)$ we have to find all nonequivalent k-dimensional representations of $\mathfrak{n}/\mathfrak{a}$.

(4) Once a representation A of the factor algebra $\mathfrak{n}/\mathfrak{a}$ on the Abelian ideal $\mathfrak{a}$ is given

$$A\colon \mathfrak{n}/\mathfrak{a} \to \mathfrak{gl}(\mathfrak{a}), \quad A(e_a)e_i = \sum_{j=1}^{k} (A_a)_i^j e_j, \quad a = k+1, \dots, n,\ i = 1, \dots, k \tag{8.10}$$

the Lie brackets

$$[e_i, e_a] = -A(e_a)e_i = -\sum_{j=1}^{k} (A_a)_i^j e_j$$

are fixed. The Lie brackets between e_a and e_b, $a, b = k+1, \dots, n$ have in general the form

$$[e_a, e_b] = \sum_{c=k+1}^{n} B_{ab}{}^c e_c + \sum_{i=1}^{k} C_{ab}{}^i e_i. \tag{8.11}$$

Equivalently, we can express them in terms of skew symmetric bilinear maps

$$B\colon \mathfrak{n}/\mathfrak{a} \times \mathfrak{n}/\mathfrak{a} \to \mathfrak{n}/\mathfrak{a}, \quad C\colon \mathfrak{n}/\mathfrak{a} \times \mathfrak{n}/\mathfrak{a} \to \mathfrak{a}$$

as

$$[e_a, e_b] = B(e_a, e_b) + C(e_a, e_b). \tag{8.12}$$

The map B represents the given Lie bracket on the factor algebra $\mathfrak{n}/\mathfrak{a}$,

$$B(\ ,\ ) = [\ ,\ ] \bmod \mathfrak{a}$$

and therefore is known. The Jacobi identities (e_a, e_b, e_c) imply that the map $C\colon \mathfrak{n}/\mathfrak{a} \times \mathfrak{n}/\mathfrak{a} \to \mathfrak{a}$ is a 2-cocycle for the representation A of the algebra $\mathfrak{n}/\mathfrak{a}$,

$$\begin{aligned} A(x)C(y,z) + A(y)C(z,x) + A(z)C(x,y) \\ + C\big(x, B(y,z)\big) + C\big(y, B(z,x)\big) + C\big(z, B(x,y)\big) = 0 \end{aligned} \tag{8.13}$$

(cf. Section 2.4). Any change of basis of the form

$$e_a \to e'_a = e_a + \sum_{i=1}^{k} M_a^i e_i \tag{8.14}$$

defines an operator $M\colon \mathfrak{n}/\mathfrak{a} \to \mathfrak{a}$, $M(e_a) = \sum_{i=1}^{k} M_a^i e_i$. In the language of Chevalley cohomology the map M defines also the 2-coboundary $\mathrm{d}M\colon \mathfrak{n}/\mathfrak{a} \times \mathfrak{n}/\mathfrak{a} \to \mathfrak{a}$

$$\mathrm{d}M(x,y) = A(x)M(y) - A(y)M(x) - M\big(B(x,y)\big).$$

The change of basis (8.14) modifies the Lie brackets (8.11). It leaves invariant the map B in (8.12), (i.e., its matrix $B_{ab}{}^c$ in (8.11)) but induces a change of the map C which can be expressed as

$$C \to C' = C + \mathrm{d}M. \tag{8.15}$$

Therefore, nilpotent algebras with the same algebraic structure of $\mathfrak{n}/\mathfrak{a}$ and representation A whose maps B differ by a coboundary are necessarily isomorphic. The classes of 2-cocycles modulo 2-coboundaries form the second cohomology group $H^2(\mathfrak{n}/\mathfrak{a}, \mathfrak{a}; A)$. Therefore, the cohomology theory can be employed in the identification of algebras which are not related by any transformation of the form (8.14), provided that the 2nd cohomology group $H^2(\mathfrak{n}/\mathfrak{a}, \mathfrak{a}; A)$ can be calculated.

On the other hand, a change of basis of the form (8.14) is not the only transformation of e_i, e_a which is allowed. In addition we can take linear combinations of the elements e_a induced by any automorphism of $\mathfrak{n}/\mathfrak{a}$ and also change the basis of $\mathfrak{a}$ in a way that commutes with the representation A. Last but not least, neither the representation A nor the maximal Abelian ideal of the given algebra $\mathfrak{n}$ is in general unique (several nonequivalent maximal Abelian ideals often exist, sometimes of different dimension). This leads to further possible isomorphisms between the nilpotent algebras found by the above procedure. That means that cohomology arguments alone are not sufficient to eliminate all the isomorphisms between the arising algebras.

Other classification methods have been devised, e.g., the Skjelbred – Sund construction of central extensions of nilpotent algebras based also on the cohomology theory of nilpotent Lie algebras [112]. It uses the center of $\mathfrak{n}$ instead of $\mathfrak{a}$, i.e., constructs all n-dimensional nilpotent algebras inductively from the knowledge of all lower dimensional ones. This method was successfully employed by M.P. Gong in his classification of 7-dimensional nilpotent algebras [51] which provides an independent check and several minor corrections[1] to the classification [110] performed by C. Seeley by a combination of other methods.

We mention that the number of indecomposable nonisomorphic nilpotent Lie algebras grows dramatically with dimension. Namely, in dimension 2 there is no such algebra, in dimensions 3 and 4 there is only one each, in dimension 5 there are 6 algebras. In dimension 6 the numbers start to differ depending on the field $\mathbb{F} = \mathbb{C}, \mathbb{R}$; there are 20 complex algebras and 24 real algebras. In dimension 7 continuous 1-parametric families of nonisomorphic algebras appear. The list of [51] which is probably the most reliable one presents the algebras organized into 119 classes over complex numbers and 24 additional classes over real numbers. A computerized classification of 9-dimensional nilpotent algebras with 7-dimensional maximal Abelian ideal in [127] claims to have found 24 168 classes of nonisomorphic nilpotent algebras.

A complete list of nilpotent Lie algebras up to dimension 6 is given in the respective chapters devoted to Lie algebras of different dimensions in Part 4. The maximal Abelian ideals $\mathfrak{a}$, which were used in the classification of nilpotent algebras, are either

[1] The list [110] almost coincides with [51] up to four corrections, only one substantial (the Lie brackets do not define an algebra), the others being restrictions on the parameter ranges or corrections of typographical errors.

- uniquely specified as an ideal in the characteristic series or a centralizer of such, or
- specified as a member of a class of equivalent ideals, described by their specific properties, which can be brought to the form presented in Part 4 by a suitable automorphism of $\mathfrak{n}$ (in this case we use the symbol $\sim$).

When there are several nonequivalent choices of the maximal Abelian ideal $\mathfrak{a}$, usually the one allowing its complement in $\mathfrak{n}$ to be an Abelian subalgebra was preferred if one exists. "s.a." is a shorthand notation for "subalgebra." The ideals employed are as follows:

- One-dimensional nilpotent Lie algebra

  $\mathfrak{n}_{1,1}$: $\mathfrak{a} = \mathfrak{n}_{1,1}$

- Three-dimensional nilpotent Lie algebra

  $\mathfrak{n}_{3,1}$: $\mathfrak{a} \sim \{\text{any 2-dim. s.a. of } \mathfrak{n}_{3,1}\}$

- Four-dimensional nilpotent Lie algebra

  $\mathfrak{n}_{4,1}$: $\mathfrak{a} = \operatorname{cent}\big(D(\mathfrak{n}_{4,1})\big) = \operatorname{cent}\big(\mathfrak{z}^2(\mathfrak{n}_{4,1})\big)$

- Five-dimensional nilpotent Lie algebras

  $\mathfrak{n}_{5,1}$: $\mathfrak{a} = \text{unique 4-dim. Abelian subalgebra of } \mathfrak{n}_{5,1}$
  $\mathfrak{n}_{5,2}$: $\mathfrak{a} = D(\mathfrak{n}_{5,2}) = \mathfrak{z}^2(\mathfrak{n}_{5,2})$
  $\mathfrak{n}_{5,3}$: $\mathfrak{a} \sim \{\text{any 3-dim. Abelian s.a. of } \mathfrak{n}_{5,3}\}$
  $\mathfrak{n}_{5,4}$: $\mathfrak{a} = \mathfrak{z}^2(\mathfrak{n}_{5,4})$
  $\mathfrak{n}_{5,5}$: $\mathfrak{a} = \operatorname{cent}\big(D(\mathfrak{n}_{5,5})\big) = \operatorname{cent}\big(\mathfrak{z}^3(\mathfrak{n}_{5,5})\big)$
  $\mathfrak{n}_{5,6}$: $\mathfrak{a} = D(\mathfrak{n}_{5,6}) = \mathfrak{z}^3(\mathfrak{n}_{5,6})$

- Six-dimensional nilpotent Lie algebras

  $\mathfrak{n}_{6,1}$: $\mathfrak{a} \sim \{\text{any 4-dim. Abelian s.a. of } \mathfrak{n}_{6,1}\}$
  $\mathfrak{n}_{6,2}$: $\mathfrak{a} \sim \{\text{any 4-dim. Abelian s.a. of } \mathfrak{n}_{6,2} \text{ such that } [\mathfrak{a}, \mathfrak{n}_{6,2}] = D(\mathfrak{n}_{6,2})\}$
  $\mathfrak{n}_{6,3}$: $\mathfrak{a} \sim \{\text{any 4-dim. Abelian s.a. of } \mathfrak{n}_{6,3}\}$
  $\mathfrak{n}_{6,4}$: $\mathfrak{a} = \operatorname{cent}\big(D(\mathfrak{n}_{6,4})\big) = \mathfrak{z}^2(\mathfrak{n}_{6,4})$
  $\mathfrak{n}_{6,5}$: $\mathfrak{a} = \operatorname{cent}\big(D(\mathfrak{n}_{6,5})\big) = \mathfrak{z}^2(\mathfrak{n}_{6,5})$
  $\mathfrak{n}_{6,6}$: $\mathfrak{a} = \operatorname{cent}\big(D(\mathfrak{n}_{6,6})\big) = \mathfrak{z}^2(\mathfrak{n}_{6,6})$
  $\mathfrak{n}_{6,7}$: $\mathfrak{a} = \operatorname{cent}\big(D(\mathfrak{n}_{6,7})\big) = \operatorname{cent}\big(\mathfrak{z}^2(\mathfrak{n}_{6,7})\big)$
  $\mathfrak{n}_{6,8}$: $\mathfrak{a} = \mathfrak{z}^2(\mathfrak{n}_{6,8})$
  $\mathfrak{n}_{6,9}$: $\mathfrak{a} = \mathfrak{z}^2(\mathfrak{n}_{6,9})$
  $\mathfrak{n}_{6,10}$: $\mathfrak{a} = D(\mathfrak{n}_{6,10}) = \mathfrak{z}^3(\mathfrak{n}_{6,10})$
  $\mathfrak{n}_{6,11}$: $\mathfrak{a} \sim \big\{\text{any 4-dim. Abelian s.a. of } \operatorname{cent}\big(D(\mathfrak{n}_{6,11})\big)\big\}$
  $\mathfrak{n}_{6,12}$: $\mathfrak{a} = \operatorname{cent}\big(D(\mathfrak{n}_{6,12})\big) = \operatorname{cent}\big(\mathfrak{z}^2(\mathfrak{n}_{6,12})\big)$
  $\mathfrak{n}_{6,13}$: $\mathfrak{a} = \operatorname{cent}\big(D(\mathfrak{n}_{6,13})\big) = \operatorname{cent}\big(\mathfrak{z}^2(\mathfrak{n}_{6,13})\big)$
  $\mathfrak{n}_{6,14}$: $\mathfrak{a} = D(\mathfrak{n}_{6,14}) = \mathfrak{z}^3(\mathfrak{n}_{6,14})$
  $\mathfrak{n}_{6,15}$: $\mathfrak{a} = \operatorname{cent}\big(D(\mathfrak{n}_{6,15})\big) = \mathfrak{z}^3(\mathfrak{n}_{6,15})$
  $\mathfrak{n}_{6,16}$: $\mathfrak{a} = \mathfrak{z}^3(\mathfrak{n}_{6,16})$
  $\mathfrak{n}_{6,17}$: $\mathfrak{a} = \operatorname{cent}\big(D(\mathfrak{n}_{6,17})\big)$
  $\mathfrak{n}_{6,18}$: $\mathfrak{a} = (\mathfrak{n}_{6,18})^3 = \mathfrak{z}^3(\mathfrak{n}_{6,18})$
  $\mathfrak{n}_{6,19}$: $\mathfrak{a} = (\mathfrak{n}_{6,19})^3 = \mathfrak{z}^3(\mathfrak{n}_{6,19})$

$\mathfrak{n}_{6,20}$: $\mathfrak{a} = \operatorname{cent}\big(D(\mathfrak{n}_{6,20})\big) = \operatorname{cent}\big(\mathfrak{z}^4(\mathfrak{n}_{6,20})\big)$
$\mathfrak{n}_{6,21}$: $\mathfrak{a} = D(\mathfrak{n}_{6,21}) = \mathfrak{z}^4(\mathfrak{n}_{6,21})$
$\mathfrak{n}_{6,22}$: $\mathfrak{a} = D(\mathfrak{n}_{6,22}) = \mathfrak{z}^4(\mathfrak{n}_{6,22})$

The identification of different nilpotent Lie algebras in the lists in Part 4 can be accomplished easily over the field of complex numbers by determination of dimensions US, CS and DS of ideals in the characteristic series and the dimension of the algebra of derivations of $\mathfrak{n}$ when the sets US, CS and DS coincide. Over the field of real numbers the identification is complicated by the existence of several different real forms of some of the complex algebras. Their differentiation is of course not possible by the above mentioned "dimensional" invariants. It turns out that these nonisomorphic real forms can be distinguished by the signatures of the Killing forms of their algebras of derivations. (The Killing form of a nilpotent Lie algebra itself automatically vanishes and consequently is useless for any identification.)

CHAPTER 9

Solvable Lie Algebras and Their Nilradicals

9.1. General structure of a solvable Lie algebra

Let us recall that every solvable Lie algebra $\mathfrak{s}$ contains a unique maximal nilpotent ideal called the *nilradical* $\mathfrak{n} = \mathrm{NR}(\mathfrak{s})$. For a solvable Lie algebra $\mathfrak{s}$ the dimension of the nilradical $\mathfrak{n}$ satisfies [78]

$$\dim \mathfrak{n} \geq \tfrac{1}{2} \dim \mathfrak{s} \tag{9.1}$$

(see Section 9.3 for an improved bound and its derivation).

Any solvable Lie algebra $\mathfrak{s}$ can be written as the algebraic sum of the nilradical $\mathfrak{n}$ and a complementary linear space F

$$\mathfrak{s} = F \dotplus \mathfrak{n}. \tag{9.2}$$

The derived algebra $\mathfrak{s}^2 = [\mathfrak{s}, \mathfrak{s}]$ of a solvable Lie algebra $\mathfrak{s}$ is contained in the nilradical

$$\mathfrak{s}^2 \subseteq \mathfrak{n}. \tag{9.3}$$

because all derivations map $\mathfrak{s}$ into $\mathfrak{n}$, in particular any inner derivation. An element x of $\mathfrak{s}$ is *nilpotent* if it satisfies

$$\big(\mathrm{ad}(x)\big)^k y = \big[x, \dots, \big[x, [x, y]\big] \dots\big] = 0, \quad \forall y \in \mathfrak{s}$$

for some positive integer k.

A set of elements $\{x_1, \dots, x_k\}$ of $\mathfrak{s}$ is called *linearly nilindependent* if no nontrivial linear combination of them is nilpotent.

Use will be made of the adjoint representation defined in (2.21) and of its restriction to the nilradical $\mathfrak{n}$ of $\mathfrak{s}$ denoted by $\mathrm{ad}|_{\mathfrak{n}}$. This restriction is realized by matrices $A \in \mathbb{F}^{n\times n}$ where $\mathbb{F}$ is the ground field. If x is a nilpotent element of $\mathfrak{s}$, it will be represented by a nilpotent matrix in the adjoint representation and consequently also $\mathrm{ad}|_{\mathfrak{n}}(x)$ will be nilpotent.

A set of matrices in $\mathbb{F}^{n\times n}$ will be called *linearly nilindependent* if no nontrivial combination of them is a nilpotent matrix.

From now on we shall assume that the solvable algebra $\mathfrak{s}$ is finite-dimensional with $\dim \mathfrak{s} = s$ and indecomposable, i.e., cannot be decomposed into a direct sum of two or more subalgebras. For the purpose of the classification we further assume that $\mathfrak{s}$ is nonnilpotent.

9.2. General procedure for classifying all solvable Lie algebras with a given nilradical

Let us consider a given nilpotent Lie algebra $\mathfrak{n}$ of dimension n as a nilradical of a solvable Lie algebra $\mathfrak{s}$ of dimension s. We wish to find all such (indecomposable, nonnilpotent) solvable Lie algebras. Any such solvable algebra will be called a *solvable extension* of the given nilradical $\mathfrak{n}$. Let us choose a basis $(e_1, \dots, e_n)$ of $\mathfrak{n}$ and extend it to a basis of $\mathfrak{s}$

$$(e_1, \dots, e_n, f_1, \dots, f_f), \quad \text{where } n + f = s. \tag{9.4}$$

Since the derived algebra of a solvable Lie algebra is contained in the nilradical, the commutation relations for the solvable Lie algebra $\mathfrak{s}$ in the basis (9.4) can be written as

$$[e_i, e_j] = N_{ij}{}^k e_k, \quad i, j \in \{1, \dots, n\}, \tag{9.5}$$

$$[f_\alpha, e_i] = (A_\alpha)_i^j e_j, \quad \alpha, \beta \in \{1, \dots, f\}, \tag{9.6}$$

$$[f_\alpha, f_\beta] = \gamma_{\alpha\beta}{}^i e_i, \tag{9.7}$$

where Latin subscripts and superscripts refer to the basis of the nilradical $\mathfrak{n}$, Greek ones to the basis of the complementary space F. Summation over repeated indices is to be understood. The structure constants $N_{ij}{}^k$ of the nilradical are assumed to be known (in the chosen basis $(e_1, \dots, e_n)$).

The Jacobi identities for $\{e_i, e_j, e_k\}$ are already satisfied. The remaining Jacobi identities must be imposed. The triplets $\{f_\alpha, e_i, e_j\}$, $\{f_\alpha, f_\beta, e_i\}$, $\{f_\alpha, f_\beta, f_\delta\}$ imply

$$N_{ij}{}^k (A_\alpha)_k^l + N_{jk}{}^l (A_\alpha)_i^k - N_{ik}{}^l (A_\alpha)_j^k = 0, \tag{9.8}$$

$$([A_\alpha, A_\beta])_i^j = \gamma_{\alpha\beta}{}^k N_{ik}{}^j, \tag{9.9}$$

$$\gamma_{\beta\delta}{}^i (A_\alpha)_i^k + \gamma_{\alpha\beta}{}^i (A_\delta)_i^k + \gamma_{\delta\alpha}{}^i (A_\beta)_i^k = 0, \tag{9.10}$$

respectively.

Equation (9.8) is a system of $n^2(n-1)/2$ linear homogeneous equations for n^2 unknowns, i.e., the matrix elements $(A_\alpha)_i^j$, for each value of $\alpha \in \{1, \dots, f\}$ separately (since $N_{ij}{}^k$ are known). It follows from (9.8) that the matrix A_α represents a derivation D_α of the nilradical. In the adjoint representation of the Lie algebra $\mathfrak{s}$ restricted to the nilradical $\mathfrak{n}$ we have

$$\left(D_1 = \operatorname{ad}|_{\mathfrak{n}}(f_1), \dots, D_f = \operatorname{ad}|_{\mathfrak{n}}(f_f)\right) \simeq (A_1, \dots, A_f) \tag{9.11}$$

where $D_\alpha(e_i) = (A_\alpha)_i^j e_j$. Finding all sets of matrices $\{A_\alpha\}$ satisfying (9.8) is equivalent to finding all sets of nilindependent derivations of the nilradical. They must be nilindependent since otherwise the nilradical would be larger (it would contain one or more of the elements of F). This also means that the derivations are outer ones: inner derivations of $\mathfrak{n}$ are always represented by nilpotent matrices. For simplicity, we shall change the symbol A_α used for the matrices to D_α, so that the derivation and its matrix are denoted identically.

As we have just seen, in order to have nontrivial solvable extensions our nilpotent algebra $\mathfrak{n}$ must possess at least one nonnilpotent derivation. This is not automatic. A nilpotent algebra $\mathfrak{n}$ which has only nilpotent derivations and consequently is not a nilradical of any solvable Lie algebra is called *characteristically nilpotent*. Characteristically nilpotent Lie algebras exist in any dimension $n \geq 7$ and are in fact prevalent among nilpotent algebras of higher dimensions in the sense that they form an open subset of the variety of nilpotent algebras in the Zariski topology.

The simplest example of such an algebra is

	e_1	e_2	e_3	e_4	e_5	e_6	e_7
e_1	0	0	0	0	0	0	0
e_2	0	0	0	0	0	0	e_1
e_3	0	0	0	0	0	e_1	e_2
e_4	0	0	0	0	$-e_1$	e_1	e_3
e_5	0	0	0	e_1	0	e_2	e_4
e_6	0	0	$-e_1$	$-e_1$	$-e_2$	0	e_5

which was first presented in [47].

Relations (9.9) exist for $f \geq 2$ and determine the properties of the set of matrices $(D_1, \ldots, D_f)$. In particular, if the nilradical is Abelian we have ${N_{ij}}^k = 0$ and hence the matrices commute

$$[D_\alpha, D_\beta] = 0. \tag{9.12}$$

In general the relation (9.12) does not hold. However, the space F is not uniquely defined by the solvable algebra $\mathfrak{s}$. In the next paragraph we will discuss "allowed transformations" of the basis of $\mathfrak{s}$ that will be used to classify the Lie algebras $\mathfrak{s}$ into equivalence classes. In all cases so far considered it turns out that F can be chosen so that matrices D_α commute, i.e., (9.12) holds in a suitably chosen basis.

The relations (9.10) represent a set of linear homogeneous equations for the structure constants ${\gamma_{\alpha\beta}}^i$ once the matrices D_α are known. Nontrivial equations (9.10) exist for $f \geq 3$.

A classification of all solvable Lie algebras $\mathfrak{s}$ with the same nilradical $\mathfrak{n}$ amounts to a classification of all matrices D_α and constants ${\gamma_{\alpha\beta}}^i$ satisfying (9.8), (9.9) and (9.10) under the following "allowed transformations":

(1) Redefinition of the space F

$$\tilde{f}_\alpha = f_\alpha + r_\alpha^j e_j, \qquad r_\alpha^j \in \mathbb{F}. \tag{9.13}$$

(2) Change of basis in the nilradical $\mathfrak{n}$

$$\tilde{e}_i = S_i^j e_j, \quad S \in \operatorname{Aut}(\mathfrak{n}) \subseteq \operatorname{GL}(n, \mathbb{F}). \tag{9.14}$$

Thus the matrices S form the group of automorphisms of the nilradical, expressed in the chosen basis $(e_1, \ldots, e_n)$. By definition, they leave the set of commutation relations (9.5) invariant and consequently respect all basis independent properties of the nilradical (in particular all ideals in the derived and lower and upper central series).

(3) Change of basis in F

$$\tilde{f}_i = G_i^j f_j, \quad G \in \operatorname{GL}(f, \mathbb{F}). \tag{9.15}$$

The derivations D_α are by definition 1-cocycles i.e., in $Z^1(\mathfrak{n}, \mathfrak{n}; \operatorname{ad}_\mathfrak{n})$. The inner derivations are 1-coboundaries, i.e., in $B^1(\mathfrak{n}, \mathfrak{n}; \operatorname{ad}_\mathfrak{n})$. The transformation (9.13) transforms the corresponding matrices D_α as follows

$$[\tilde{f}_\alpha, e_i] = [f_\alpha, e_i] + r_\alpha^j [e_j, e_i] = (\widetilde{D}_\alpha)_i^k e_k,$$

i.e.,

$$(\widetilde{D}_\alpha)_j^i = (D_\alpha)_j^i - r_\alpha^k {N_{jk}}^i. \tag{9.16}$$

The equivalence defined by (9.16) implies that D_α should be viewed as elements of the first cohomology group $H^1(\mathfrak{n}, \mathfrak{n}; \operatorname{ad}_\mathfrak{n})$, i.e., equivalence classes, rather than outer derivations themselves. In other words, we combine inner derivations with the outer derivations to modify the matrices D_α. The matrices D_α defined in (9.6) must commute when the nilradical $\mathfrak{n}$ is Abelian. Our experience leads us to the following conjecture

Conjecture 9.1. The matrices D_α for non-Abelian nilradicals can be transformed by the redefinition (9.16) into matrices $\widetilde{D}_\alpha$ that commute pairwise. In the cohomological formulation that means that a representative basis of the set $[F] = \mathrm{span}\{[D_\alpha], \alpha \in \{1,\ldots,f\}\}$ can be chosen to consist of commuting operators $\widetilde{D}_\alpha$.

The classification of solvable extensions has been performed for the following classes of nilpotent Lie algebras: Heisenberg algebras $\mathfrak{h}(N)$ (where $\dim \mathfrak{h}(N) = 2N+1,\ N \geq 1$) [106], Abelian Lie algebras $\mathfrak{a}_n,\ n \geq 1$ [83, 84], nilradicals of Borel subalgebras [118], in particular "triangular" Lie algebras $\mathfrak{t}(N)$, ($\dim \mathfrak{t}(N) = N(N-1)/2,\ N \geq 2$) [123, 124], naturally graded and $\mathbb{Z}$-graded nilradicals of maximal degree of nilpotency [3, 19, 116, 117] and some other special types of nilradicals [115, 133]. Moreover, all solvable algebras in dimensions up to 6 have been found in this way [79, 80, 129]. In all these cases the Conjecture 9.1 was confirmed.

Let us present several low dimensional examples in order to demonstrate the procedure.

First, let us consider the case of 3-dimensional solvable Lie algebras. In this case the restriction (9.1) shows that the dimension of the nilradical $\dim \mathrm{NR}(\mathfrak{s})$ is 2 or 3. When $\dim \mathrm{NR}(\mathfrak{s})$ is 3, the algebra is equal to its nilradical, i.e., nilpotent, and was already found before. When $\dim \mathrm{NR}(\mathfrak{s}) = 2$ we have an Abelian nilradical and the solvable algebra $\mathfrak{s}$ is determined once the action of one nonnilpotent element f_1 on the nilradical $\mathfrak{n} = \mathrm{NR}(\mathfrak{s}) = \mathrm{span}\{e_1, e_2\}$ is specified. Any change of basis in the nilradical is allowed because any regular linear map is an automorphism of $\mathfrak{n}$ and consequently the task is reduced to the classification of 2×2 nonnilpotent matrices with respect to conjugation and overall rescaling. We find the following canonical forms for the matrix D_1.

- Over the field of complex numbers the matrix D_1 has one of the following forms
$$\begin{pmatrix} 1 & 0 \\ 0 & a \end{pmatrix}, \quad \begin{pmatrix} 1 & 1 \\ 0 & 1 \end{pmatrix}$$
where the parameter a satisfies $0 < |a| \leq 1$, if $|a| = 1$ then $\arg(a) \leq \pi$.
- Over the field of real numbers the matrix D_1 has one of the following forms
$$\begin{pmatrix} 1 & 0 \\ 0 & a \end{pmatrix}, \quad \begin{pmatrix} \alpha & 1 \\ -1 & \alpha \end{pmatrix}, \quad \begin{pmatrix} 1 & 1 \\ 0 & 1 \end{pmatrix}$$
where the parameters a, α satisfy $-1 \leq a \leq 1,\ a \neq 0,\ \alpha \geq 0$.

The condition $a \neq 0$ arises from the restriction to indecomposable algebras. The matrix $\left(\begin{smallmatrix} \alpha & 1 \\ -1 & \alpha \end{smallmatrix}\right)$ is present only over the field $\mathbb{R}$ because over the field $\mathbb{C}$ it is upon rescaling conjugated to $\left(\begin{smallmatrix} 1 & 0 \\ 0 & a \end{smallmatrix}\right)$ with the choice $a = (\alpha + \mathrm{i})/(\alpha - \mathrm{i})$.

The corresponding solvable algebras as listed in Section 16.4 are

$\mathfrak{s}_{3,1}$:

	e_1	e_2	e_3
e_1	0	0	$-e_1$
e_2		0	ae_2

,

$\mathfrak{s}_{3,3}$

	e_1	e_2	e_3
e_1	0	0	$-\alpha e_1 + e_2$
e_2		0	$-e_1 - \alpha e_2$

which is isomorphic to $\mathfrak{s}_{3,1}$ over the field $\mathbb{C}$, and

$\mathfrak{s}_{3,2}$

	e_1	e_2	e_3
e_1	0	0	$-e_1$
e_2		0	$-e_1 - e_2$

.

A similar investigation can be performed in any dimension when the nilradical is Abelian and has codimension one in $\mathfrak{s}$. For more details see Section 10.4.

When the codimension of the Abelian nilradical is greater than one the situation becomes more involved — one has to classify all Abelian subalgebras of the matrix algebra $\mathfrak{gl}(\mathfrak{n})$ and then use this classification in the construction of nonisomorphic solvable algebras with the Abelian nilradical $\mathfrak{n}$, see, e.g., Section 10.5.

When the nilradical $\mathfrak{n}$ is not Abelian its Lie brackets put restrictions on automorphisms of $\mathfrak{n}$, i.e., it is no longer an arbitrary regular linear map. Consequently, not all changes of basis in the nilradical are allowed. Similarly, the space of derivations of $\mathfrak{n}$ is limited. As an example let us consider solvable Lie algebras with the Heisenberg nilradical, i.e., $\mathfrak{n} = \mathfrak{n}_{3,1}$

	e_1	e_2	e_3
e_1	0	0	0
e_2		0	e_1

.

Any derivation D and any automorphism ϕ must preserve the ideal $\mathfrak{n}^2 = \operatorname{span}\{e_1\}$ and in addition $D(e_1)$ or $\phi(e_1)$ is determined by the action of D, ϕ, respectively, on e_2, e_3. We find that an arbitrary derivation of $\mathfrak{n}$ has the matrix form

$$D = \begin{pmatrix} \operatorname{Tr} X & \vec{b} \\ 0 & X \end{pmatrix}, \quad X \in \mathbb{F}^{2,2}, \quad \vec{b} \in \mathbb{F}^2 \tag{9.17}$$

and any automorphism takes the form

$$\phi = \begin{pmatrix} \det C & \vec{d} \\ 0 & C \end{pmatrix}, \quad C \in \mathbb{F}^{2,2}, \quad \det C \neq 0, \quad \vec{d} \in \mathbb{F}^2. \tag{9.18}$$

Solvable Lie algebras with Heisenberg nilradical $\mathfrak{h}(m)$ will be considered below in Chapter 11 for an arbitrary dimension $n = 2N + 1$. The lowest dimension $n = 3$ will be treated as a special case in Section 11.3.

9.3. Upper bound on the dimension of a solvable extension of a given nilradical

Let us discuss in some detail the upper bound on the maximal number f of nonnilpotent elements f_a we can add to a given nilradical $\mathfrak{n}$. We recall that there is a simple bound

$$f \leq n$$

implied by (9.1) but as we shall see it is never saturated for non-Abelian nilradicals and can be easily improved as was shown in [113].

We start by choosing a convenient basis $\mathcal{E}$ of the nilpotent Lie algebra $\mathfrak{n}$. We first choose some complement $\mathfrak{m}_1$ of $\mathfrak{n}^2$ in $\mathfrak{n}$,

$$\mathfrak{n} = \mathfrak{n}^2 \dotplus \mathfrak{m}_1$$

and denote $m_1 = \dim \mathfrak{m}_1$. We construct a basis $\mathcal{E}_{\mathfrak{m}_1} = (e_{n-m_1+1}, \dots, e_n)$ of $\mathfrak{m}_1$. In the next step, we recall that

$$\mathfrak{n}^2 = [\mathfrak{n}, \mathfrak{n}] = [\mathfrak{m}_1 \dot{+} \mathfrak{n}^2, \mathfrak{m}_1 \dot{+} \mathfrak{n}^2] = [\mathfrak{m}_1, \mathfrak{m}_1] + \mathfrak{n}^3,$$

(the last sum is not necessarily direct). Consequently, we can choose a complement $\mathfrak{m}_2$ of $\mathfrak{n}^3$ in $\mathfrak{n}^2$ such that $\mathfrak{m}_2 \subseteq [\mathfrak{m}_1, \mathfrak{m}_1]$. We write its basis $\mathcal{E}_{\mathfrak{m}_2}$ in the form of some subset of Lie brackets of vectors in $\mathcal{E}_{\mathfrak{m}_1}$, i.e. $\mathcal{E}_{\mathfrak{m}_2} = (e_{n-m_1-m_2+1}, \dots, e_{n-m_1})$ where $m_2 = \dim \mathfrak{m}_2$ and for any $k \in \{n - m_1 - m_2 + 1, n - m_1\}$ a pair $y_k, z_k \in \mathcal{E}_{\mathfrak{m}_1}$ exists such that $e_k = [y_k, z_k]$ (y_k, z_k are not necessarily unique).

Proceeding by induction we have

$$\mathfrak{n}^k = [\mathfrak{m}_{k-1} \dot{+} \mathfrak{n}^k, \mathfrak{m}_1 \dot{+} \mathfrak{n}^2] = [\mathfrak{m}_{k-1}, \mathfrak{m}_1] + \mathfrak{n}^{k+1}$$

and we find a complement $\mathfrak{m}_k$ of $\mathfrak{n}^{k+1}$ in $\mathfrak{n}^k$,

$$\mathfrak{n}^k = \mathfrak{n}^{k+1} \dot{+} \mathfrak{m}_k,$$

$\mathfrak{m}_k \subseteq [\mathfrak{m}_{k-1}, \mathfrak{m}_1]$, $m_k = \dim \mathfrak{m}_k$ and a basis $\mathcal{E}_{\mathfrak{m}_k} = \left(e_{n+1-\sum_{i=1}^{k} m_i}, \dots, e_{n-\sum_{i=1}^{k-1} m_i}\right)$ of $\mathfrak{m}_k$ such that

$$\forall e_j \in \mathcal{E}_{\mathfrak{m}_k} \quad \exists y_j \in \mathcal{E}_{\mathfrak{m}_{k-1}}, z_j \in \mathcal{E}_{\mathfrak{m}_1}: \quad e_j = [y_j, z_j]. \tag{9.19}$$

Together the elements of the bases $\mathcal{E}_{\mathfrak{m}_k}$ form a basis $\mathcal{E} = (e_1, \dots, e_n)$ of the whole nilpotent algebra $\mathfrak{n}$. The main advantage of working in the basis $\mathcal{E}$ lies in the fact that any automorphism ϕ, or any derivation D, is fully specified once its action on the elements of the basis $\mathcal{E}_{\mathfrak{m}_1}$ of $\mathfrak{m}_1$ is known due to the definition of an automorphism, (2.25), or a derivation, (2.26), together with (9.19), respectively. In all our computations concerning solvable extensions of the given nilradicals and in the lists containing the classification of low dimensional solvable Lie algebras we shall use bases of indecomposable nilradicals satisfying (9.19). For decomposable nilradicals we use bases satisfying (9.19) up to permutation of basis elements (in order not to obscure the explicit decomposition).

The matrix of any derivation D of $\mathfrak{n}$ is by construction upper block triangular in the basis $\mathcal{E}$

$$D = \begin{pmatrix} D_{\mathfrak{m}_K\mathfrak{m}_K} & \cdots & D_{\mathfrak{m}_K\mathfrak{m}_2} & D_{\mathfrak{m}_K\mathfrak{m}_1} \\ & \ddots & \vdots & \vdots \\ & & D_{\mathfrak{m}_2\mathfrak{m}_2} & D_{\mathfrak{m}_2\mathfrak{m}_1} \\ & & & D_{\mathfrak{m}_1\mathfrak{m}_1} \end{pmatrix}. \tag{9.20}$$

Due to the relation $D([e_k, e_j]) = [D(e_k), e_j] + [e_k, D(e_j)]$ the elements of the diagonal blocks $D_{\mathfrak{m}_k\mathfrak{m}_k}$, $k = 2, \dots, K$ are linear functions of elements of $D_{\mathfrak{m}_1\mathfrak{m}_1}$. The elements in the general block $D_{\mathfrak{m}_j\mathfrak{m}_k}$, $2 \le k \le j \le K$ are linear functions of elements in the last column blocks $D_{\mathfrak{m}_1\mathfrak{m}_1}, \dots, D_{\mathfrak{m}_{j-k+1}\mathfrak{m}_1}$.

Inner derivations by definition map $\mathfrak{n}^k \to \mathfrak{n}^{k+1}$, i.e., have strictly upper triangular block structure. Therefore, any set of outer derivations $\{D_1, \dots, D_f\}$ such that $[D_j, D_k] \in \mathfrak{Inn}(\mathfrak{n})$ for all $j, k = 1, \dots, f$ must have commuting $\mathfrak{m}_1\mathfrak{m}_1$-submatrices,

$$[(D_j)_{\mathfrak{m}_1\mathfrak{m}_1}, (D_k)_{\mathfrak{m}_1\mathfrak{m}_1}] = 0. \tag{9.21}$$

An iterative consequence of the definition of a derivation (2.26)

$$D^n[x, y] = \sum_{j=0}^{n} \binom{n}{j} [D^j x, D^{n-j} y]$$

implies that the derivation D is nilpotent if and only if its submatrix $D_{\mathfrak{m}_1\mathfrak{m}_1}$ is nilpotent. This equivalence implies that derivations $D_1, \dots, D_f$ are linearly nilindependent if and only if their submatrices $(D_1)_{\mathfrak{m}_1\mathfrak{m}_1}, \dots, (D_f)_{\mathfrak{m}_1\mathfrak{m}_1}$ are linearly nilindependent. Together with (9.21) this means that the number f of linearly nilindependent outer derivations of the given nilpotent algebra $\mathfrak{n}$ commuting to inner derivations is bounded from above by the maximal number of linearly nilindependent commuting matrices of dimension $m_1 \times m_1$. This number is $m_1 = \dim \mathfrak{n} - \dim \mathfrak{n}^2$. Thus we have

Proposition 9.2. *Let $\mathfrak{n}$ be a nilpotent Lie algebra and $\mathfrak{s}$ a solvable Lie algebra with the nilradical $\mathfrak{n}$. Let $\dim \mathfrak{n} = n$, $\dim \mathfrak{s} = n + f$. Then f satisfies*

$$f \leq \dim \mathfrak{n} - \dim \mathfrak{n}^2. \tag{9.22}$$

The bound (9.22) is saturated for many classes of nilpotent Lie algebras whose solvable extensions were investigated—e.g., Abelian [84], (reproduced in Chapter 10), naturally graded filiform $\mathfrak{n}_{n,1}, \mathcal{Q}_n$ [3, 116] (partly reproduced in Section 13.1), a decomposable central extension of $\mathfrak{n}_{n,1}$ in [133] and triangular [123] (reproduced in Section 12.3).

On the other hand, it is obvious that even the improved bound (9.22) cannot give a precise estimate of the maximal dimension of a solvable extension in all cases. In particular, we have always $\dim \mathfrak{n} - \dim \mathfrak{n}^2 \geq 2$, i.e., characteristically nilpotent Lie algebras cannot be easily detected using (9.22). Similarly, the bound (9.22) is not saturated in the case of Heisenberg nilradicals $\mathfrak{h}$ [106] (reproduced in Chapter 11) where the maximal number of nonnilpotent elements is in fact equal to $(\dim \mathfrak{h} + 1)/2 < \dim \mathfrak{h} - 1$.

We notice that a different improvement of the bound (9.1) was found in [81]

$$f \leq \dim \mathfrak{n} - \dim C(\mathfrak{s}). \tag{9.23}$$

It turns out that the estimate (9.22) is in most cases more restrictive than (9.23). In addition, (9.22) does not involve a priori knowledge of the algebra $\mathfrak{s}$ whereas (9.23) requires it.

9.4. Particular classes of nilradicals and their solvable extensions

In the following chapters we shall consider several classes of nilradicals and present some results concerning their solvable extensions.

The nilpotent algebras whose solvable extensions have been classified with various degrees of detail, can be divided into three major classes:

- Algebras with low degree of nilpotency.

The algebras investigated so far in this class are the Abelian and Heisenberg algebras (in arbitrary finite dimensions). These algebras possess large algebras of derivations that have well-understood properties. E.g., for an Abelian nilradical, any linear transformation is a derivation and any regular linear map is an automorphism. Consequently, the construction of solvable extensions is reduced to the consideration of Abelian subalgebras in $\mathfrak{gl}(n)$ and their equivalence. Similarly, for Heisenberg algebras $\mathfrak{h}(n)$, the task is reduced to the study of Abelian subalgebras of $\mathfrak{sp}(2n)$.

- Nilradicals of Borel subalgebras of simple Lie algebras.

Nilpotent algebras in this class have a very particular structure given by the corresponding root diagram. Consequently, all derivations of such algebras can be found in explicit form using cohomological arguments. This was done by G.F. Leger and E.M. Luks in [67]. A prime example of a nilradical in this class is the algebra of strictly upper triangular matrices.

- Algebras with a high degree of nilpotency.

The structure of Lie brackets of such algebras usually significantly restricts the algebra of derivations. Therefore the algebras of derivations can often be written down explicitly in arbitrary dimension and similarly for the automorphisms. Consequently, many explicit lists of solvable algebras with nilradicals in this class are known.

9.5. Vector fields realizing bases of the coadjoint representation of a solvable Lie algebra

When calculating invariants of the coadjoint representation of a solvable Lie algebra $\mathfrak{s}$ we use the vector fields (3.1). When we use the basis (9.4) of $\mathfrak{s}$, i.e., $(e_1, \ldots, e_n, f_1, \ldots, f_f)$, in which the Lie brackets are expressed as in (9.5) – (9.7), the corresponding vector fields can be written as

$$
\begin{aligned}
\widehat{E}_i &= N_{ji}{}^k e_k \frac{\partial}{\partial e_j} + (D_\alpha)_i^k e_k \frac{\partial}{\partial f_\alpha}, \qquad & i = 1, \ldots, n, \\
\widehat{F}_\alpha &= -(D_\alpha)_i^k e_k \frac{\partial}{\partial e_i} + \gamma_{\beta\alpha}{}^k e_k \frac{\partial}{\partial f_\beta}, \qquad & \alpha = 1, \ldots, f
\end{aligned}
\tag{9.24}
$$

(summation over repeated indices $i, j, k = 1, \ldots, n$ and $\beta = 1, \ldots, f$ applies). As in Section 3.2 the operators $\widehat{E}_i, \widehat{F}_\alpha$ act on smooth functions on the space $\mathfrak{s}^*$ dual to the Lie algebra $\mathfrak{s}$ with coordinates $(e_1, \ldots, e_n, f_1, \ldots, f_f)$, i.e., on functions of $n+p$ variables

$$I(e_1, \ldots, e_n, f_1, \ldots, f_f).$$

The operators $\widehat{E}_i, \widehat{F}_\alpha$ define an infinite dimensional representation of the Lie algebra $\mathfrak{s}$ on the space $C^\infty(\mathfrak{s}^*)$ of smooth functions on $\mathfrak{s}^*$, i.e., satisfy the same commutation relations as the corresponding elements e_i, f_α in (9.5) – (9.7).

CHAPTER 10

Solvable Lie Algebras with Abelian Nilradicals

The simplest possible structure of the nilradical is Abelian. Thus we start our discussion of solvable extensions of various nilradicals by considering the case of an Abelian nilradical.

The results presented in this chapter were obtained in [84]. Our presentation follows the general outline of Chapter 9.

10.1. Basic structural theorems

Since the nilradical is Abelian, the structure constants ${N_{ij}}^k$ satisfy ${N_{ij}}^k = 0$. Equation (9.8) is satisfied trivially. From (9.9) we see that the matrices D_α all commute as in (9.12).

Theorem 10.1. *Let $\mathfrak{s}$ be a finite-dimensional solvable nonnilpotent Lie algebra over the field $\mathbb{F}$ and let its nilradical $\mathfrak{n}$ of dimension n be Abelian. We can choose a basis of $\mathfrak{s}$ such that the commutation relations are*

$$([e_1, f_\alpha], \ldots, [e_n, f_\alpha]) = (e_1, \ldots, e_n) \cdot D_\alpha, \quad 1 \leq \alpha \leq f \leq n,$$

$$D_\alpha = \mathrm{ad}_{\mathfrak{n}}(f_\alpha) \in \mathbb{F}^{n \times n}, \tag{10.1}$$

$$[e_i, e_k] = 0, \tag{10.2}$$

$$[f_\alpha, f_\beta] = {\gamma_{\alpha\beta}}^j e_j, \quad {\gamma_{\alpha\beta}}^j \in \mathbb{F}. \tag{10.3}$$

The matrices D_α are linearly nilindependent and commute pairwise

$$[D_\alpha, D_\beta] = 0, \quad 1 \leq \alpha, \beta \leq f. \tag{10.4}$$

For $f \geq 3$ the matrices D_α and the constants ${\gamma_{\alpha\beta}}^j$ satisfy

$$\begin{gathered} {\gamma_{\alpha\beta}}^j (D_\delta)_j^k + {\gamma_{\delta\alpha}}^j (D_\beta)_j^k + {\gamma_{\beta\delta}}^j (D_\alpha)_j^k = 0 \\ 1 \leq k \leq n, \quad 1 \leq \alpha < \beta < \delta \leq f. \end{gathered} \tag{10.5}$$

A classification of the Lie algebras $\mathfrak{s}$ thus amounts to a classification of the matrices D_α and constants ${\gamma_{\alpha\beta}}^j$ under the transformations (9.13) – (9.15).

Proof. Equations (10.1) – (10.3) simply express the results about the dimension of the nilradical of a solvable Lie algebra (9.1), and the structure of a solvable extension (9.5) – (9.7) in the particular case when $\mathfrak{n}$ is Abelian. Equation (10.4) is a consequence of the Jacobi identifies (9.9). Similarly, (10.5) is the same as (9.10). The transformations (9.13) – (9.15) leave the relations (10.4) – (10.5) invariant, but transform the matrices D_α and constants ${\gamma_{\alpha\beta}}^j$ in (10.1) – (10.3). This completes the proof. □

In more detail, if we put

$$e_i' = S_i^k e_k, \quad f_\alpha' = G_\alpha^\beta f_\beta + r_\alpha^a e_a \tag{10.6}$$

we obtain the commutation relations (10.1) – (10.3) with D_α and ${\gamma_{\alpha\beta}}^j$ replaced by

$$D'_\alpha = G_\alpha^\beta (S D_\beta S^{-1}) \tag{10.7}$$

$${\gamma'_{\alpha\beta}}^j = \left(G_\alpha^\mu G_\beta^\nu {\gamma_{\mu\nu}}^a + r_\alpha^b G_\beta^\nu (D_\nu)_b^a - r_\beta^b G_\alpha^\mu (D_\mu)_b^a \right) (S^{-1})_a^j. \tag{10.8}$$

We mention that the transformation formulas for D_α, ${\gamma_{\alpha\beta}}^j$ would be more complicated for a non-Abelian nilradical which possesses nonvanishing inner derivations.

Theorem 10.1 provides us with a general classification procedure for obtaining all solvable Lie algebras $\mathfrak{s}$ of the considered type for given values of $s = \dim \mathfrak{s}$ and $n = \dim \mathfrak{n}$. The procedure is:

(1) Classify all Abelian subalgebras $A_f(n)$ of $\mathfrak{gl}(n, \mathbb{F})$ of dimension $f = s - n$, containing no nilpotent elements, into conjugacy classes under the action of $\mathrm{GL}(\mathfrak{n}, \mathbb{F})$. Choose a representative of each conjugacy class, i.e., use the transformation (10.6) to transform the matrices D_α, $\alpha = 1, \dots, f$ into some chosen canonical form.

For $f = 1$ there is just one matrix D and it can be transformed to its Jordan canonical form. For $f = n$ the matrices $(D_1, \dots, D_f)$ form a basis of a Cartan subalgebra of $\mathfrak{gl}(n, \mathbb{F})$. As shown below (Theorem 10.6) the algebra $\mathfrak{s}$ will be decomposable for $n > 2$, $\mathbb{F} = \mathbb{C}$ and $n > 4$, $\mathbb{F} = \mathbb{R}$.

For $1 < f < n$, $4 \leq n$, the problem of classifying the Abelian subalgebras $A_f(n) \subset \mathfrak{gl}(n, \mathbb{F})$ is more difficult. Use should be made of known results on maximal Abelian subalgebras of $\mathfrak{gl}(n, \mathbb{F})$ [57, 74, 122] (see below).

(2) Determine the structure constants ${\gamma_{\alpha\beta}}^j$ of (10.3). For $f = 1$ the question does not arise. For $f \geq 2$ we can simplify them using the transformation (10.6). We have always at our disposal the transformation (10.6) with $S = \mathbf{1}_{n\times n}$, $G = \mathbf{1}_{f\times f}$, i.e., with arbitrary constants r_α^a. That leads to equivalent structure constants ${\gamma'_{\alpha\beta}}^j$

$${\gamma'_{\alpha\beta}}^j = {\gamma_{\alpha\beta}}^j + r_\alpha^a (D_\beta)_a^j - r_\beta^a (D_\alpha)_a^j. \tag{10.9}$$

For some sets of derivations D_α the more general formula (10.6), rather than (10.9), can be used to further simplify and standardize ${\gamma_{\alpha\beta}}^j$. One must however require that the matrices S, G in (10.6) leave the already standardized matrices D_α invariant.

From a slightly different viewpoint, cohomological considerations along the lines of Section 8.2, (8.10) with $B \equiv 0$ can also be employed. The cohomology group encountered here is $H^2(\mathfrak{s}/\mathfrak{n}, \mathfrak{n}; A)$ where A stands for the representation of the Abelian algebra $\mathfrak{s}/\mathfrak{n}$ by matrices D_α on the vector space $\mathfrak{n}$. When the cohomology group $H^2(\mathfrak{s}/\mathfrak{n}, \mathfrak{n}; A)$ is trivial we can always set ${\gamma_{\alpha\beta}}^j = 0$ for all α, β and j.

(3) Weed out decomposable Lie algebras among the constructed solvable Lie algebras.

We see from Theorem 10.1 and the discussion above that we cannot expect to have a nice closed form result for solvable Lie algebras of arbitrary dimensions. We shall present further partial results that together make the classification of solvable Lie algebras easy, once the dimensions n and f are fixed. We shall concentrate on the case $\mathbb{F} = \mathbb{C}$ but we will also point out the differences that occur for the field $\mathbb{F} = \mathbb{R}$.

From (10.7) we see that a good classification procedure is to first use the basis transformations (10.6) to transform the matrices $D_1, \dots, D_f$ to some standard form. The constants r_α^j can then be used to simplify the structure constants ${\gamma_{\alpha\beta}}^j$.

In particular, for given α, β the structure constants $\gamma_{\alpha\beta}{}^j$ can be always annulled when the sum of image spaces of the derivations D_α and D_β contains the whole nilradical $\mathfrak{n}$.

The matrices D_α, $\alpha = 1, \ldots, f$, form an Abelian subalgebra $A_f(n) \subset \mathfrak{gl}(n, \mathbb{F})$, containing no nilpotent matrices. Each set $A_f(n)$ is contained in at least one maximal Abelian subalgebra (MASA) of $\mathfrak{gl}(n, \mathbb{F})$. A sizable literature exists on MASAs of the classical Lie algebras. For a review of classical results, including the Kravchuk signatures and Kravchuk normal forms of MASAs of $\mathfrak{sl}(n, \mathbb{C})$, we refer to D.A. Suprunenko and R.I. Tyshkevich [122]. More recent results on MASAs of other classical Lie algebras can be found in [57, 74, 137].

The pertinent results for our purposes are:

(1) A MASA of $\mathfrak{gl}(n, \mathbb{F})$ can always be written as

$$\mathrm{MASA}\big(\mathfrak{gl}(n, \mathbb{F})\big) = \mathbb{F}I_n \oplus \mathrm{MASA}\big(\mathfrak{sl}(n, \mathbb{F})\big).$$

(2) A MASA of $\mathfrak{sl}(n, \mathbb{F})$ as a set of matrices can be either indecomposable, or decomposable into a direct sum of indecomposable ones.

(3) An indecomposable MASA of $\mathfrak{sl}(n, \mathbb{C})$ is always a maximal Abelian nilpotent subalgebra (MANS) [37, 74, 90, 122, 137]. A MANS is represented by nilpotent matrices in any finite dimensional representation. A MANS is characterized by its Kravchuk signature

$$(\lambda\,\mu\,\nu), \quad 1 \leq \lambda, \quad 1 \leq \nu, \quad 0 \leq \mu, \quad \lambda + \mu + \nu = n, \quad \lambda, \mu, \nu \in \mathbb{Z}. \tag{10.10}$$

All elements X of a MANS of $\mathfrak{sl}(n, \mathbb{C})$ can be simultaneously written in the Kravchuk normal form

$$X = \begin{pmatrix} 0_\lambda & P & Q \\ 0 & R & S \\ 0 & 0 & 0_\nu \end{pmatrix}, \quad P \in \mathbb{C}^{\lambda,\mu},\ Q \in \mathbb{C}^{\lambda,\nu},\ R \in \mathbb{C}^{\mu,\mu},\ S \in \mathbb{C}^{\mu,\nu}. \tag{10.11}$$

where the matrix Q is free, i.e., contains $\lambda \cdot \nu$ arbitrary parameters. The matrices R and S are completely determined by the matrix P; in addition, the matrices R form an ANS of $\mathfrak{sl}(\mu, \mathbb{C})$

$$R = \begin{pmatrix} 0 & & * \\ & \ddots & \\ 0 & & 0 \end{pmatrix}. \tag{10.12}$$

The matrix P may be free-rowed, i.e., the elements of its first row are free (μ arbitrary parameters) and all the others are determined by the first row. Alternatively for $k \geq 2$, non-free-rowed MANS may exist where the first row of P contains some zeroes and new parameters occur in lower rows. The symbols 0_λ and 0_ν denote square null matrices of the indicated dimension.

(4) An indecomposable MASA of $\mathfrak{sl}(n, \mathbb{R})$ can be either absolutely indecomposable or indecomposable, but not absolutely indecomposable. The absolutely indecomposable ones are MANS and can be written as in (10.12), but with real entries. The nonabsolutely indecomposable MASAs of $\mathfrak{sl}(n, \mathbb{R})$, become decomposable

after complexification. They exist only for n even and have the form [137]

$$\mathbb{R}\begin{pmatrix} 0 & -1 & & & & & \\ 1 & 0 & & & & & \\ & & 0 & -1 & & & \\ & & 1 & 0 & & & \\ & & & & \ddots & & \\ & & & & & 0 & -1 \\ & & & & & 1 & 0 \end{pmatrix} \oplus \mathrm{MANS}(\mathfrak{sl}(r,\mathbb{C})),$$

where $r = n/2$ and $\mathfrak{sl}(r,\mathbb{C})$ is represented by the matrices

$$\begin{pmatrix} \alpha_{11} & \dots & \alpha_{1r} \\ \vdots & & \vdots \\ \alpha_{r1} & \dots & \alpha_{rr} \end{pmatrix} \quad \alpha_{ij} = \begin{pmatrix} a_{ij} & -b_{ij} \\ b_{ij} & a_{ij} \end{pmatrix}, \quad a_{ij}, b_{ij} \in \mathbb{R}, \quad \sum_{i=1}^{r} a_{ii} = 0.$$

The corresponding MANS of $\mathfrak{sl}(r,\mathbb{C})$ can again be written in Kravchuk normal form.

Let us now return to the commuting matrices D_α of (10.1) – (10.3) and (10.7). They form an Abelian subalgebra of $\mathfrak{gl}(n,\mathbb{F})$ and hence a subalgebra of some MASA. This MASA cannot be a MANS; as a matter of fact it contains no nilpotent elements at all. Hence all matrices D_α can be simultaneously brought to the same block diagonal form. Moreover each block can be brought to a triangular form.

We thus arrive at the following theorem.

Theorem 10.2. *Let $\mathfrak{s}$ be a finite-dimensional solvable Lie algebra over $\mathbb{C}$ with an Abelian nilradical $\mathfrak{n}$. The matrices D_α in (10.1) – (10.3) in Theorem 10.1 can be simultaneously transformed into a block diagonal form*

$$D_\alpha = \begin{pmatrix} T^1_\alpha(a^1_\alpha) & & & & & \\ & \ddots & & & & \\ & & T^p_\alpha(a^p_\alpha) & & & \\ & & & T^{p+1}_\alpha(0) & & \\ & & & & \ddots & \\ & & & & & T^{p+q}_\alpha(0) \end{pmatrix}, \tag{10.13}$$

$$T^j_\alpha(a^j_\alpha) = \begin{pmatrix} a^j_\alpha & & * \\ & \ddots & \\ 0 & & a^j_\alpha \end{pmatrix} \in \mathbb{C}^{n_j \times n_j}, \quad a^j_\alpha \begin{cases} = \delta^j_\alpha, & 1 \le j \le f, \\ \in \mathbb{C}, & f+1 \le j \le p, \\ = 0, & p+1 \le j \le p+q, \end{cases} \tag{10.14}$$

$$\sum_{j=1}^{p+q} n_j = n, \quad 1 \le f \le p \le n, \quad 1 \le \alpha \le f, \quad p+q \le n, \tag{10.15}$$

$$n_1 \ge n_2 \ge \dots \ge n_p \ge 1, \quad n_{p+1} \ge n_{p+2} \ge \dots \ge n_{p+q} \ge 0. \tag{10.16}$$

For each j, the off-diagonal parts of the corresponding matrices $T^j_\alpha(a^j_\alpha)$, i.e., $\{T^j_\alpha(a^j_\alpha) - a^j_\alpha \mathbf{1}_{n_j \times n_j} \mid \alpha = 1, \dots, f\}$, span an Abelian nilpotent subalgebra of $\mathfrak{sl}(n_j,\mathbb{C})$.

In order to obtain all complex solvable Lie algebras of the considered type, we must consider all partitions of n satisfying (10.16). For each partition we must construct all inequivalent Abelian indecomposable subalgebras of $\mathfrak{sl}(n_j,\mathbb{C})$, $1 \le j \le p+q$.

Remark 10.3. (1) Notice that the Abelian subalgebras of $\mathfrak{sl}(n_j,\mathbb{C})$ need not be maximal. They must however be indecomposable, otherwise they would also appear in some other, finer partition of n.

(2) The situation for $\mathbb{F}=\mathbb{R}$ is conceptually quite similar, however the usual complications arise. Thus, in addition to the nonnilpotent and nilpotent blocks $T_j^\alpha(a_\alpha^j)$ and $T_k^\alpha(0)$, respectively, a further type of nonnilpotent block can occur, namely

$$\widetilde{T}_j^\alpha(b_j^\alpha,c_j^\alpha)=\begin{pmatrix} b_\alpha^j & -c_\alpha^j & & & \\ c_\alpha^j & b_\alpha^j & & * & \\ & & \ddots & & \\ & 0 & & b_\alpha^j & -c_\alpha^j \\ & & & c_\alpha^j & b_\alpha^j \end{pmatrix} \tag{10.17}$$

with $c_\alpha^j\neq 0$ for at least one value of α. Each block $\widetilde{T}_\alpha^j(b_\alpha^j,c_\alpha^j)$ provides either one, or two, nonnilpotent matrices D_α. As an example consider $f=2$, $n=4$. We can have one real 4×4 Jordan block (10.17)

$$D_1=\begin{pmatrix} 0 & -1 & \tilde{b}_1 & \tilde{c}_1 \\ 1 & 0 & -\tilde{c}_1 & \tilde{b}_1 \\ & & 0 & -1 \\ & & 1 & 0 \end{pmatrix},\quad D_2=\begin{pmatrix} 1 & 0 & \tilde{b}_2 & \tilde{c}_2 \\ 0 & 1 & -\tilde{c}_2 & \tilde{b}_2 \\ & & 1 & 0 \\ & & 0 & 1 \end{pmatrix}$$

or two real 2×2 Jordan blocks, for instance

$$D_1=\begin{pmatrix} b_1 & -1 & & \\ 1 & b_1 & & \\ & & b_2 & 0 \\ & & 0 & b_2 \end{pmatrix},\quad D_2=\begin{pmatrix} 0 & 0 & & \\ 0 & 0 & & \\ & & b_3 & -1 \\ & & 1 & b_3 \end{pmatrix}$$

or a combination of one real 2×2 Jordan block (10.17) with blocks $T_j^\alpha(a_\alpha^j)$ and $T_k^\alpha(0)$.

In general the matrices D_α will have the form

$$D_\alpha=\begin{pmatrix} T_\alpha^1(a_\alpha^1) & & & & & & & \\ & \ddots & & & & & & \\ & & T_\alpha^p(a_\alpha^p) & & & & & \\ & & & \widetilde{T}_\alpha^{p+1}(b_\alpha^{p+1},c_\alpha^{p+1}) & & & & \\ & & & & \ddots & & & \\ & & & & & \widetilde{T}_\alpha^{p+q}(b_\alpha^{p+t},c_\alpha^{p+t}) & & \\ & & & & & & T_\alpha^{p+t+1}(0) & \\ & & & & & & & \ddots \\ & & & & & & & & T_\alpha^{p+q+t}(0) \end{pmatrix}$$

with $T_\alpha^j(a_\alpha^j)$ as in (10.14), $\widetilde{T}_\alpha^{p+i}(b_\alpha^{p+i},c_\alpha^{p+i})$ as in (10.17). The linear nilindependence of $D_1,\ldots,D_f$ is assured by setting

$$a_\alpha^j=\delta_\alpha^j,\quad 1\leq j\leq \min(p,f),\qquad 1\leq\alpha\leq f\leq p+2t$$

and by appropriately specifying the entries $(b_\alpha^{p+j},c_\alpha^{p+j})$, $1\leq j\leq t$.

Theorem 10.2 goes quite far towards a classification of the solvable algebras $\mathfrak{s}$ with Abelian nilradical $\mathfrak{n}$, as can be seen from the following example.

Example 10.4 ($\mathbb{F} = \mathbb{C}$, $\dim \mathfrak{s} = s = 8$, $\dim \mathfrak{n} = n = 5$, $f = 3$). We have 3 matrices $D_1, D_2, D_3 \in \mathbb{C}^{5\times 5}$, hence we need at least 3 blocks to assure linear nilindependence. The allowed partitions, values of p and q and corresponding matrices are

(1) $5 = 3 + 1 + 1$ ($p = 3$, $q = 0$)

$$D_1 = \begin{pmatrix} 1 & a_1 & b_1 & & \\ & 1 & c_1 & & \\ & & 1 & & \\ & & & 0 & \\ & & & & 0 \end{pmatrix}, \quad D_2 = \begin{pmatrix} 0 & a_2 & b_2 & & \\ & 0 & c_2 & & \\ & & 0 & & \\ & & & 1 & \\ & & & & 0 \end{pmatrix}, \quad D_3 = \begin{pmatrix} 0 & a_3 & b_3 & & \\ & 0 & c_3 & & \\ & & 0 & & \\ & & & 0 & \\ & & & & 1 \end{pmatrix}.$$

Commutativity requires that the three-dimensional block should represent a MASA of $\mathfrak{gl}(3, \mathbb{C})$. There are 3 possible Kravchuk signatures, for which we have

$$\begin{array}{lll} (2\,0\,1) & a_\alpha = 0, & \\ (1\,0\,2) & c_\alpha = 0, & \alpha = 1, 2, 3, \\ (1\,1\,1) & a_\alpha = c_\alpha. & \end{array} \tag{10.18}$$

We mention that here, and in some cases below, further simplifications are possible. Take, e.g., the Kravchuk signature $(1\,1\,1)$, i.e., $a_\alpha = c_\alpha$. By a change of basis in the nilradical we can transform (a_1, b_1) into $(1, 0)$, $(0, 1)$, or $(0, 0)$. For $a_1 = 1$, $b_1 = 0$ no further simplifications are possible. For $a_1 = 0$, $b_1 = 1$ we can transform (a_2, b_2) into $(a_2, 0)$, $(0, b_2)$ or $(0, 0)$ for $a_2 \neq 0$; $a_2 = 0$, $b_2 \neq 0$; or $a_2 = b_2 = 0$, respectively. In the last case we can take (a_3, b_3) into $(1, 0)$, $(0, 1)$, or $(0, 0)$, as the case may be. Thus, the number and range of parameters can be greatly restricted at the price of splitting each case into many subcases. We shall skip this type of discussion below.

(2) $5 = 2 + 2 + 1$ ($p = 3$, $q = 0$)

$$D_1 = \begin{pmatrix} 1 & a_1 & & & \\ & 1 & & & \\ & & 0 & b_1 & \\ & & & 0 & \\ & & & & 0 \end{pmatrix}, \quad D_2 = \begin{pmatrix} 0 & a_2 & & & \\ & 0 & & & \\ & & 1 & b_2 & \\ & & & 1 & \\ & & & & 0 \end{pmatrix}, \quad D_3 = \begin{pmatrix} 0 & a_3 & & & \\ & 0 & 0 & & \\ & & 0 & b_3 & \\ & & & 0 & \\ & & & & 1 \end{pmatrix}.$$

(3) $5 = 2 + 1 + 1 + 1$

$$D_1 = \begin{pmatrix} 1 & a_1 & & & \\ & 1 & & & \\ & & 0 & & \\ & & & 0 & \\ & & & & b_1 \end{pmatrix}, \quad D_2 = \begin{pmatrix} 0 & a_2 & & & \\ & 0 & & & \\ & & 1 & & \\ & & & 0 & \\ & & & & b_2 \end{pmatrix}, \quad D_3 = \begin{pmatrix} 0 & a_3 & & & \\ & 0 & & & \\ & & 0 & & \\ & & & 1 & \\ & & & & b_3 \end{pmatrix},$$

$$\begin{array}{lll} p = 3, & q = 1 & \text{for } (b_1, b_2, b_3) = (0, 0, 0), \\ p = 4, & q = 0 & \text{for } (b_1, b_2, b_3) \neq (0, 0, 0). \end{array}$$

(4) $5 = 5 \times 1$

$$D_1 = \begin{pmatrix} 1 &&&& \\ & 0 &&& \\ && 0 && \\ &&& a_1 & \\ &&&& b_1 \end{pmatrix}, \quad D_2 = \begin{pmatrix} 0 &&&& \\ & 1 &&& \\ && 0 && \\ &&& a_2 & \\ &&&& b_2 \end{pmatrix}, \quad D_3 = \begin{pmatrix} 0 &&&& \\ & 0 &&& \\ && 1 && \\ &&& a_3 & \\ &&&& b_3 \end{pmatrix},$$

$$\begin{aligned} &p = 3, && q = 2 && \text{for } a_i = b_i = 0, \\ &p = 4, && q = 1 && \text{for } (a_1, a_2, a_3) \neq (0,0,0),\ b_i = 0, \\ &p = 5, && q = 0 && \text{for } (a_1, a_2, a_3) \neq (0,0,0),\ (b_1, b_2, b_3) \neq (0,0,0). \end{aligned}$$

All 4 above case also occur for $\mathbb{F} = \mathbb{R}$, but further ones also exist, for example

(5) $5 = 3 + 2$ ($p = 1$, $q = 0$, $t = 1$)

$$D_1 = \begin{pmatrix} 1 & a_1 & b_1 && \\ & 1 & c_1 && \\ && 1 && \\ &&& 0 & \\ &&&& 0 \end{pmatrix}, \quad D_2 = \begin{pmatrix} 0 & a_2 & b_2 && \\ & 0 & c_2 && \\ && 0 && \\ &&& 0 & -1 \\ &&& 1 & 0 \end{pmatrix}, \quad D_3 = \begin{pmatrix} 0 & a_3 & b_3 && \\ & 0 & c_3 && \\ && 0 && \\ &&& 1 & 0 \\ &&& 0 & 1 \end{pmatrix}.$$

(Equation (10.18) holds for a_i, b_i, c_i.)

(6) $5 = 1 + 2 + 2$ ($p = 2$, $q = 0$, $t = 1$)

$$D_1 = \begin{pmatrix} 1 &&&& \\ & 0 & a_1 && \\ & 0 & 0 && \\ &&& b_1 & \\ &&&& b_1 \end{pmatrix}, \quad D_2 = \begin{pmatrix} 0 &&&& \\ & 1 & a_2 && \\ & 0 & 1 && \\ &&& b_2 & \\ &&&& b_2 \end{pmatrix}, \quad D_3 = \begin{pmatrix} 0 &&&& \\ & 0 & a_3 && \\ & 0 & 0 && \\ &&& b_3 & -1 \\ &&& 1 & b_3 \end{pmatrix}.$$

(7) $5 = 1 + 2 + 2$ ($p = 1$, $q = 1$, $t = 1$)

$$D_1 = \begin{pmatrix} 1 &&&& \\ & 0 & 0 && \\ & 0 & 0 && \\ &&& 0 & 1 \\ &&& 0 & 0 \end{pmatrix}, \quad D_2 = \begin{pmatrix} 0 &&&& \\ & 0 & -1 && \\ & 1 & 0 && \\ &&& 0 & a_2 \\ &&& 0 & 0 \end{pmatrix}, \quad D_3 = \begin{pmatrix} 0 &&&& \\ & 1 & 0 && \\ & 0 & 1 && \\ &&& 0 & a_3 \\ &&& 0 & 0 \end{pmatrix}.$$

(8) $5 = 1 + 2 + 2$ ($p = 1$, $q = 0$, $t = 2$)

$$D_1 = \begin{pmatrix} 1 &&&& \\ & 0 & 0 && \\ & 0 & 0 && \\ &&& a_1 & -b_1 \\ &&& b_1 & a_1 \end{pmatrix}, \quad D_2 = \begin{pmatrix} 0 &&&& \\ & 0 & -1 && \\ & 1 & 0 && \\ &&& a_2 & -b_2 \\ &&& b_2 & a_2 \end{pmatrix}, \quad D_3 = \begin{pmatrix} 0 &&&& \\ & 1 & 0 && \\ & 0 & 1 && \\ &&& a_3 & -b_3 \\ &&& b_3 & a_3 \end{pmatrix},$$

$$(b_1, b_2, b_3) \neq (0,0,0).$$

(9) $5 = 1 + 2 + 2$ ($p = 1$, $q = 0$, $t = 2$)

$$D_1 = \begin{pmatrix} 1 &&&& \\ & 0 & 0 && \\ & 0 & 0 && \\ &&& b_1 & 0 \\ &&& 0 & b_1 \end{pmatrix}, \quad D_2 = \begin{pmatrix} 0 &&&& \\ & 0 & -1 && \\ & 1 & 0 && \\ &&& b_2 & 0 \\ &&& 0 & b_2 \end{pmatrix}, \quad D_3 = \begin{pmatrix} 0 &&&& \\ & 0 & 0 && \\ & 0 & 0 && \\ &&& b_3 & -1 \\ &&& 1 & b_3 \end{pmatrix}.$$

(10) $5 = 1 + 2 + 1 + 1$

$$D_1 = \begin{pmatrix} 1 & & & & \\ & 0 & 0 & & \\ & 0 & 0 & & \\ & & & a_1 & \\ & & & & b_1 \end{pmatrix}, \quad D_2 = \begin{pmatrix} 0 & & & & \\ & 0 & -1 & & \\ & 1 & 0 & & \\ & & & a_2 & \\ & & & & b_2 \end{pmatrix}, \quad D_3 = \begin{pmatrix} 0 & & & & \\ & 1 & 0 & & \\ & 0 & 1 & & \\ & & & a_3 & \\ & & & & b_3 \end{pmatrix},$$

$p = 1, \quad q = 2, \quad t = 1$ for $a_i = b_i = 0$,
$p = 2, \quad q = 1, \quad t = 1$ for $(a_1, a_2, a_3) \neq (0, 0, 0)$, $b_i = 0$,
$p = 3, \quad q = 0, \quad t = 1$ for $(a_1, a_2, a_3) \neq (0, 0, 0)$, $(b_1, b_2, b_3) \neq (0, 0, 0)$.

10.2. Decomposability properties of the solvable Lie algebras

So far, nothing guarantees that the Lie algebras $\mathfrak{s}$ described in Theorems 10.1 and 10.2 are indecomposable. Indeed, in general, they may be decomposable into direct sums of lower-dimensional Lie algebras, either solvable with Abelian nilradicals, or Abelian

$$\mathfrak{s} = \mathfrak{s}_1 \oplus \mathfrak{s}_2 \oplus \cdots \oplus \mathfrak{s}_k. \tag{10.19}$$

As described in Chapter 5, this decomposition may be central, which happens when some of the factors are Abelian, or noncentral.

Proposition 10.5. *Let the algebra $\mathfrak{s}$ of Theorems 10.1 and 10.2 be such that the set of matrices D_α contains q_0 one-dimensional zero blocks:*

$$T_{p+t+j}(0) = 0, \quad q - q_0 + 1 \leq j \leq q \tag{10.20}$$

and let

$$q_0 > \frac{f(f-1)}{2}. \tag{10.21}$$

Then the algebra is decomposable.

PROOF. The existence of the q_0 zero blocks implies that the center $C(\mathfrak{s})$ contains the corresponding q_0 elements e_j, $f - q_0 + 1 \leq j \leq f$. These elements e_j do not figure in $[\mathfrak{s}, \mathfrak{n}]$ in (10.1). The only way they can be contained in $\mathfrak{s}^2$ is on the right-hand side of relations (10.3). Only $f(f-1)/2$ such relations exist, hence at most that many linearly independent elements of $\mathfrak{n}$ figure in (10.3). If (10.20)–(10.21) hold then the center $C(\mathfrak{s})$ contains at least

$$q_1 = q_0 - \frac{f(f-1)}{2} > 0$$

linearly independent elements, not contained in $\mathfrak{s}^2$. By Theorem 5.7, the algebra is decomposable. □

A noncentral decomposition into two (or, successively, more) solvable Lie algebras occurs if the matrices D_α of (10.13) can be split into two sets satisfying

$$D_\alpha = \begin{pmatrix} D_\alpha^1 & 0 \\ 0 & 0 \end{pmatrix}, \quad D_\beta = \begin{pmatrix} 0 & 0 \\ 0 & D_\beta^2 \end{pmatrix},$$

$1 \leq \alpha \leq f_0 < f, \quad f_0 + 1 \leq \beta \leq f, \quad D_\alpha^1 \in \mathbb{F}^{n_1 \times n_1}, \quad D_\beta^1 \in \mathbb{F}^{n_2 \times n_2}, \quad n_1 + n_2 = n.$

The two sets of elements

$$S_1 = \{f_1, \ldots, f_{f_0}, e_1, \ldots, e_{n_1}\}, \quad S_2 = \{f_{f_0+1}, \ldots, f_f, e_{n_1+1}, \ldots, e_{n_1+n_2}\}$$

will form two mutually commuting Lie algebras if we also have

$$[f_\alpha, f_\beta] = 0, \quad 1 \le \alpha \le f_0, \quad f_0 + 1 \le \beta \le f.$$

10.2.1. Nilradicals of minimal dimension. Let us consider the case when

$$n = \dim \mathfrak{n} = \frac{1}{2} \dim \mathfrak{s} = \frac{s}{2}, \quad \text{i.e., } n = f.$$

In this case the matrices D_α, $\alpha = 1, \dots, n$ form an n-dimensional Abelian subalgebra of $\mathfrak{gl}(n, \mathbb{F})$ containing no nilpotent elements. This is only possible if the algebra $\{D_1, \dots, D_n\}$ is actually a Cartan subalgebra of $\mathfrak{gl}(n, \mathbb{F})$ (a maximal Abelian self-normalizing subalgebra, i.e., a MASA containing only semisimple elements).

Consider first the case of $\mathbb{F} = \mathbb{C}$. Over the field of complex numbers all Cartan subalgebras of a semisimple Lie algebra are mutually conjugate. In particular, the Cartan subalgebra of $\mathfrak{gl}(n, \mathbb{C})$ can be represented by n diagonal matrices and we can put

$$(D_\alpha)^i_k = \delta_{\alpha i}\delta_{\alpha k}, \quad [e_i, f_\alpha] = \delta_{i\alpha} e_i.$$

From relations (10.5) we obtain ${\gamma_{\alpha\beta}}^j = 0$ for $j \ne \alpha, \beta$. From (10.8) with $G = \mathbf{1}_{f\times f}$, $S = \mathbf{1}_{n\times n}$, $r^\beta_\alpha = -{\gamma_{\alpha\beta}}^\beta$ (no summation) we see that we can set

$${\gamma_{\alpha\beta}}^j = 0, \quad 1 \le \alpha, \beta, j \le n.$$

Thus the algebra $\mathfrak{s}$ is indecomposable for $f = n = 1$ and is decomposable into a direct sum of two-dimensional algebras for $n = f \ge 2$.

Now consider the algebra $\mathfrak{s}$ over the field of real numbers $\mathbb{F} = \mathbb{R}$. Cartan subalgebras of the real simple Lie algebras were classified by B. Kostant [63] and M. Sugiura [121]. In particular $\mathfrak{gl}(n, \mathbb{R})$ has $[n/2] + 1$ inequivalent classes of Cartan subalgebras. Each Cartan subalgebra can be represented by the matrices:

$$C_{n_2} = \begin{pmatrix} a_1 & & & & & & & \\ & \ddots & & & & & & \\ & & a_{n_1} & & & & & \\ & & & b_1 & -c_1 & & & \\ & & & c_1 & b_1 & & & \\ & & & & & \ddots & & \\ & & & & & & b_{n_2} & -c_{n_2} \\ & & & & & & c_{n_2} & b_{n_2} \end{pmatrix}, \quad 0 \le n_2 \le \left[\frac{n}{2}\right], \tag{10.22}$$

$$n_1 + 2n_2 = n.$$

From (10.22) we see that we can choose the matrices D_α to satisfy

$$(D_\alpha)^i_k = \begin{cases} \delta_{\alpha i}\delta_{\alpha k}, & 1 \le \alpha \le n_1; \\ b_\alpha(\delta_{\alpha i}\delta_{\alpha k} + \delta_{i\alpha+1}\delta_{k\alpha+1}), & \alpha = n_1+1, n_1+3, \dots, n_1+2n_2-1; \\ c_\alpha(\delta_{i\alpha-1}\delta_{k\alpha} - \delta_{i\alpha}\delta_{k\alpha-1}), & \alpha = n_1+2, n_1+4, \dots, n_1+2n_2. \end{cases} \tag{10.23}$$

From relations (10.5) and (10.8) we see that we can set ${\gamma_{\alpha\beta}}^j = 0$ for all values of α, β, j.

Let us sum up the results as a theorem.

Theorem 10.6. *Precisely two indecomposable solvable real Lie algebras $\mathfrak{s}$ with Abelian nilradical $\mathfrak{n}$ satisfying*

$$s = \dim \mathfrak{s} = 2 \dim \mathfrak{n} = 2n \tag{10.24}$$

exist, namely:

$$n = 1, \qquad [e, f] = e \tag{10.25}$$

$$\begin{aligned} n = 2, \qquad &[e_1, f_1] = e_1 \qquad &[e_1, f_2] = e_2 \\ &[e_2, f_1] = e_2 \qquad &[e_2, f_2] = -e_1 \\ &[f_1, f_2] = 0 \qquad &[e_1, e_2] = 0. \end{aligned} \tag{10.26}$$

Precisely one indecomposable complex solvable Lie algebra satisfying (10.24) *exists; it is given by the commutation relations* (10.25).

Every other solvable Lie algebra satisfying (10.24) *is decomposable into a direct sum of the Lie algebras* (10.25) *for* $\mathbb{F} = \mathbb{C}$, *and into a direct sum of the Lie algebras of* (10.25) *and* (10.26) *for* $\mathbb{F} = \mathbb{R}$.

10.3. Solvable Lie algebras with centers of maximal dimension

Consider again a solvable Lie algebra $\mathfrak{s}$ with an Abelian nilradical $\mathfrak{n}$. An important characteristic of $\mathfrak{s}$ is the dimension of its center $C(\mathfrak{s})$. Here we shall establish the maximal possible dimension

$$d_M = \max[\dim C(\mathfrak{s})], \tag{10.27}$$

compatible with the requirement that the algebra $\mathfrak{s}$ must be indecomposable.

The indecomposability together with (9.3) implies

$$C(\mathfrak{s}) \subseteq \mathfrak{s}^2 \subseteq \mathfrak{n}.$$

We also have $C(\mathfrak{s}) \neq \mathfrak{n}$, otherwise $\mathfrak{s}$ would be nilpotent.

Let us define a matrix $A \in \mathbb{F}^{fn \times n}$ obtained by stacking together all the matrices D_α of Theorems 10.1 and 10.2

$$A = \begin{pmatrix} D_1 \\ D_2 \\ \vdots \\ D_f \end{pmatrix} \in \mathbb{F}^{fn \times n}. \tag{10.28}$$

In view of (10.1) – (10.3) the center $C(\mathfrak{s})$ of $\mathfrak{s}$ is equal to the intersection of kernels of all derivations D_α, $1 \leq \alpha \leq f$. Therefore, $C(\mathfrak{s})$ is the nullspace of the matrix A. In order to maximize the dimension of $C(\mathfrak{s})$ we must hence maximize the dimension of the nullspace of A, i.e., minimize the rank of A.

If we are working over $\mathbb{C}$ we may assume that the matrices D_α are in the form (10.13). Obviously, only the nilpotent blocks $T_{p+j}(0)$ contribute to the computation of the nullspace.

The rank of A is bounded from below by $\sum_{j=1}^{p} n_j$. Since we are interested in the case of minimal rank of A we can assume without loss of generality that all the blocks $T^j_\alpha(a^j_\alpha)$, $j \leq p$ are one-dimensional. Equations (10.13) – (10.14) now imply

$$(D_\alpha)^i_k = \delta_{\alpha i}\delta_{\alpha k}, \quad 1 \leq i, k \leq n \text{ where } i \leq f \text{ or } k \leq f. \tag{10.29}$$

We investigate the possible structures of the remaining blocks which minimize the rank of A and preserve the indecomposability of $\mathfrak{s}$. We distinguish two cases depending on the relation between the dimensions n and f.

(1) $n - f \leq f(f-1)/2$. In order to maximize the dimension of the nullspace of A in this case we can set all the remaining blocks in D_α equal to zero

$$(D_\alpha)^i_k = 0, \quad f+1 \leq i \leq n, \ 1 \leq k \leq n \tag{10.30}$$

and put

$$[f_\alpha, f_\beta] = \tilde{e}_{\alpha\beta} \tag{10.31}$$

where $\tilde{e}_{\alpha\beta}$, $1 \leq \alpha < \beta \leq f$ span the whole vector space spanned by $e_{f+1}, \dots, e_n$.

In this case we have

$$C(\mathfrak{s}) = \operatorname{span}\{e_{f+1}, \dots, e_n\}$$

and

$$d_M = n - f. \tag{10.32}$$

Notice that the Jacobi identities (10.5) are satisfied identically and we have

$$\operatorname{span}\{[f_\alpha, f_\beta] \mid 1 \leq \alpha < \beta \leq f\} = C(\mathfrak{s}).$$

(2) $n-f > f(f-1)/2$. In this case we can only assume that $f(f-1)/2$ columns of A consist entirely of zeros which we assume without the loss of generality to be the last ones. If there were more vanishing columns in A, the corresponding nullspace would not be spanned by the commutators $[f_\alpha, f_\beta]$ and the algebra would necessarily be decomposable. To avoid decomposability the commutation relations between f_α, f_β must take the form (10.31) with

$$\{\tilde{e}_{\alpha\beta} \mid 1 \leq \alpha < \beta \leq f\} = \operatorname{span}\{e_{n-f(f+1)/2+1}, \dots, e_n\}. \tag{10.33}$$

On dimensional grounds, $\tilde{e}_{\alpha\beta}$ must be linearly independent.

That leaves the square block in the matrices D_α between the rows $f+1$ and $n - f(f-1)/2$ undetermined. Its dimension is

$$N_0 = n - \frac{f(f+1)}{2}. \tag{10.34}$$

Let us introduce two integer constants q_0, ν such that

$$N_0 = (f+1)q_0 + \nu, \quad 0 \leq \nu \leq f, \quad 0 \leq q_0, \quad q_0, \nu \in \mathbb{Z}. \tag{10.35}$$

We can now maximize the dimension of the nullspace of A by choosing the remaining blocks in D_α to be of the type $T^{f+j}(0)$, satisfying

$$T^{f+j}(0) \in \mathbb{F}^{(f+1)\times(f+1)}, \quad j = 1, \dots, q_0, \quad T^{f+q_0+1}(0) \in \mathbb{F}^{\nu\times\nu}, \quad \nu \geqq 2;$$

the precise structure of the blocks will be specified below.

As we shall see this is the structure which allows to maximize the dimension of the nullspace of A while keeping the algebra indecomposable. If the blocks $T^{f+j}(0)$ were of a larger size we could not accomplish the indecomposability in view of (10.40).

For $\nu = 1$ we must add one further nonzero 1×1 block on the diagonal of D_α instead of $T^{f+q_0+1}(0)$.

We thus have for $\nu \geq 2$

$$D_\alpha = \begin{pmatrix} E^\alpha & & & & & \\ & T^1_\alpha(0) & & & & \\ & & \ddots & & & \\ & & & T^{q_0}_\alpha(0) & & \\ & & & & T^{q_0+1}_\alpha(0) & \\ & & & & & \mathbf{0}_{d\times d} \end{pmatrix} \tag{10.36}$$

$$E^\alpha \in \mathbb{C}^{f\times f}, \quad (E^\alpha)_{ik} = \delta_{i\alpha}\delta_{k\alpha}, \quad d = \frac{f(f-1)}{2}. \tag{10.37}$$

In (10.36) the blocks $T^j_\alpha(0), T^{q_0+1}_\alpha(0)$ take the following form which maximizes the nullspace

$$T^j_\alpha(0) = \begin{pmatrix} 0 & \dots & 0 & b^{(j)}_{\alpha\,1} \\ & \ddots & \vdots & \vdots \\ & & 0 & b^{(j)}_{\alpha\,f} \\ & & & 0 \end{pmatrix} \in \mathbb{C}^{(f+1)\times(f+1)}, \quad 1 \le j \le q_0, \tag{10.38}$$

$$T^{q_0+1}_\alpha(0) = \begin{pmatrix} 0 & \dots & 0 & b^{(q_0+1)}_{\alpha\,1} \\ & \ddots & & \vdots \\ & & 0 & b^{(q_0+1)}_{\alpha\,\nu-1} \\ & & & 0 \end{pmatrix} \in \mathbb{C}^{\nu\times\nu}. \tag{10.39}$$

In this way, each block $T^j_\alpha(0)$ corresponds to a f-dimensional subspace of the nullspace of A.

The condition that the algebra $\mathfrak{s}$ be indecomposable, i.e., $C(\mathfrak{s}) \subset \mathfrak{s}^2$, requires that the matrices

$$M^{(j)} = \begin{pmatrix} b^{(j)}_{11} & \dots & b^{(j)}_{f1} \\ \vdots & & \vdots \\ b^{(j)}_{1f} & \dots & b^{(j)}_{ff} \end{pmatrix}, \quad j = 1, \dots, q_0, \qquad M^{(q_0+1)} = \begin{pmatrix} b^{(q_0+1)}_{11} & \dots & b^{(q_0+1)}_{f1} \\ \vdots & & \vdots \\ b^{(q_0+1)}_{1\,\nu-1} & \dots & b^{(q_0+1)}_{f\,\nu-1} \end{pmatrix}$$

all have maximal rank

$$\operatorname{rank} M^{(j)} = f, \quad j = 1, \dots, q_0, \qquad \operatorname{rank} M^{(q_0+1)} = \nu - 1 < f. \tag{10.40}$$

It is easy to verify that the Jacobi identities (10.5) are satisfied. The dimension of the center is

$$\dim C(\mathfrak{s}) = \begin{cases} \dfrac{f(f-1)}{2} + f q_0 + \nu - 1, & \text{for } \nu \ge 1, \\ \dfrac{f(f-1)}{2} + f q_0, & \text{for } \nu = 0. \end{cases}$$

When we eliminate q_0 using (10.34) and (10.35) we obtain

$$\dim C(\mathfrak{s}) = \begin{cases} \dfrac{1}{2(f+1)}\big(2fn - f(f+1) - 2(f+1-\nu)\big), & \nu \ge 1, \\ \dfrac{1}{2(f+1)}\big(2fn - f(f+1)\big), & \nu = 0. \end{cases}$$

The matrices D_α of (10.36) – (10.38) can be subjected to a simultaneous similarity transformation $D'_\alpha = G^{-1} D_\alpha G$. We choose G in the form

$$G = \begin{pmatrix} \mathbf{1}_{f\times f} & & & & & \\ & G_1 & & & & \\ & & \ddots & & & \\ & & & G_{q_0} & & \\ & & & & G_{(q_0+1)} & \\ & & & & & \mathbf{1}_{f(f-1)/2\times f(f-1)/2} \end{pmatrix}, \tag{10.41}$$

$$G_j = \begin{pmatrix} M^{(j)} & \\ & 1 \end{pmatrix}, \quad j = 1, \dots, q_0, \qquad G_{q_0+1} = \begin{pmatrix} \widetilde{M}^{(q_0+1)} & \\ & 1 \end{pmatrix},$$

where $\widetilde{M}^{(q_0+1)} \in \mathbb{F}^{(\nu-1)\times(\nu-1)}$ is a square matrix containing $(\nu-1)$ linearly independent columns of $M^{(q_0+1)}$. Thus we transform the rows in the matrix (10.38) into ones satisfying

$$b^{(j)}_{\alpha\, i} = \delta^i_\alpha, \qquad \alpha = 1, \ldots, f, \ j = 1, \ldots, q_0 \tag{10.42}$$
$$b^{(q_0+1)}_{\alpha\, i} = \delta^i_\alpha, \qquad \alpha = 1, \ldots, \nu - 1.$$

The results of this section can now be summed up as a theorem.

Theorem 10.7. *Let $\mathfrak{s}$ be a solvable nonnilpotent indecomposable Lie algebra of dimension $s = n + f$ with an Abelian nilradical $\mathfrak{n}$ of dimension n with $f < n$. Let $C(\mathfrak{s})$ be the center of $\mathfrak{s}$. Define a nonnegative integer $\nu \leq f$ by the following relation*

$$\nu = \left(n - \frac{f(f+1)}{2}\right) \pmod{f+1}.$$

The maximal possible dimension d_M of the center $C(\mathfrak{s})$ is given by

$$d_M = \begin{cases} n - f & \text{if } n - f \leq \dfrac{f(f-1)}{2}; \\ \dfrac{1}{2(f+1)}[2fn - f(f+1)] & \text{if } n - f > \dfrac{f(f-1)}{2},\ \nu = 0; \\ \dfrac{1}{2(f+1)}[2fn - f(f+1) + 2(\nu - f - 1)] & \text{if } n - f > \dfrac{f(f-1)}{2},\ \nu \geq 1. \end{cases} \tag{10.43}$$

The algebra $\mathfrak{s}$, over the fields $\mathbb{F} = \mathbb{C}$, or $\mathbb{F} = \mathbb{R}$, can be realized as in Theorems 10.1 *and* 10.2, *with the matrices D_α realized as in* (10.36) – (10.39), *satisfying* (10.42). *For $\nu = 1$ the nilpotent block of dimension $\nu \times \nu$ is replaced by $a^{q_0+1}_\alpha$ on the diagonal of D_α with $(a^{q_0+1}_1, \ldots, a^{q_0+1}_f) \neq (0, 0, \ldots, 0)$. The structure constants ${\gamma_{\alpha\beta}}^j$ in* (10.3) *must be such that the commutation relations $[f_\alpha, f_\beta]$ generate the entire subspace* (10.33).

Remark 10.8. (1) The construction of the matrices D_α, described above, is always possible and guarantees the maximal dimension of $C(\mathfrak{s})$. It is however not necessarily unique. What is unique is the rank of the matrix A, i.e., the value d_M in (10.43).

(2) Over the field $\mathbb{F} = \mathbb{R}$ the linear nilindependence of the matrices D_α can be arranged as in (10.36) – (10.38). Alternatively, in the left top corner of D_α we can replace any pair

$$\left\{\begin{pmatrix} 1 & \\ & 0 \end{pmatrix}, \begin{pmatrix} 0 & \\ & 1 \end{pmatrix}\right\} \quad \text{by} \quad \left\{\begin{pmatrix} 1 & 0 \\ 0 & 1 \end{pmatrix}, \begin{pmatrix} 0 & -1 \\ 1 & 0 \end{pmatrix}\right\}$$

to obtain a different Lie algebra $\mathfrak{s}$ with the same center $C(\mathfrak{s})$ (these algebras are inequivalent over $\mathbb{R}$, equivalent over $\mathbb{C}$).

Let us consider an example with $s = 15$, $f = 3$, $n = 12$. We then have $N_0 = 6$, $q_0 = 1$, $\nu = 2$. We have

$$D_1 = \begin{pmatrix}
1 &&&&&&&&&&& \\
& 0 &&&&&&&&&& \\
&& 0 &&&&&&&&& \\
&&& 0 & 0 & 0 & 1 &&&&& \\
&&& 0 & 0 & 0 & 0 &&&&& \\
&&& 0 & 0 & 0 & 0 &&&&& \\
&&& 0 & 0 & 0 & 0 &&&&& \\
&&&&&&& 0 & a_1 &&& \\
&&&&&&& 0 & 0 &&& \\
&&&&&&&&& 0 && \\
&&&&&&&&&& 0 & \\
&&&&&&&&&&& 0
\end{pmatrix},$$

$$D_2 = \begin{pmatrix}
0 &&&&&&&&&&& \\
& 1 &&&&&&&&&& \\
&& 0 &&&&&&&&& \\
&&& 0 & 0 & 0 & 0 &&&&& \\
&&& 0 & 0 & 0 & 1 &&&&& \\
&&& 0 & 0 & 0 & 0 &&&&& \\
&&& 0 & 0 & 0 & 0 &&&&& \\
&&&&&&& 0 & a_2 &&& \\
&&&&&&& 0 & 0 &&& \\
&&&&&&&&& 0 && \\
&&&&&&&&&& 0 & \\
&&&&&&&&&&& 0
\end{pmatrix},$$

$$D_3 = \begin{pmatrix}
0 &&&&&&&&&&& \\
& 0 &&&&&&&&&& \\
&& 1 &&&&&&&&& \\
&&& 0 & 0 & 0 & 0 &&&&& \\
&&& 0 & 0 & 0 & 0 &&&&& \\
&&& 0 & 0 & 0 & 1 &&&&& \\
&&& 0 & 0 & 0 & 0 &&&&& \\
&&&&&&& 0 & a_3 &&& \\
&&&&&&& 0 & 0 &&& \\
&&&&&&&&& 0 && \\
&&&&&&&&&& 0 & \\
&&&&&&&&&&& 0
\end{pmatrix},$$

$$(a_1, a_2, a_3) = (1, 1, a) \text{ or } (1, 0, 0) \quad \text{(up to permutations)}$$

and condition (10.33) is satisfied by putting

$$[f_1, f_2] = e_{10}, \quad [f_2, f_3] = e_{11}, \quad [f_3, f_1] = e_{12}.$$

10.4. Solvable Lie algebras with one nonnilpotent element and an n-dimensional Abelian nilradical

Any such algebra has commutation relations

$$\begin{aligned}
[f_1, e_i] &= \sum_{k=1}^{n} D_i^k e_k, & 1 \le i \le n, \\
[e_i, e_j] &= 0, & 1 \le i, j \le n.
\end{aligned}$$

Two algebras given by $D, \tilde{D}$ are isomorphic if and only if we have

$$\tilde{D} = \lambda G D G^{-1}, \quad \lambda \in \mathbb{F}, \lambda \neq 0, G \in \mathrm{SL}(n, \mathbb{F}).$$

The matrix D can be transformed into Jordan canonical form over the respective field $\mathbb{F}$. We shall first put the blocks that are indecomposable over the field $\mathbb{C}$ and order them by decreasing size. In the case of the field $\mathbb{R}$ we first put the blocks that are indecomposable over both $\mathbb{R}$ and $\mathbb{C}$ and below them blocks that are indecomposable over $\mathbb{R}$ but not absolutely decomposable (i.e., they become decomposable after field extension). These blocks are also ordered by decreasing size. The algebras given by D and $\tilde{D}$ that are related by

(1) permutations of blocks of equal size and Jordan type, and/or
(2) multiplication by a nonzero constant

are isomorphic.

We can have the following cases for matrix D:

(1) Diagonalizable over $\mathbb{F}$. In this case

(a) Order eigenvalues $|a_1| \ge |a_2| \ge \cdots \ge |a_n| > 0$ (if $a_n = 0$ then the algebra is decomposable).

(b) Divide by the largest one and get

$$D = \begin{pmatrix} 1 & & & \\ & a_2 & & \\ & & \ddots & \\ & & & a_n \end{pmatrix}.$$

If $1 > |a_2| > \cdots > |a_n| > 0$ then D is unique.

(c) If we have $|a_2| = \cdots = |a_k| = 1$, $k \le n$ set $a_j = \mathrm{e}^{\mathrm{i}\phi_j}$, $0 \le \phi_j < 2\pi$, $1 \le j \le k$ and order

$$0 = \phi_1 \le \phi_2 \le \cdots \le \phi_k. \tag{10.44}$$

Two algebras given by $D, \tilde{D}$ of such form are isomorphic if the numbers ϕ_1, $\phi_2, \ldots, \phi_k$ can be transformed into each other by an overall change of phase $\phi_j \to \phi_j + \alpha \bmod 2\pi$, $j = 1, \ldots, k$ accompanied by a suitable permutation of indices j so that (10.44) holds.

(d) Order all the remaining elements by increasing phases if their magnitudes are equal.

(2) Real matrix D diagonalizable over $\mathbb{C}$ but not over $\mathbb{R}$. We put

$$D = \begin{pmatrix} \alpha_1 & & & & & & & & \\ & \ddots & & & & & & & \\ & & \alpha_k & & & & & & \\ & & & \beta_1 & \gamma_1 & & & & \\ & & & -\gamma_1 & \beta_1 & & & & \\ & & & & & \ddots & & & \\ & & & & & & \beta_l & \gamma_l \\ & & & & & & -\gamma_l & \beta_l \end{pmatrix}.$$

where $k + 2l = n$, $\gamma_i > 0$, $\alpha_i \neq 0$ ($\gamma_i = 0$ is not allowed by the assumed block structure and the sign of γ_i can be changed by the permutation of the elements in the nilradical; if any of the constants α_i satisfies $\alpha_i = 0$ then the algebra is decomposable).

Order the blocks and choose the overall multiple so that $\alpha_1 = |\alpha_1| \geq \alpha_2 \geq \dots \geq \alpha_k > -|\alpha_1|$, $1 = \gamma_1 \geq \gamma_2 \geq \dots \geq \gamma_l > 0$. If $\gamma_i = \gamma_{i+1}$ then order $\beta_i > \beta_{i+1}$.

(3) Nondiagonalizable over $\mathbb{C}$

(a) In the case of algebras over $\mathbb{C}$ proceed as follows

(i) Order Jordan blocks by size

$$D = \begin{pmatrix} D_1 & & & \\ & D_2 & & \\ & & \ddots & \\ & & & D_k \end{pmatrix},$$

where $\dim D_i = n_i$, $n_i \geq n_{i+1}$, $\sum_{i=1}^{k} n_i = n$, and the blocks are of the form

$$D_i = \begin{pmatrix} a_i & 1 & 0 & \dots & 0 \\ & a_i & 1 & \dots & 0 \\ & & \ddots & \ddots & \\ & & & a_i & 1 \\ & & & & a_i \end{pmatrix}.$$

If $n_i = 1$ then $a_i \neq 0$ (otherwise the algebra would be decomposable). If all $n_i \geq 2$ then at least one $a_i \neq 0$ (otherwise the algebra would be nilpotent). If $a_i = 0$ for some i then the characteristic series are different from the generic case.

(ii) Let $n_1 = \dots = n_j > n_{j+1}$. Order $|a_1| \geq |a_2| \geq \dots \geq |a_j|$ and apply the rules of the diagonalizable case if $|a_i| = |a_{i+1}|$.

(iii) Order the remaining blocks of the same size by decreasing absolute value of a_i. In the case of equal absolute values order by increasing phases.

(b) In the case of algebras over $\mathbb{R}$ the block diagonal form of D

$$D = \begin{pmatrix} D_1 & & & \\ & D_2 & & \\ & & \ddots & \\ & & & D_k \end{pmatrix},$$

may contain two types of blocks.

Type I

$$D_{\mathrm{I}} = \begin{pmatrix} \alpha & 1 & 0 & \dots & 0 \\ & \alpha & 1 & \dots & 0 \\ & & \ddots & \ddots & \\ & & & \alpha & 1 \\ & & & & \alpha \end{pmatrix}.$$

Type I blocks with $\alpha = 0$ are allowed if $\dim D_I \geq 2$ but the characteristic series for algebras with $\alpha = 0$ differ from those with $\alpha \neq 0$.

Type II

$$D_{\mathrm{II}} = \begin{pmatrix} \beta & \gamma & 1 & 0 & \dots & 0 \\ -\gamma & \beta & 0 & 1 & \dots & 0 \\ & \ddots & \ddots & \ddots & \ddots & \\ & & \beta & \gamma & 1 & 0 \\ & & -\gamma & \beta & 0 & 1 \\ & & & & \beta & \gamma \\ & & & & -\gamma & \beta \end{pmatrix}, \quad \gamma > 0.$$

(i) Put type I blocks first (if any occur). If all the blocks are of type I then at least one α_i must satisfy $\alpha_i \neq 0$. Order both types of blocks by decreasing size.
  - Order blocks of type I of same size by decreasing value of α_i.
  - Order blocks of type II of same size by decreasing value of γ_i. If $\gamma_i = \gamma_{i+1}$ order them by decreasing value of β_i.

(ii) If the first block is of type I set $\alpha_1 = 1$ using the overall multiplying factor.

(iii) If the first block is of type II set $\gamma_1 = 1$ and $\beta_1 \geq 0$. If $\beta_1 = 0$ then set the first nonzero β_i positive. If more than one block of Type II occurs with $n_i = n_{i+1}$ and $\gamma_i = \gamma_{i+1}$ then order them by decreasing values of β_i.

10.5. Solvable Lie algebras with two nonnilpotent elements and n-dimensional Abelian nilradical

Any such algebra $\mathfrak{g}$ has commutation relations

$$\begin{aligned} [f_1, e_i] &= \sum_{k=1}^{n} (D_1)_i^k e_k, \quad 1 \leq i \leq n, \\ [f_2, e_i] &= \sum_{k=1}^{n} (D_2)_i^k e_k, \quad 1 \leq i \leq n, \\ [e_i, e_j] &= 0, \qquad\qquad\quad 1 \leq i, j \leq n, \\ [f_1, f_2] &\in C(\mathfrak{g}), \end{aligned}$$

where the matrices $D_1, D_2 \in \mathbb{F}^{n \times n}$ are linearly nilindependent and commute. Two algebras given by D_1, D_2 and $\widetilde{D}_1, \widetilde{D}_2$ are isomorphic if and only if we have

$$\widetilde{D}_1 = G(x_1 D_1 + x_2 D_2)G^{-1}, \quad \widetilde{D}_2 = G(y_1 D_1 + y_2 D_2)G^{-1} \tag{10.45}$$

where

$$\begin{pmatrix} x_1 & x_2 \\ y_1 & y_2 \end{pmatrix} \in \mathrm{GL}(2,\mathbb{F}), \quad G \in \mathrm{SL}(n,\mathbb{F}).$$

Using the $\mathrm{SL}(n,\mathbb{F})$ part of the equivalence (10.45) we can put simultaneously D_1 and D_2 into block diagonal form

$$D_1 = \begin{pmatrix} A_1 & & & \\ & A_2 & & \\ & & \ddots & \\ & & & A_k \end{pmatrix}, \quad D_2 = \begin{pmatrix} B_1 & & & \\ & B_2 & & \\ & & \ddots & \\ & & & B_k \end{pmatrix} \tag{10.46}$$

where $\dim A_i = \dim B_i = n_i$, $n_1 \geq n_2 \geq \cdots \geq n_k$, $\sum_{i=1}^k n_i = n$ and $\mathrm{span}\{A_i, B_i\}$ are indecomposable Abelian subalgebras of $\mathfrak{gl}(n_i,\mathbb{F})$ meaning that A_i and B_i cannot be simultaneously block diagonalized into smaller blocks. There remain some $\mathrm{SL}(n-2,\mathbb{F})$ transformations at our disposal, namely

(1) Permutations of the same size blocks (simultaneous in both D_1 and D_2), and
(2) Changes of basis inside the blocks, i.e., conjugation by G of the form

$$G \in \mathrm{GL}(n_1,\mathbb{F}) \times \cdots \times \mathrm{GL}(n_k,\mathbb{F}).$$

This amounts to a classification of two-dimensional Abelian subalgebras of $\mathfrak{gl}(n_i,\mathbb{F})$ for $1 \leq i \leq k$.

Therefore the form (10.46) still doesn't fix the representant of the equivalence class with respect to (10.45) uniquely. For example, let us consider the case of D_1 and D_2 both diagonal,

$$D_1 = \begin{pmatrix} a_1 & & & \\ & a_2 & & \\ & & \ddots & \\ & & & a_n \end{pmatrix}, \quad D_2 = \begin{pmatrix} b_1 & & & \\ & b_2 & & \\ & & \ddots & \\ & & & b_n \end{pmatrix}.$$

By permutation of the basis elements of the nilradical together with the $\mathrm{GL}(2,\mathbb{F})$ part of the equivalence (10.45) we may transform D_1, D_2 into the form

$$D_1 = \begin{pmatrix} a_1 & & & & \\ & \ddots & & & \\ & & a_{n-2} & & \\ & & & 0 & \\ & & & & 1 \end{pmatrix}, \quad D_2 = \begin{pmatrix} b_1 & & & & \\ & \ddots & & & \\ & & b_{n-2} & & \\ & & & 1 & \\ & & & & 0 \end{pmatrix} \tag{10.47}$$

with

$$|a_1| \leq |a_2| \leq \cdots \leq |a_{n-2}| \leq 1, \quad |b_i| \leq |a_{n-2}|.$$

If $a_1 = b_1 = 0$ then by indecomposability we find

$$[f_1, f_2] = ce_1, \ c \neq 0$$

which by an obvious rescaling of e_1 can be put in the form

$$[f_1, f_2] = e_1.$$

If $|a_1| + |b_1| \neq 0$ then $C(\mathfrak{g}) = 0$ and we have $[f_1, f_2] = 0$.

The choice of D_1, D_2 in the form of (10.47) fixes all the continuous symmetries in the equivalence (10.45). Unfortunately, in general there are still some discrete symmetries present which further complicate the choice of the representative of the equivalence class.

For example if $n = 4$ then we may use

$$\begin{pmatrix} x_1 & x_2 \\ y_1 & y_2 \end{pmatrix} = \frac{1}{\Delta}\begin{pmatrix} b_2 & -a_2 \\ -b_1 & a_1 \end{pmatrix}, \quad \Delta = a_1 b_2 - a_2 b_1$$

accompanied by a permutation of the basis elements in the nilradical to show that the algebras given by

$$D_1 = \begin{pmatrix} a_1 & & & \\ & a_2 & & \\ & & 0 & \\ & & & 1 \end{pmatrix}, \quad D_2 = \begin{pmatrix} b_1 & & & \\ & b_2 & & \\ & & 1 & \\ & & & 0 \end{pmatrix}$$

and

$$D_1 = \begin{pmatrix} -b_2/\Delta & & & \\ & a_2/\Delta & & \\ & & 0 & \\ & & & 1 \end{pmatrix}, \quad D_2 = \begin{pmatrix} b_1/\Delta & & & \\ & -a_1/\Delta & & \\ & & 1 & \\ & & & 0 \end{pmatrix}$$

are isomorphic. Consequently we may impose an additional constraint on our parameters so that

$$|\Delta| = |a_1 b_2 - a_2 b_1| \leq 1.$$

This fixes the choice of D_1, D_2 uniquely if sharp inequalities hold in all the inequalities encountered. If there is any equality, e.g., $|\Delta| = 1$, we have several choices of D_1, D_2 allowed for isomorphic algebras and one of them must be chosen.

10.6. Generalized Casimir invariants of solvable Lie algebras with Abelian nilradicals

Let us now apply the methods of Chapter 3 to calculate the generalized Casimir invariants of solvable Lie algebras $\mathfrak{s}$ with Abelian nilradicals. The Lie brackets of these algebras were given in Theorem 10.1. The vector fields (9.24) of Section 9.5 in this case are

$$\begin{aligned} \widehat{E}_i &= (D_\alpha)_i^k e_k \frac{\partial}{\partial f_\alpha}, & i &= 1, \dots, n, \\ \widehat{F}_\alpha &= -(D_\alpha)_i^k e_k \frac{\partial}{\partial e_i} + {\gamma_{\beta\alpha}}^k e_k \frac{\partial}{\partial f_\beta}, & \alpha &= 1, \dots, f. \end{aligned}$$

The function

$$I(e_1, \dots, e_n, f_1, \dots, f_f)$$

will be an invariant of the coadjoint representation of $\mathfrak{s}$ if it satisfies the following linear first order partial differential equations

$$\widehat{E}_i I = 0, \quad i = 1, \dots, n, \tag{10.48}$$

$$\widehat{F}_\alpha I = 0, \quad \alpha = 1, \dots, f. \tag{10.49}$$

Our aim is to find a complete set of elementary solutions of equations (10.49)–(10.48), i.e., a functionally independent set of generalized Casimir invariants.

10.6.1. General form of the generalized Casimir invariants and their number.

Theorem 10.9. *The solvable Lie algebra* $\mathfrak{s}$ *over the field* $\mathbb{C}$ *satisfying the commutation relations* (10.1) – (10.3) *has exactly*

$$m = n - f \tag{10.50}$$

functionally independent generalized Casimir invariants. They depend only on the variables e_i*, corresponding to the elements of the Abelian nilradical* $\mathfrak{n}$,

$$I_j = I_j(e_1, \ldots, e_n), \quad j = 1, \ldots, m. \tag{10.51}$$

PROOF. The general invariant $I(e_1, \ldots, e_n, f_1, \ldots, f_f)$ must satisfy (10.49), (10.48), in particular

$$\widehat{E}_i I = \sum_{\alpha=1}^{f} \sum_{k=1}^{n} (D_\alpha)_i^k \, e_k \frac{\partial}{\partial f_\alpha} I = 0, \quad i = 1, \ldots, n. \tag{10.52}$$

The matrices D_α have the block diagonal form of Theorem 10.2, Equation (10.13). Consequently, particular vector fields in (10.52) take the form

$$\begin{aligned} \widehat{E}_1 &= -e_1 \frac{\partial}{\partial f_1}, \quad \widehat{E}_{n_1+1} = -e_{n_1+1} \frac{\partial}{\partial f_2}, \ldots, \\ \widehat{E}_{n_1+n_2+\cdots+n_{f-1}+1} &= -e_{n_1+n_2+\cdots+n_{f-1}+1} \frac{\partial}{\partial f_f}. \end{aligned} \tag{10.53}$$

The form of the vector fields in (10.53) implies that the invariants of the algebra $\mathfrak{s}$, and hence in particular the elementary invariants I_j, do not depend on $f_1, \ldots, f_f$, as indicated in (10.51).

Now consider (10.49) which simplifies to

$$\widehat{F}_\alpha I(e_1, \ldots, e_n) = \sum_{i,k=1}^{n} (D_\alpha)_i^k \, e_k \frac{\partial}{\partial e_i} I = 0, \quad \alpha = 1, \ldots, f. \tag{10.54}$$

Equation (10.54) can be rewritten in matrix form as

$$M \begin{pmatrix} \partial I/\partial e_1 \\ \vdots \\ \partial I/\partial e_n \end{pmatrix} = \begin{pmatrix} 0 \\ \vdots \\ 0 \end{pmatrix}. \tag{10.55}$$

From (10.54) we see that the matrix M has maximal rank at generic points of the space $\mathfrak{n}^*$ with the coordinates $(e_1, \ldots, e_n)$. Thus, we have f independent equations for a function of n variables. The number of independent solutions is hence $n - f$, as stated in (10.50).

This completes the proof. □

Notice that the structure constants $\gamma_{\alpha\beta}{}^j$ of (10.3) play no role in the calculation of the Casimir invariants.

Theorem 10.9 is valid also over the field $\mathbb{F} = \mathbb{R}$. The proof is quite analogous, but involves the usual complications due to the fact that the field $\mathbb{R}$ is not algebraically closed and hence the normal forms of commuting matrices can be more complicated. We omit the details of the proof.

The generalized Casimir invariants (10.51) are thus a complete set of $n - f$ functionally independent solutions of (10.55) (or equivalently of (10.54)). The actual form of the invariants depends on the dimension f of the factor algebra $\mathfrak{s}/\mathfrak{n}$ and on the specific form of the matrices D_α. The invariants are of course basis dependent.

Let us now consider some special cases in appropriate bases. They bring out all the characteristic properties of the general case.

10.6.2. Diagonal structure matrices. The simplest case, for any values of f and n, occurs when all the structure matrices D_α are diagonal (for an appropriate choice of the basis in the nilradical $\mathfrak{n}$). By linear combinations of the elements $f_\alpha \subset F$ we can transform the structure matrices to the form

$$D_1 = \begin{pmatrix} 1 & & & & & & \\ & 0 & & & & & \\ & & \ddots & & & & \\ & & & 0 & & & \\ & & & & a_{11} & & \\ & & & & & \ddots & \\ & & & & & & a_{1n-f} \end{pmatrix}, \dots, D_f = \begin{pmatrix} 0 & & & & & & \\ & \ddots & & & & & \\ & & 0 & & & & \\ & & & 1 & & & \\ & & & & a_{f1} & & \\ & & & & & \ddots & \\ & & & & & & a_{fn-f} \end{pmatrix}. \tag{10.56}$$

The $n - f$ invariants can be chosen to be

$$I_k = e_{f+k}\, e_1^{-a_{1k}}\, e_2^{-a_{2k}} \cdots e_f^{-a_{fk}}, \quad k = 1, \dots, n-f. \tag{10.57}$$

Thus, I_k are all rational if the eigenvalues $a_{\alpha k}$ are rational.

10.6.3. Casimir invariants in the case $f = 1$. Let us consider the Lie algebra $\mathfrak{s}$ over the field $\mathbb{C}$ for the case $f = 1$. The algebra $\mathfrak{s}$ has a basis $(e_1, \dots, e_n, f_1)$, where f_1 is the only element of the basis not contained in the nilradical. The commutation relations in this case are

$$[f_1, e_i] = \sum_{k=1}^{n} D_i^k\, e_k, \quad [e_i, e_k] = 0. \tag{10.58}$$

In view of Theorem 10.9 any invariant of such a Lie algebra $\mathfrak{s}$ will depend only on $e_1, \dots, e_n$ and will be a solution of a single 1st order linear PDE

$$\sum_{i,k=1}^{n} D_i^k\, e_k \frac{\partial}{\partial e_i} I(e_1, \dots, e_n) = 0. \tag{10.59}$$

The solution of (10.59) is precisely the task that was studied in Section 3.2.2. The vector field

$$\widehat{F}_1 = \sum_{i,k=1}^{n} D_i^k\, e_k \frac{\partial}{\partial e_i}$$

which represents the nonnilpotent element f_1, has the form (3.11) after identification of variables $e_j \equiv x_j$ and the matrix D once reduced to its Jordan canonical form coincides with the matrix F of (3.18). The complete set of elementary invariants of the algebra $\mathfrak{s} = \operatorname{span}\{e_1, \dots, e_n, f_1\}$ is therefore given by Theorem 3.4 upon replacing the variables x_j by e_j.

10.6.4. Casimir invariants for low-dimensional nilradicals. In the previous Section 10.6.3 we considered the case of the largest possible nilradical of a solvable nonnilpotent Lie algebra, namely the case $f = 1$. Now we shall consider the opposite situation, when the factor algebra $\mathfrak{s}/\mathfrak{n}$ is large, namely $f = n$, $n - 1$ and $n - 2$.

The case $f = n$ is only possible (over $\mathbb{C}$) if the algebra $\mathfrak{s}$ satisfies $\dim \mathfrak{s} = 2$, or is decomposable into a direct sum of two-dimensional solvable Lie algebras as was shown in Theorem 10.6. The algebra $\mathfrak{s}$ in this case has no Casimir invariants at all (see (10.50) of Theorem 10.9).

The case $f = n-1$. We have $n-1$ linearly nilindependent commuting matrices $D_1, \dots, D_{n-1} \in \mathbb{C}^{n\times n}$. Hence, they form a subalgebra of a decomposable MASA of $\mathfrak{gl}(n, \mathbb{C})$. Only 2 decompositions of n are allowed by the condition of linear nilindependence, namely n one-dimensional blocks, i.e., all matrices D_i are diagonalizable, or 1 two-dimensional block and $(n-2)$ one-dimensional ones.

In this case we have the following result.

Theorem 10.10. *Let $\mathfrak{s}$ be a complex indecomposable solvable Lie algebra of dimension $2n-1$ with an Abelian nilradical of dimension n. Then $\mathfrak{s}$ has precisely one generalized Casimir invariant that, in an appropriate basis, has one of the following two forms*

(1) *If the matrices D_α of (10.1) are simultaneously diagonalizable, we have*

$$I = \frac{1}{e_n}(e_1^{a_1} e_2^{a_2} \cdots e_{n-1}^{a_{n-1}}). \tag{10.60}$$

(2) *If the matrices D_α are not simultaneously diagonalizable, we have*

$$I = \frac{e_2}{e_1} - \ln(e_1^{b_1} e_3^{b_2} \cdots e_n^{b_{n-1}}). \tag{10.61}$$

In the first case the constants $a_\alpha \in \mathbb{C}$ are eigenvalues of the structure matrices D_α as in (10.56). In the second case the constants $b_\alpha \in \mathbb{C}$ are the off-diagonal elements of the structure matrices.

Proof. It suffices to notice that in the first case all matrices D_α can be written as

$$D_\alpha = \operatorname{diag}(\delta_{\alpha 1}, \dots, \delta_{\alpha n-1}, a_\alpha), \quad a_j \in \mathbb{C}. \tag{10.62}$$

Reading off the operators $\widehat{F}_\alpha$ of (10.48) in this case, we see that they all annihilate the invariant (10.60).

In the second case we choose

$$D_\alpha = \operatorname{diag}\left(\begin{pmatrix} \delta_{\alpha 1} & 0 \\ b_\alpha & \delta_{\alpha 1} \end{pmatrix}, \delta_{\alpha 2}, \dots, \delta_{\alpha n-1}\right), \quad b_\alpha \in \mathbb{C},\ \alpha = 1, \dots, n-1. \tag{10.63}$$

Again, all the corresponding operators $\widehat{F}_\alpha$ will annihilate the "logarithmic" invariant I of (10.61). □

The case $f = n-2$. The matrices $D_\alpha \in \mathbb{C}^{n\times n}$ again form a decomposable MASA of $\mathfrak{gl}(n, \mathbb{C})$ and four different decompositions are allowed. Let us state the results as a theorem.

Theorem 10.11. *Let $\mathfrak{s}$ be a complex indecomposable solvable Lie algebra of dimension $2n-2$ with an Abelian nilradical of dimension n. The structure* *matrices* $\{D_\alpha\}$ *and invariants* $\{I_1, I_2\}$ *have, in the appropriate basis, one of the following forms*

(1) *Decomposition* $n = n \times 1$

$$D_\alpha = \operatorname{diag}(\delta_{\alpha 1} \dots \delta_{\alpha n-2}, a_\alpha, b_\alpha), \quad a_\alpha, b_\alpha \in \mathbb{C}, \ \alpha = 1, \dots, n-2. \tag{10.64}$$

Then

$$I_1 = \frac{1}{e_{n-1}}(e_1^{a_1} e_2^{a_2} \cdots e_{n-2}^{a_{n-2}}), \quad I_2 = \frac{1}{e_n}(e_1^{b_1} e_2^{b_2} \cdots e_{n-2}^{b_{n-2}}). \tag{10.65}$$

(2) *Decomposition* $n = 2 + (n-2) \times 1$

$$D_\alpha = \operatorname{diag}\left[\begin{pmatrix} c_\alpha & 0 \\ y_\alpha & c_\alpha \end{pmatrix}, b_{\alpha 3}, \dots, b_{\alpha n-1}, a_\alpha\right], \tag{10.66}$$

$$c_1 = \begin{cases} 1 \\ 0 \end{cases}, \quad c_\alpha = 0, \quad 2 \le \alpha \le n-2, \quad a_\alpha \in \mathbb{C}, \tag{10.67}$$

$$b_{\alpha j} = \delta_{j \alpha+1} \quad j = 3, \dots, n-1 \text{ for } c_1 = 1, \tag{10.68}$$

$$b_{\alpha j} = \delta_{j \alpha+1} \quad j = 3, \dots, n \text{ for } c_1 = 0, \tag{10.69}$$

$y_\alpha \in \mathbb{C}$, *at least one* y_α *satisfies* $y_\alpha \neq 0$. *For* $c_1 = 1$ *we have*

$$\begin{aligned} I_1 &= \frac{1}{e_n}(e_1^{a_1} e_3^{a_2} \cdots e_{n-1}^{a_{n-2}}), \\ I_2 &= \frac{e_2}{e_1} - \ln(e_1^{y_1} e_3^{y_2} \cdots e_{n-1}^{y_{n-2}}). \end{aligned} \tag{10.70}$$

For $c_1 = 0$ *we have*

$$\begin{aligned} I_1 &= e_1, \\ I_2 &= e_2 - e_1 \ln(e_3^{y_1} e_4^{y_2} \cdots e_n^{y_{n-2}}), \end{aligned} \tag{10.71}$$

(3) *Decomposition* $n = 2 + 2 + (n-4) \times 1$

$$D_\alpha = \operatorname{diag}\left[\begin{pmatrix} \delta_{\alpha 1} & 0 \\ y_\alpha & \delta_{\alpha 1} \end{pmatrix}, \begin{pmatrix} \delta_{\alpha 2} & 0 \\ z_\alpha & \delta_{\alpha 2} \end{pmatrix}, \delta_{\alpha 3}, \dots, \delta_{\alpha n-2}\right] \tag{10.72}$$

$y_\alpha, z_\alpha \in \mathbb{C}$, *at least one* y_α *and one* z_β *satisfies* $y_\alpha \neq 0$, $z_\beta \neq 0$. *In this case the invariants are*

$$I_1 = \frac{e_2}{e_1} - \ln(e_1^{y_1} e_3^{y_2} e_5^{y_3} \cdots e_n^{y_{n-2}}) \tag{10.73}$$

$$I_2 = \frac{e_4}{e_3} - \ln(e_1^{z_1} e_3^{z_2} e_5^{z_3} \cdots e_n^{z_{n-2}}). \tag{10.74}$$

(4) *Decomposition* $n = 3 + (n-3) \times 1$

$$D_\alpha = \operatorname{diag}\left[\begin{pmatrix} \delta_{\alpha 1} & & \\ u_\alpha & \delta_{\alpha 1} & \\ y_\alpha & z_\alpha & \delta_{\alpha 1} \end{pmatrix}, \delta_{\alpha 2}, \dots, \delta_{\alpha n-2}\right] \tag{10.75}$$

The complex constants u_α, y_α *and* z_α *satisfy one of the following conditions*

- $u_\alpha = z_\alpha$ *for all* α, $u_\alpha = z_\alpha \neq 0$ *for at least one value of* α.
- $u_\alpha = 0$ *for all* α, $y_\beta \neq 0$, $z_\gamma \neq 0$ *for at least one value of* β *and* γ.
- $z_\alpha = 0$ *for all* $\alpha, y_\beta \neq 0$, $u_\gamma \neq 0$ *for at least one value of* β *and* γ.

The invariants in this basis are

$$I_1 = \frac{e_2}{e_1} - \ln e_1^{u_1} e_4^{u_2} \cdots e_n^{u_{n-2}} \tag{10.76}$$

$$\begin{aligned} I_2 = y_{n-2} \frac{e_2}{e_1} - u_{n-2} \frac{e_3}{e_1} + \frac{1}{2} z_{n-2} \left(\frac{e_2}{e_1}\right)^2 + \\ + \ln\left[e_1^{(u_{n-2}y_1 - u_1 y_{n-2})} e_4^{(u_{n-2}y_2 - u_2 y_{n-2})} \cdots e_{n-1}^{(u_{n-2}y_{n-3} - u_{n-3}y_{n-2})}\right]. \end{aligned} \tag{10.77}$$

PROOF. The proof is the same in all cases. The operators $\widehat{F}_\alpha$ of (10.48) are read off from the matrices D_α. It is then easy to verify that we have $\widehat{F}_\alpha I_1 = \widehat{F}_\alpha I_2 = 0$ for all α in all cases. It is obvious from the form of I_1 and I_2 that they are functionally independent. The conditions on the constants u_α, y_α and z_α in the fourth case are necessary to assure commutativity of the matrices D_α. They correspond to Kravchuk signatures (1 1 1), (2 0 1) and (1 0 2), respectively. □

10.6.5. Summary. The main results of this section can be summed up as follows:

(1) A solvable Lie algebra $\mathfrak{s}$, over the field $\mathbb{C}$, of dimension $n + f$, with an Abelian nilradical of dimension n, has precisely $n - f$ generalized Casimir invariants. All of them are functions of the variables e_i, representing the elements of the nilradical in $(\mathfrak{s}^*)^*$ (Theorem 10.9).

(2) The generalized Casimir invariants in general involve logarithms of polynomials in e_i, as well as rational and irrational functions of e_i (Theorems 3.4, 10.10 and 10.11).

(3) Explicit expressions for the generalized Casimir operators were given for the cases when the dimension of the nilradical is equal, or close to its minimal, or alternatively to itsmaximal possible value.

(4) The generalized Casimir operators are rational functions only when all the structure matrices D_α of (10.1) are diagonal and the eigenvalues satisfy certain rationality conditions. Logarithmic expressions occur as soon as any of the structure matrices contains nontrivial Jordan blocks.

To illustrate the complications arising over the field of real numbers, consider the four-dimensional Lie algebra $\mathfrak{s}_{4,5}$ of Section 17.2, expressed as in Theorem 10.1 with

$$D_1 = \begin{pmatrix} \alpha & & \\ & \beta & 1 \\ & -1 & \beta \end{pmatrix} \qquad \alpha, \beta \in \mathbb{R} \tag{10.78}$$

Obviously, D_1 is diagonalizable over $\mathbb{C}$, but not over $\mathbb{R}$. In agreement with Theorem 10.9 (valid also for $\mathbb{F} = \mathbb{R}$) we have two invariants, both depending on (e_1, e_2, e_3) only. However, their form is:

$$I_1 = \frac{(e_2^2 + e_3^2)^\alpha}{e_1^{2\beta}}, \qquad I_2 = \ln e_1 + \alpha \arctan \frac{e_2}{e_3} \tag{10.79}$$

(or, equivalently, $\tilde{I}_2 = e_1 \exp(\alpha \arctan e_2/e_3)$, as presented in Section 17.2). Thus, in addition to rational and irrational functions and logarithms, we obtain inverse trigonometric functions.

In Part 4 of this book we classify all complex and real Lie algebras of dimension $\dim \mathfrak{s} \leq 6$ and present a basis for the (generalized) Casimir invariants in each case.

CHAPTER 11

Solvable Lie Algebras with Heisenberg Nilradical

The purpose of this chapter is to construct all indecomposable solvable Lie algebras $\mathfrak{s}$ that have the $(2m+1)$-dimensional Heisenberg algebra $\mathfrak{h}(m)$ as their nilradical. The main results presented in this chapter were obtained in [106].

11.1. The Heisenberg relations and the Heisenberg algebra

The Heisenberg algebras $\mathfrak{h}(m)$ with their basis written conventionally in the form

$$(h, p_1, \ldots, p_m, q_1, \ldots, q_m) \tag{11.1}$$

is of primordial importance in quantum mechanics. In this context the extension $\mathfrak{s}$ of the algebra $\mathfrak{h}(m)$ by further operators $f_1, \ldots, f_f$ is a question of the algebra of quantum mechanical observables.

The Heisenberg relations

$$[\hat{x}_a, \hat{p}_b] = \mathrm{i}\hbar\delta_{ab}, \quad 1 \le a, b \le m \tag{11.2}$$

underlie all of quantum mechanics. In standard quantum theory $\hat{x}_a$ are the coordinates of a particle and $\hat{p}_a = -\mathrm{i}\hbar\partial_{x_a}$ are the components of its linear momentum operator. An abstract theory of the Heisenberg relations exists and is called "umbral calculus" [38, 103]. Other realizations exist, for instance $\hat{p}_a$ can be represented by difference operators ∇_a and $\hat{x}_a$ by coordinates on a lattice multiplied by an appropriate shift operator [40, 69, 70]. This can be used to develop quantum theory on a lattice while preserving continuous symmetries like Galilei, Poincaré or conformal invariance. Standard quantum mechanics and quantum mechanics on a lattice can be viewed as the theory of polynomials or convergent power series in the operators realizing the Lie brackets (11.2).

The Heisenberg algebra $\mathfrak{h}(m)$ is also a subalgebra of the quantum mechanical Galilei algebra [67, 132] (or extended Galilei algebra). The operators $\hat{p}_a$ in this interpretation generate space translations, $\hat{q}_a$ generate Galilei boosts and h changes the phase of the wavefunction. The algebra $\mathfrak{h}(m)$ is also a subalgebra of the symmetry algebra of the heat equation, [86], of the nonlinear Schrödinger equation with any nonlinearity $F(|\psi|)$ depending only on the absolute value of the wavefunction ψ and of many other partial differential equations occurring in nonrelativistic physical theories [49, 76]. In this context, extensions of the Heisenberg algebra $\mathfrak{h}(m)$ are part of a study of physical theories with symmetries going beyond translations and Galilei boosts.

An algebra that plays an important role in the microscopic theory of collective motions in nuclei is a semidirect product of the symplectic Lie algebra with the Heisenberg algebra as an ideal [96, 97, 104, 105]

$$\mathfrak{wsp}(2m, \mathbb{R}) = \mathfrak{sp}(2m, \mathbb{R}) \ni \mathfrak{h}(m). \tag{11.3}$$

As we shall show below, all solvable indecomposable Lie algebras $\mathfrak{s}$ obtained as extensions of the Heisenberg algebra $\mathfrak{h}(m)$ are closely related to subalgebras of $\mathfrak{wsp}(2m,\mathbb{R})$ (or its complex version $\mathfrak{wsp}(2m,\mathbb{C})$).

11.2. Classification of solvable Lie algebras with nilradical $\mathfrak{h}(m)$

11.2.1. Basic classification theorem. Let us consider the Heisenberg algebra $\mathfrak{h}(m)$ in its standard basis $(h, p_1, \ldots, p_m, q_1, \ldots, q_m)$ with commutation relations

$$\begin{gathered}[p_i, q_j] = \delta_{ij} h, \qquad [p_i, p_j] = [q_i, q_j] = 0, \\ [p_i, h] = [q_i, h] = 0, \quad 1 \le i, j \le m.\end{gathered} \tag{11.4}$$

For future convenience let us denote

$$\vec{p} = (p_1, \ldots, p_m), \quad \vec{q} = (q_1, \ldots, q_m). \tag{11.5}$$

We wish to extend this algebra to an indecomposable solvable Lie algebra $\mathfrak{s}$ of dimension $2m+1+f$ having $\mathfrak{h}(m)$ as its nilradical. This means we wish to add f further linearly nilindependent elements to $\mathfrak{h}(m)$. Let us denote them $f_1, \ldots, f_f$.

The commutation relations of $\mathfrak{s}$ involving the new elements f_α will have the form

$$\begin{gathered}([f_\alpha, h], [f_\alpha, \vec{p}], [f_\alpha, \vec{q}]) = (h, \vec{p}, \vec{q}) \cdot \begin{pmatrix} 2a_\alpha & \rho_{\alpha 1} & \rho_{\alpha 2} \\ \sigma_{\alpha 1}^T & a_\alpha \mathbf{1}_{m\times m} + A_\alpha & B_\alpha \\ \sigma_{\alpha 2}^T & C_\alpha & a_\alpha \mathbf{1}_{m\times m} + D_\alpha \end{pmatrix}, \\ a_\alpha \in \mathbb{F}, \quad \rho_{\alpha 1}, \rho_{\alpha 2}, \sigma_{\alpha 1}, \sigma_{\alpha 2} \in \mathbb{F}^{1\times m}, \quad A_\alpha, B_\alpha, C_\alpha, D_\alpha \in \mathbb{F}^{m\times m}.\end{gathered} \tag{11.6}$$

The superscript T denotes transposition and the constants a_α were split off from A_α and D_α for future convenience. Further, we have

$$[f_\alpha, f_\beta] = r_{\alpha\beta} h + \mu_{\alpha\beta}{}^i p_i + \nu_{\alpha\beta}{}^i q_i \tag{11.7}$$

for some constants $r_{\alpha\beta}, \mu_{\alpha\beta}{}^i, \nu_{\alpha\beta}{}^i \in \mathbb{F}$.

First, we proceed to simplify the Lie brackets (11.6) as much as is possible. We perform a change of basis

$$f'_\alpha = f_\alpha + \rho_{\alpha 1,i} q_i - \rho_{\alpha 2,i} p_i.$$

This allows us to set $\rho_{\alpha 1,i}$ and $\rho_{\alpha 2,i}$ to zero in (11.6).

Let us now impose the Jacobi identities. From the triplets (f_α, p_i, h) and (f_α, q_i, h) we obtain that $\sigma_{\alpha 1} = \sigma_{\alpha 2} = 0$ in (11.6). From the triplets (f_α, p_i, p_j), (f_α, p_i, q_j) and (f_α, q_i, q_j) we find that the matrices $A_\alpha, B_\alpha, C_\alpha, D_\alpha \in \mathbb{F}^{m\times m}$ must satisfy

$$D_\alpha = -A_\alpha^T, \quad B_\alpha = B_\alpha^T, \quad C_\alpha = C_\alpha^T. \tag{11.8}$$

It follows that the matrices

$$X_\alpha = \begin{pmatrix} A_\alpha & B_\alpha \\ C_\alpha & -A_\alpha^T \end{pmatrix} \tag{11.9}$$

belong to the symplectic Lie algebra $\mathfrak{sp}(2m,\mathbb{F})$, i.e., the satisfy.

$$X_\alpha K + K X_\alpha^T = 0, \quad K = \begin{pmatrix} 0 & \mathbf{1}_{m\times m} \\ -\mathbf{1}_{m\times m} & 0 \end{pmatrix}. \tag{11.10}$$

By taking suitable linear combinations of the elements f_α we can arrange to have $a_1 = 1$ or 0 and $a_2 = \cdots = a_f = 0$.

The Jacobi identities for the triplets (f_α, f_β, p_i) and (f_α, f_β, q_i) imply that ${\mu_{\alpha\beta}}^i = {\nu_{\alpha\beta}}^i = 0$. Further they imply that the matrices X_α must commute. For $a_1 = 0$ they must also be linearly nilindependent, otherwise the nilradical would be larger than $\mathfrak{h}(m)$. For $a_1 = 1$ the matrices $X_2, \dots, X_f$ must be linearly nilindependent for the same reason, though X_1 may be nilpotent or even vanish. This imposes a restriction on the number of elements f that can be added. Indeed, the number of commuting linearly nilindependent matrices $X_\alpha \in \mathfrak{sp}(2m, \mathbb{F})$ is less or equal to the rank m of $\mathfrak{sp}(2m, \mathbb{F})$ [90].

The remaining Jacobi identities for $(f_\alpha, f_\beta, f_\gamma)$ (for $f \geq 3$) imply that for $a_1 = 1$ we have $r_{\alpha\beta} = 0$ whenever $\alpha \neq 1$ and $\beta \neq 1$. Redefining f_α as $f'_\alpha = f_\alpha - r_{1\alpha} h$ we obtain $r_{1\alpha} = 0$ as well. For $a_1 = 0$ the Jacobi identities $(f_\alpha, f_\beta, f_\gamma)$ do not restrict the values of $r_{\alpha\beta}$.

The obtained commutation relations can be further simplified by transformations that respect the commutation relations in the nilradical $\mathfrak{h}(m)$ and the simplifications already achieved. We put

$$\vec{\xi} = (\vec{p}, \vec{q}), \quad \vec{\xi'} = \vec{\xi} \cdot G. \tag{11.11}$$

The commutation relations (11.4) are written as

$$[\xi_a, \xi_b] = K_{ab} h, \quad 1 \leq a < b \leq 2m. \tag{11.12}$$

The transformation G in (11.11) must then satisfy

$$G^T \cdot K \cdot G = K \tag{11.13}$$

with K as in (11.10), i.e., $G \in \mathrm{Sp}(2m, \mathbb{F})$ belongs to the symplectic Lie group.

A further allowed transformation is the scaling

$$p'_i = \lambda p_i, \quad q'_i = \lambda q_i, \quad h' = \lambda^2 h, \qquad \lambda \in \mathbb{F},\ \lambda \neq 0. \tag{11.14}$$

We have thus proven the following result.

Theorem 11.1. *Every indecomposable solvable Lie algebra $\mathfrak{s}$ (over the field $\mathbb{F} = \mathbb{C}$ or $\mathbb{F} = \mathbb{R}$), containing the Heisenberg algebra $\mathfrak{h}(m)$ as its nilradical, can be written in a canonical basis $(h, p_1, \dots, p_m, q_1, \dots, q_m, f_1, \dots, f_f)$ with the commutation relations* (11.4)*, supplemented by*

$$\begin{gathered} ([f_\alpha, h], [f_\alpha, \vec{\xi}]) = (h, \vec{\xi}) \cdot M_\alpha, \\ M_\alpha = \begin{pmatrix} 2a_\alpha & 0 \\ 0 & a_\alpha \mathbf{1}_{2m\times 2m} + X_\alpha \end{pmatrix}, \\ [f_\alpha, f_\beta] = r_{\alpha\beta} h, \quad \alpha, \beta = 1, \dots, f. \end{gathered} \tag{11.15}$$

The vector $\vec{\xi}$ is defined as

$$\xi = (p_1, \dots, p_m, q_1, \dots, q_m).$$

The constants a_α satisfy

$$a_1 \in \{0, 1\}, \qquad a_2 = \dots = a_f = 0.$$

The matrices $X_1, \dots, X_f \in \mathfrak{sp}(2m, \mathbb{F})$ satisfy (11.10) *and*

$$[X_\alpha, X_\beta] = 0. \tag{11.16}$$

For $a_1 = 0$, or $a_1 = 1$ the sets $\{X_1, \dots, X_f\}$, or $\{X_2, \dots, X_f\}$ are linearly nilindependent, respectively.

The constants $r_{\alpha\beta}$ satisfy

$$r_{\alpha\beta} = 0, \qquad 1 \le \alpha, \beta \le f, \qquad \textit{for } a_1 = 1,$$
$$r_{\alpha\beta} = -r_{\beta\alpha} \in \mathbb{F}, \qquad 1 \le \alpha, \beta \le f, \qquad \textit{for } a_1 = 0. \tag{11.17}$$

The dimension of the solvable Lie algebra $\mathfrak{s}$ is

$$\dim \mathfrak{s} = 2m + 1 + f, \quad 0 < f \le m + 1. \tag{11.18}$$

The maximal value $f = m + 1$ is achieved precisely if we have $a_1 = 1$, $X_1 = 0$ and $\operatorname{span}\{X_2, \dots, X_{m+1}\}$ *is a Cartan subalgebra of $\mathfrak{sp}(2m, \mathbb{F})$. We then also have $r_{\alpha\beta} = 0$ for all α, β.*

Two algebras $\mathfrak{s}$ and $\mathfrak{s}'$ characterized by $(a_1, X_\alpha, r_{\alpha\beta})$ and $(a_1', X_\alpha', r_{\alpha\beta}')$ are equivalent if, after an allowed transformation we have

$$a_1' = a_1, \quad X_\alpha' = X_\alpha, \quad r_{\alpha\beta}' = r_{\alpha\beta}. \tag{11.19}$$

Allowed transformations are the symplectic transformations (11.11), *the scaling* (11.14) *and linear combinations of f_α, respecting the form of* (11.15).

A classification of the solvable algebras $\mathfrak{s}$ thus reduces to a classification of Abelian subalgebras of $\mathfrak{sp}(2m, \mathbb{F})$ containing no nilpotent elements. Elements of $\mathfrak{sp}(2m, \mathbb{C})$ and $\mathfrak{sp}(2m, \mathbb{R})$ have been classified [12, 42], as have maximal Abelian subalgebras [90].

11.3. The lowest dimensional case $m = 1$

The Heisenberg algebra $\mathfrak{h}(1)$ is three dimensional with basis (h, p, q). Following Theorem 11.1, we can extend $\mathfrak{h}(1)$ either by one element f_1, or by two commuting elements f_1, f_2. Let us consider these two cases separately.

(1) $\dim \mathfrak{s} = 4$ $(f = 1)$. The algebra $\mathfrak{sp}(2, \mathbb{C})$ has two types of elements, nilpotent and nonnilpotent. The algebra $\mathfrak{sp}(2, \mathbb{R})$ has three types: nilpotent, compact and noncompact. The algebra $\mathfrak{s}$ is completely characterized by one matrix M, figuring in the commutation relation (11.15).

For $\mathbb{F} = \mathbb{C}$ precisely three possibilities occur, namely

$$M^{(1)} = \begin{pmatrix} 2 & & \\ & 1+b & 0 \\ & 0 & 1-b \end{pmatrix}, \quad b \ge 0,$$
$$M^{(2)} = \begin{pmatrix} 2 & & \\ & 1 & 1 \\ & 0 & 1 \end{pmatrix}, \qquad M^{(3)} = \begin{pmatrix} 0 & & \\ & 1 & 0 \\ & 0 & -1 \end{pmatrix}. \tag{11.20}$$

For $\mathbb{F} = \mathbb{R}$ two additional inequivalent possibilities occur, characterized by the matrices

$$M^{(4)} = \begin{pmatrix} 2 & & \\ & 1 & -\alpha \\ & \alpha & 1 \end{pmatrix}, \quad \alpha > 0, \qquad M^{(5)} = \begin{pmatrix} 0 & & \\ & 0 & -1 \\ & 1 & 0 \end{pmatrix}. \tag{11.21}$$

Over $\mathbb{C}$, $M^{(4)}$ is equivalent to $M^{(1)}$, $M^{(5)}$ to $M^{(3)}$.

The corresponding algebras are presented in Part 4, Section 17.3. $M^{(1)}$ gives rise to two classes of algebras therein, namely $\mathfrak{s}_{4,8}$ and $\mathfrak{s}_{4,11}$ which differ by the dimensions of ideals in the characteristic series.

(2) $\dim \mathfrak{s} = 5$ ($f = 2$). Since in this case we must have $a_1 = 1$ on dimensional grounds, we have

$$[f_1, f_2] = 0 \tag{11.22}$$

and the action of f_1 and f_2 on $\mathfrak{h}(1)$ is characterized by a pair of matrices M_1, M_2. Up to a change of basis there is only one possibility over $\mathbb{F} = \mathbb{C}$

$$M_1 = \begin{pmatrix} 2 & & \\ & 1 & 0 \\ & 0 & 1 \end{pmatrix}, \quad M_2 = \begin{pmatrix} 0 & & \\ & 1 & 0 \\ & 0 & -1 \end{pmatrix}. \tag{11.23}$$

Over $\mathbb{F} = \mathbb{R}$ two inequivalent possibilities occur, namely (11.23) and

$$M_1 = \begin{pmatrix} 2 & & \\ & 1 & 0 \\ & 0 & 1 \end{pmatrix}, \quad M_2 = \begin{pmatrix} 0 & & \\ & 0 & -1 \\ & 1 & 0 \end{pmatrix}. \tag{11.24}$$

The corresponding algebras are presented in Part 4, Section 18.6.

11.4. The case $m = 2$

The Heisenberg algebra $\mathfrak{h}(2)$ is five-dimensional; we have $0 < f \leq 3$, i.e., we can add one, two or three elements f_α.

(1) $\dim \mathfrak{s} = 6$. Construction of these algebras amounts to the determination of canonical forms of elements of $\mathfrak{sp}(4, \mathbb{C}/\mathbb{R})$ modulo scaling. This was done in [42] and the resulting algebras are presented in Section 19.7 in Part 4. They were also presented in [106, Table A1].

For $\mathbb{F} = \mathbb{C}$, the list contains 12 classes of such algebras. For $\mathbb{F} = \mathbb{R}$, there are 25 classes in the list.

(2) $\dim \mathfrak{s} = 7$. These algebras require the classification of 2-dimensional Abelian subalgebras of $\mathfrak{sp}(4, \mathbb{C}/\mathbb{R})$. The resulting matrices A_α were presented in [106, Table A2]. Since the results are quite lengthy we do not reproduce them here.

(3) $\dim \mathfrak{s} = 8$. We must have $a_1 = 1$, $a_2 = a_3 = 0$. The commutation relations for f_1, f_2, f_3 therefore read

$$[f_\alpha, f_\beta] = 0, \quad 1 \leq \alpha, \beta \leq 3.$$

The action of f_α on h, p_1, p_2, q_1, q_2 is given by three matrices M_α as in (11.15). For $\mathbb{F} = \mathbb{C}$ the Cartan subalgebra of $\mathfrak{sp}(4, \mathbb{C})$ is unique and we have

$$M_1^{(1)} = \begin{pmatrix} 2 & & & & \\ & 1 & & & \\ & & 1 & & \\ & & & 1 & \\ & & & & 1 \end{pmatrix}, \quad M_2^{(1)} = \begin{pmatrix} 0 & & & & \\ & 1 & & & \\ & & 0 & & \\ & & & -1 & \\ & & & & 0 \end{pmatrix}, \quad M_3^{(1)} = \begin{pmatrix} 0 & & & & \\ & 0 & & & \\ & & 1 & & \\ & & & 0 & \\ & & & & -1 \end{pmatrix}. \tag{11.25}$$

For $\mathfrak{sp}(4, \mathbb{R})$ we have four inequivalent Cartan subalgebras [90], so in addition to the algebra $\mathfrak{s}$ characterized by the matrices (11.25) we have three more cases, characterized by

$$M_1^{(2)} = \begin{pmatrix} 2 &&&& \\ & 1 &&& \\ && 1 && \\ &&& 1 & \\ &&&& 1 \end{pmatrix}, \quad M_2^{(2)} = \begin{pmatrix} 0 &&&& \\ & 1 &&& \\ && 1 && \\ &&& -1 & \\ &&&& -1 \end{pmatrix}, \quad M_3^{(2)} = \begin{pmatrix} 0 &&&& \\ & 0 & 1 && \\ & -1 & 0 && \\ &&& 0 & 1 \\ &&& -1 & 0 \end{pmatrix}, \tag{11.26}$$

$$M_1^{(3)} = \begin{pmatrix} 2 &&&& \\ & 1 &&& \\ && 1 && \\ &&& 1 & \\ &&&& 1 \end{pmatrix}, \quad M_2^{(3)} = \begin{pmatrix} 0 &&&& \\ & 1 &&& \\ && 0 && \\ &&& -1 & \\ &&&& 0 \end{pmatrix}, \quad M_3^{(3)} = \begin{pmatrix} 0 &&&& \\ & 0 & 0 & 0 & 0 \\ & 0 & 0 & 0 & 1 \\ & 0 & 0 & 0 & 0 \\ & 0 & -1 & 0 & 0 \end{pmatrix}, \tag{11.27}$$

$$M_1^{(4)} = \begin{pmatrix} 2 &&&& \\ & 1 &&& \\ && 1 && \\ &&& 1 & \\ &&&& 1 \end{pmatrix}, \quad M_2^{(4)} = \begin{pmatrix} 0 &&&& \\ & 0 & 0 & 1 & 0 \\ & 0 & 0 & 0 & 0 \\ & -1 & 0 & 0 & 0 \\ & 0 & 0 & 0 & 0 \end{pmatrix}, \quad M_3^{(4)} = \begin{pmatrix} 0 &&&& \\ & 0 & 0 & 0 & 0 \\ & 0 & 0 & 0 & 1 \\ & 0 & 0 & 0 & 0 \\ & 0 & -1 & 0 & 0 \end{pmatrix}. \tag{11.28}$$

The low-dimensional cases $m = 1$ and $m = 2$ show that explicit detailed classifications become rather long and cumbersome for $2 \leq f \leq m$ already for $m = 2$ and we go no further here. On the other hand for $f = m+1$ the results are very simple: $a_1 = 1$, $r_{\alpha\beta} = 0$, $X_1 = 0$ and the matrices X_α, $2 \leq \alpha \leq m+1$ form a Cartan subalgebra of $\mathfrak{sp}(2m, \mathbb{F})$. There is only one conjugacy class of Cartan subalgebras for $\mathfrak{sp}(2m, \mathbb{C})$ (as for any complex simple Lie algebra). The real algebra $\mathfrak{sp}(2m, \mathbb{R})$ has $(m+2)^2/4$ different classes of Cartan subalgebras for m even and $(m+1)(m+3)/4$ for m odd [54, 63, 90, 121]. For $f = 1$ the results are also quite simple both for $\mathbb{F} = \mathbb{C}$ and $\mathbb{F} = \mathbb{R}$ and reduce to a classification of elements of $\mathfrak{sp}(2m, \mathbb{F})$ [42].

11.5. Generalized Casimir invariants

We shall realize the coadjoint representation of $\mathfrak{s}$ in a space of differentiable functions of $2m + 1 + f$ variables which we denote

$$(h, p_1, \ldots, p_m, q_1, \ldots, q_m, f_1, \ldots, f_f).$$

The algebra $\mathfrak{s}$ of Theorem 11.1 will be realized by differential operators, the form of which can be read off from the Lie brackets (11.4) and (11.15). Namely, we have

$$\begin{aligned} \widehat{H} &= -2a_1 h \frac{\partial}{\partial f_1}, \\ \widehat{P}_i &= h\frac{\partial}{\partial q_i} - \left((A_\alpha)_i^k p_k + (C_\alpha)_i^k q_k\right)\frac{\partial}{\partial f_\alpha} - a_1 p_i \frac{\partial}{\partial f_1}, \\ \widehat{Q}_i &= -h\frac{\partial}{\partial p_i} - \left((B_\alpha)_i^k p_k - (A_\alpha)_k^i q_k\right)\frac{\partial}{\partial f_\alpha} - a_1 q_i \frac{\partial}{\partial f_1}, \\ \widehat{F}_\alpha &= 2a_\alpha h\frac{\partial}{\partial h} + \left((A_\alpha)_i^k p_k + (C_\alpha)_i^k q_k + a_\alpha p_i\right)\frac{\partial}{\partial p_i} \\ &\qquad + \left((B_\alpha)_i^k p_k - (A_\alpha)_k^i q_k + a_\alpha q_i\right)\frac{\partial}{\partial q_i} + r_{\alpha\beta} h \frac{\partial}{\partial f_\beta}, \\ &\quad i, k = 1, \ldots, m, \qquad \alpha = 1, \ldots, f, \qquad a_1 = 0, 1, \quad a_\alpha = 0, \quad \alpha \geq 2, \\ &\quad r_{\alpha\beta} = 0 \quad \text{for } a_1 = 1 \end{aligned} \tag{11.29}$$

(summation over repeated indices is understood). The matrices A_α, $B_\alpha = B_\alpha^T$ and $C_\alpha = C_\alpha^T$ are the same as in the Lie brackets (11.15). The invariants of $\mathfrak{s}$ are obtained as functionally independent solutions of the system of first-order linear partial differential equations

$$(11.30)\quad \widehat{H}I = 0, \quad \widehat{P}_i I = 0, \quad \widehat{Q}_i I = 0, \quad \widehat{F}_\alpha I = 0, \qquad i = 1, \dots, m, \quad \alpha = 1, \dots, f$$

where $I = I(h, p_1, \dots, p_m, q_1, \dots, q_m, f_1, \dots, f_f)$.

Let us treat the two cases $a_1 = 1$ and $a_1 = 0$ separately.

Theorem 11.2. *An indecomposable solvable Lie algebra $\mathfrak{s}$ with the Lie brackets* (11.4) *and* (11.15) *and $a_1 = 1$ (and hence $r_{\alpha\beta} = 0$) has precisely $(f-1)$ generalized Casimir invariants. They can be written in the form*

$$(11.31)\quad I_{\alpha-1} = \frac{1}{h}\left(2hf_\alpha + 2(A_\alpha)^i_j p_i q_j + (C_\alpha)^i_j q_i q_j - (B_\alpha)^i_j p_i p_j\right), \quad \alpha = 2, \dots, f$$

(repeated indices $i, j = 1, \dots, m$ are summed over).

Proof. We have $a_1 = 1$. Applying $\widehat{H}$ to the function

$$I(h, p_1, \dots, p_m, q_1, \dots, q_m, f_1, \dots, f_f)$$

we find that I is independent of f_1. Applying the remaining operators corresponding to the nilradical $\mathfrak{h}(m)$, namely $\widehat{P}_i$ and $\widehat{Q}_i$ of (11.29), successively to I and using the method of characteristics, we find that I can depend only on h and Z_α, $2 \leq \alpha \leq f$ where

$$(11.32)\qquad Z_\alpha = 2hf_\alpha + 2(A_\alpha)^i_j p_i q_j + (C_\alpha)^i_j q_i q_j - (B_\alpha)^i_j p_i p_j.$$

Let us now apply the operators $\widehat{F}_\alpha$ $(\alpha \geq 2)$ to $I(h, Z_2, \dots, Z_f)$. We have

$$(11.33)\quad \widehat{F}_\alpha Z_\beta = 2(A_\alpha A_\beta - A_\beta A_\alpha + B_\alpha C_\beta - B_\beta C_\alpha)^i_k\, q_i p_k + 2(A_\beta B_\alpha - A_\alpha B_\beta)^i_k\, p_i p_k + 2(C_\alpha A_\beta - C_\beta A_\alpha)^i_k\, q_i q_k$$

where $\alpha, \beta \geq 2$ and matrix multiplication of A_α with A_β etc. is assumed. The commutativity conditions (11.16) imply that the first term in (11.33) vanishes. The symmetric parts of the second and third term also vanish as a result of the same commutativity relations. Finally, the antisymmetric parts, when contracted with the symmetric tensor $p_i p_j$ ($q_i q_j$, respectively) also vanish. Hence we have $\widehat{F}_\alpha Z_\beta = 0$ for $\alpha, \beta \geq 2$. Finally, we apply $\widehat{F}_1$, and obtain

$$(11.34)\qquad \widehat{F}_1 I(h, Z_2, \dots, Z_f) = 2\left(h\frac{\partial I}{\partial h} + \sum_{\beta=2}^{f} Z_\beta \frac{\partial I}{\partial Z_\beta}\right) = 0.$$

From (11.34) we conclude that the solution I of (11.30) is an arbitrary function of the invariants (11.31) and this proves the theorem. □

Theorem 11.3. *An indecomposable solvable Lie algebra $\mathfrak{s}$ with the Lie brackets* (11.4) *and* (11.15) *and $a_1 = 0$ has $N = (f + 1 - \operatorname{rank} R)$ functionally independent Casimir invariants*

$$(11.35)\qquad I_1 = h, \qquad I_\mu = \sum_{\alpha=1}^{f} a_{\mu\alpha} Z_\alpha, \quad 2 \leq \mu \leq N$$

where

$$R = (r_{\alpha\beta})_{\alpha,\beta=1}^{f}, \quad R^T = -R$$

and

$$Z_\alpha = 2hf_\alpha + 2(A_\alpha)^i_j p_i q_j + (C_\alpha)^i_j q_i q_j - (B_\alpha)^i_j p_i p_j, \quad 1 \leq \alpha \leq f \tag{11.36}$$

(Z_α are the same as in (11.32), the only difference is in the range of index α).

The constants $a_{\mu\alpha}$ are determined from the condition that the column vectors a_μ span the nullspace of the matrix R

$$R \cdot a_\mu = 0, \quad 2 \leq \mu \leq N. \tag{11.37}$$

PROOF. From the form of the operators (11.29) with $a_1 = 0$ we see directly that h is an invariant. The equations $\widehat{P}_i I = 0$ and $\widehat{Q}_i I = 0$ imply

$$I = I(h, Z_1, \ldots, Z_f)$$

with Z_α as in (11.36), and $1 \leq \alpha \leq f$. Following the same reasoning as in the proof of Theorem 11.2 we find that

$$\widehat{F}_\alpha Z_\beta = 2h^2 r_{\alpha\beta}. \tag{11.38}$$

It follows that elementary solutions of (11.30) are linear combinations (11.35) of Z_α with constant coefficients $a_{\mu\alpha}$ such that $\sum_{\beta=1}^{f} r_{\alpha\beta} a_{\mu\beta} = 0$. This completes the proof. □

11.5.1. Examples. To illustrate Theorem 11.2, let us consider the eight-dimensional Lie algebra $\mathfrak{s}$ characterized by the matrices of (11.25). Applying the theorem we obtain

$$I_1 = \frac{1}{h}(hf_2 + p_1 q_1), \quad I_2 = \frac{1}{h}(hf_3 + p_2 q_2). \tag{11.39}$$

To illustrate Theorem 11.3, let us consider the case of the algebras $\mathfrak{s}$ with m arbitrary and $f = 3$. The matrix R in this case is

$$R = \begin{pmatrix} 0 & r_{12} & r_{13} \\ -r_{12} & 0 & r_{23} \\ -r_{13} & -r_{23} & 0 \end{pmatrix} \tag{11.40}$$

and its rank is either 0 or 2. For $\operatorname{rank} R = 0$ we have $r_{12} = r_{13} = r_{23} = 0$ and we obtain 4 invariants

$$h, \quad Z_1, \quad Z_2, \quad Z_3 \tag{11.41}$$

with Z_α as in (11.36).

When $\operatorname{rank} R = 2$ the two invariants are

$$I_1 = h, \quad I_2 = r_{23}Z_1 - r_{13}Z_2 + r_{12}Z_3. \tag{11.42}$$

As a further illustration of Theorem 11.3, consider the Lie algebras $\mathfrak{s}$ with m arbitrary and $f = 4$. The 4×4 matrix antisymmetric matrix R can have rank 0, 2 or 4. The invariants are respectively

$$\operatorname{rank} R = 0 \qquad h, Z_1, Z_2, Z_3, Z_4, \tag{11.43}$$

$$\operatorname{rank} R = 2 \qquad h, r_{23}Z_1 - r_{13}Z_2 + r_{12}Z_3, r_{24}Z_1 - r_{14}Z_2 + r_{12}Z_4, \tag{11.44}$$

$$\operatorname{rank} R = 4 \qquad h. \tag{11.45}$$

Notice that the invariants presented in (11.44) are valid for a generic antisymmetric rank-2 matrix R and vanish for some particular choices of R; when that happens a different set of solutions of(11.37) must be substituted into (11.35). An example of such a nongeneric case is obtained if we have $r_{34} \neq 0$, $r_{12} = r_{13} = r_{14} = r_{23} = r_{24} = 0$. Equations (11.38) imply

$$\frac{\partial I}{\partial Z_3} = \frac{\partial I}{\partial Z_4} = 0$$

so the Casimir invariants are h, Z_1 and Z_2.

CHAPTER 12

Solvable Lie Algebras with Borel Nilradicals

Let us now consider an infinite class of nilpotent Lie algebras which have a close relationship with simple Lie algebras. Every complex simple Lie algebra $\mathfrak{g}$ has a uniquely defined (up to equivalence) maximal solvable subalgebra, its Borel subalgebra $\mathfrak{b}(\mathfrak{g})$. In turn, the Borel subalgebra (like every solvable Lie algebra) has a unique maximal nilpotent ideal, its nilradical $\mathrm{NR}\big(\mathfrak{b}(\mathfrak{g})\big)$. We shall call these nilpotent algebras "Borel nilradicals." For each simple complex Lie algebra $\mathfrak{g}$ the Borel nilradical is the Lie algebra consisting of all positive root spaces. It is of course the nilradical only of the Borel subalgebra $\mathfrak{b}(\mathfrak{g})$, not of $\mathfrak{g}$ itself ($\mathfrak{g}$ being simple has no ideals at all).

For real Lie algebras the maximal solvable subalgebra is in general not unique. The maximal solvable subalgebras of all real classical algebras were classified in [88, 89, 93]. For the split real form of the complex simple Lie algebra there exists a distinguished maximal solvable subalgebra that has the same basis as the Borel subalgebra but considered over $\mathbb{R}$ instead of $\mathbb{C}$. We shall also call their nilradicals "Borel nilradicals" (over $\mathbb{R}$), sometimes adding the adjective "split" to highlight its relation to the split real form.

We shall treat all simple Lie algebras simultaneously (the classical and exceptional complex ones and the split real ones) using bases for the nilradicals provided by the positive roots. This simultaneous treatment is made possible by the fact that all outer derivations of these nilradicals are known, due to the work of G. F. Leger and E. M. Luks [67]. Our analysis of solvable extensions of Borel nilradicals based on this knowledge was originally presented in [118] which we shall closely follow here.

An example of a Borel nilradical is the algebra of strictly upper triangular matrices $\mathfrak{t}(l+1)$, the Borel nilradical of $\mathrm{A}_l \equiv \mathfrak{sl}(l+1,\mathbb{F})$. This case was considered in [123].

12.1. Outer derivations of nilradicals of Borel subalgebras

Let $\mathfrak{g}$ be a simple complex Lie algebra, $\mathfrak{g}_0$ its Cartan subalgebra, $l = \operatorname{rank}\mathfrak{g} = \dim\mathfrak{g}_0$. As in Section 2.3 let us denote by Δ the set of all roots, by Δ^+ the set of all positive roots and by $\Delta^S = \{\alpha_1, \dots, \alpha_l\}$ the set of simple roots. Let $\mathfrak{g}_\lambda$ denote the root subspace of the root λ, spanned by the root vector e_λ. We use the notation $\langle\ ,\ \rangle$ for the scalar product induced by the Killing form on $\mathfrak{g}$ on the real vector subspace of $(\mathfrak{g}_0)^*$ spanned by the simple roots. Let S_β denote the Weyl reflection with respect to the root β,

$$S_\beta(\alpha) = \alpha - 2\frac{\langle\alpha,\beta\rangle}{\langle\beta,\beta\rangle}\beta, \quad \alpha \in \Delta.$$

As recalled in Section 2.3 every (semi)simple complex Lie algebra $\mathfrak{g}$ contains a unique (up to isomorphisms) maximal solvable subalgebra, its *Borel subalgebra* $\mathfrak{b}(\mathfrak{g})$. It contains the Cartan subalgebra and all positive root vectors

$$\mathfrak{b}(\mathfrak{g}) = \mathfrak{g}_0 \dotplus (\dotplus\{\mathfrak{g}_\lambda \mid \lambda \in \Delta^+\}). \tag{12.1}$$

The properties of root systems imply that the Borel subalgebra is indeed a solvable subalgebra of $\mathfrak{g}$ with the nilradical

$$\mathrm{NR}\big(\mathfrak{b}(\mathfrak{g})\big) = \dotplus\{\mathfrak{g}_\lambda \mid \lambda \in \Delta^+\}. \tag{12.2}$$

For the sake of brevity we shall call the nilpotent Lie algebra $\mathrm{NR}\big(\mathfrak{b}(\mathfrak{g})\big)$ the *Borel nilradical*.

It is always possible to realize the simple Lie algebra $\mathfrak{g}$ in such a manner that its Borel subalgebra $\mathfrak{b}(\mathfrak{g})$ is represented by upper triangular matrices. For $\mathfrak{sl}(l+1,\mathbb{C})$, $\mathfrak{b}(\mathfrak{sl}(l+1,\mathbb{C}))$ is simply the set of all traceless $(l+1)\times(l+1)$ upper triangular matrices and its nilradical is the set of traceless strictly upper triangular matrices. The same holds true also for its *split*, i.e., maximally noncompact, *real form* $\mathfrak{sl}(l+1,\mathbb{R})$. All other series $\mathrm{B}_l, \mathrm{C}_l, \mathrm{D}_l$ of classical complex Lie algebras and their split real forms can be realized by matrices $X \in \mathbb{C}^{n\times n}$ or $X \in \mathbb{R}^{n\times n}$ satisfying the equation

$$XK + KX^T = 0 \tag{12.3}$$

where the matrix K is chosen as follows

$$\begin{aligned} &\mathrm{B}_l = \mathfrak{so}(2l+1,\mathbb{C}),\ \mathfrak{so}(l+1,l) && K = J_{2l+1}, \\ &\mathrm{C}_l = \mathfrak{sp}(2l,\mathbb{C}),\ \mathfrak{sp}(2l,\mathbb{R}) && K = \begin{pmatrix} 0 & J_l \\ -J_l & 0 \end{pmatrix}, \\ &\mathrm{D}_l = \mathfrak{so}(2l,\mathbb{C}),\ \mathfrak{so}(l,l) && K = J_{2l}, \end{aligned} \tag{12.4}$$

where

$$J_m = \begin{pmatrix} & & & 1 \\ & & 1 & \\ & \iddots & & \\ 1 & & & \end{pmatrix} \in \mathbb{F}^{m\times m}.$$

The corresponding Borel subalgebra $\mathfrak{b}(\mathfrak{g})$ is represented by upper triangular matrices satisfying the condition (12.3). The nilradical $\mathrm{NR}\big(\mathfrak{b}(\mathfrak{g})\big)$ is represented by strictly upper triangular matrices satisfying the condition (12.3).

Let α_i, $i = 1,\dots,l = \operatorname{rank}\mathfrak{g}$ denote the simple roots, $\Delta^S = \{\alpha_i\}_{i=1}^l$ and let

$$\mathfrak{g}_m = \dotplus\Big\{\mathfrak{g}_\lambda \,\Big|\, \lambda = \sum_{i=1}^l m_i\alpha_i \in \Delta^+, \sum_{i=1}^l m_i \geq m\Big\}. \tag{12.5}$$

The root subspaces $\mathfrak{g}_\alpha = \operatorname{span}\{e_\alpha\}$, $\alpha \in \Delta^S$ generate the entire $\mathrm{NR}\big(\mathfrak{b}(\mathfrak{g})\big) = \dotplus\{\mathfrak{g}_\lambda \mid \lambda \in \Delta^+\}$ through commutators

$$[\mathfrak{g}_\lambda, \mathfrak{g}_\mu] = \mathfrak{g}_{\lambda+\mu} \quad \text{whenever } \lambda, \mu, \lambda+\mu \in \Delta^+$$

and this implies that the ideals in the lower central series of the nilradical $\mathrm{NR}\big(\mathfrak{b}(\mathfrak{g})\big)$ of the Borel subalgebra are

$$\Big(\mathrm{NR}\big(\mathfrak{b}(\mathfrak{g})\big)\Big)^m = \mathfrak{g}_m.$$

The center $\mathfrak{z}$ of $\mathrm{NR}(\mathfrak{b}(\mathfrak{g}))$ is one-dimensional and is spanned by e_λ where λ is the *highest root* of $\mathfrak{g}$, i.e., the unique root such that no root of the form $\lambda + \alpha_j$ exists. The center $\mathfrak{z}$ coincides with the last nonvanishing ideal in the lower central series.

All derivations of the nilradical $\mathfrak{n} = \mathrm{NR}(\mathfrak{b}(\mathfrak{g}))$ were found in [67] and the result is summed up in Proposition 12.1.

Proposition 12.1. *Let $\mathfrak{g}$ be a complex simple Lie algebra of rank l, $\mathfrak{g}_0$ its Cartan subalgebra, $\Delta^S = \{\alpha_1, \ldots, \alpha_l\}$ the set of simple roots and $\mathfrak{n} = \mathrm{NR}(\mathfrak{b}(\mathfrak{g}))$. The algebra $\mathfrak{Der}(\mathfrak{n})$ of derivations of the nilradical $\mathfrak{n} = \mathrm{NR}(\mathfrak{b}(\mathfrak{g}))$ of the Borel subalgebra of a complex simple Lie algebra $\mathfrak{g}$ satisfies*

- $\mathfrak{Der}(\mathfrak{n}) = \mathfrak{Out}(\mathfrak{n}) \dotplus \mathfrak{Inn}(\mathfrak{n})$,
- $\mathfrak{Inn}(\mathfrak{n}) = \{\mathrm{ad}\, x \mid x \in \mathfrak{n}\}$ *is the ideal of inner derivations,*
- $\dim \mathfrak{Out}(\mathfrak{n}) = 2l$,
- $\mathfrak{Out}(\mathfrak{n}) = \mathrm{span}\{D_i, \widetilde{D}_i \mid i = 1, \ldots, l\}$ *where the derivations $D_i, \widetilde{D}_i$ are defined as follows. The derivations D_i act diagonally in the basis of $\mathfrak{n}$ consisting of positive root vectors e_α, $\alpha \in \Delta^+$*

$$D_i(e_\alpha) = m_i e_\alpha, \quad \alpha = \sum_{j=1}^{l} m_j \alpha_j \in \Delta^+.$$

$\widetilde{D}_i$ are nilpotent outer derivations which act on simple root vectors as

$$\widetilde{D}_i(e_\beta) = \begin{cases} e_\gamma, \ \textit{where}\ \gamma = S_{\alpha_i}(\lambda), & \textit{if}\ \beta = \alpha_i, \\ 0, & \textit{if}\ \beta = \alpha_j,\ j \neq i \end{cases} \tag{12.6}$$

where λ is the highest root of $\mathfrak{g}$. The action of $\widetilde{D}_i$ on e_α, $\alpha \in \Delta^+ \backslash \Delta^S$ follows from the definition of a derivation (2.26).

For future reference, we list the highest roots corresponding to all simple Lie algebras in Table 12.1.

For the sake of brevity, we shall write $S_i(\lambda)$ instead of $S_{\alpha_i}(\lambda)$ and introduce nonnegative integer constants s_i

$$S_i(\lambda) = \lambda - s_i \alpha_i, \quad i = 1 \ldots, l \ (\text{no summation}).$$

We notice that for $\mathfrak{g} = \mathrm{A}_l$ only two constants s_i, namely s_1 and s_l, are nonvanishing and equal to one; for all other simple algebras only one s_i is nonvanishing and turns

TABLE 12.1. The highest roots of simple Lie algebras.

A_l:	$\lambda = \sum_{j=1}^{l} \alpha_i$
B_l:	$\lambda = \alpha_1 + 2\sum_{j=2}^{l} \alpha_i$
C_l:	$\lambda = 2\sum_{j=1}^{l-1} \alpha_i + \alpha_l$
D_l:	$\lambda = \alpha_1 + 2\sum_{j=2}^{l-2} \alpha_i + \alpha_{l-1} + \alpha_l$
E_6:	$\lambda = \alpha_1 + 2\alpha_2 + 3\alpha_3 + 2\alpha_4 + \alpha_5 + 2\alpha_6$
E_7:	$\lambda = 2\alpha_1 + 3\alpha_2 + 4\alpha_3 + 3\alpha_4 + 2\alpha_5 + \alpha_6 + 2\alpha_7$
E_8:	$\lambda = 2\alpha_1 + 4\alpha_2 + 6\alpha_3 + 5\alpha_4 + 4\alpha_5 + 3\alpha_6 + 2\alpha_7 + 3\alpha_8$
F_4:	$\lambda = 2\alpha_1 + 3\alpha_2 + 4\alpha_3 + 2\alpha_4$
G_2:	$\lambda = 2\alpha_1 + 3\alpha_2$

TABLE 12.2. Dimensions of the nilradicals $\mathrm{NR}(\mathfrak{b}(\mathfrak{g}))$ and nonvanishing constants s_i in the equation $S_i(\lambda) = \lambda - s_i\alpha_i$.

A_l	$\dim \mathrm{NR}(\mathfrak{b}(\mathrm{A}_l)) = \dfrac{l(l+1)}{2}$	$s_1 = s_l = 1,$
B_l	$\dim \mathrm{NR}(\mathfrak{b}(\mathrm{B}_l)) = l^2$	$s_2 = 1,$
C_l	$\dim \mathrm{NR}(\mathfrak{b}(\mathrm{C}_l)) = l^2$	$s_1 = 2,$
D_l	$\dim \mathrm{NR}(\mathfrak{b}(\mathrm{D}_l)) = l(l-1)$	$s_2 = 1,$
E_6	$\dim \mathrm{NR}(\mathfrak{b}(\mathrm{E}_6)) = 36$	$s_6 = 1,$
E_7	$\dim \mathrm{NR}(\mathfrak{b}(\mathrm{E}_7)) = 63$	$s_1 = 1,$
E_8	$\dim \mathrm{NR}(\mathfrak{b}(\mathrm{E}_8)) = 120$	$s_7 = 1,$
F_4	$\dim \mathrm{NR}(\mathfrak{b}(\mathrm{F}_4)) = 24$	$s_1 = 1,$
G_2	$\dim \mathrm{NR}(\mathfrak{b}(\mathrm{G}_2)) = 6$	$s_1 = 1.$

out to be equal to 1 or 2, see Table 12.2. We also list the dimensions of the nilradicals $\mathrm{NR}(\mathfrak{b}(\mathfrak{g}))$ in Table 12.2.

Let us determine how $\widetilde{D}_i$ act on the vectors e_β when $\beta \in \Delta^+ \backslash \Delta^S$. We proceed by induction, using the Leibniz property of derivations (2.26). First, we evaluate

$$\widetilde{D}_i([e_{\alpha_j}, e_{\alpha_k}]) = [\widetilde{D}_i(e_{\alpha_j}), e_{\alpha_k}] + [e_{\alpha_j}, \widetilde{D}_i(e_{\alpha_k})].$$

When both $i \neq j$ and $i \neq k$ the result is 0 by definition of $\widetilde{D}_i$. Without loss of generality we take $i = j$ and $i \neq k$ and obtain

$$\widetilde{D}_i([e_{\alpha_i}, e_{\alpha_k}]) = [\widetilde{D}_i(e_{\alpha_i}), e_{\alpha_k}] = [e_{S_i(\lambda)}, e_{\alpha_k}].$$

Were the result nonvanishing, we would have $S_i(\lambda) + \alpha_k \in \Delta^+$, i.e., also

$$S_i(S_i(\lambda) + \alpha_k) = \lambda + \alpha_k - 2\frac{\langle \alpha_k, \alpha_i \rangle}{\langle \alpha_i, \alpha_i \rangle}\alpha_i \in \Delta^+.$$

Since $2\langle \alpha_k, \alpha_i \rangle / \langle \alpha_i, \alpha_i \rangle$ is an off-diagonal element of the Cartan matrix, it is a nonpositive integer. That means we would have a root larger than λ, contradicting the maximality of the highest root λ. Therefore,

$$\widetilde{D}_i([e_{\alpha_j}, e_{\alpha_k}]) = 0$$

for any $j, k = 1, \ldots, l$. Next, let us assume that $\widetilde{D}_i(e_\beta) = 0$ for all roots $\beta = \sum_{j=1}^{l} m_j \alpha_j$ such that $2 \leq \sum_{j=1}^{l} m_j \leq M$ and consider $\widetilde{D}_i([e_{\alpha_j}, e_\beta])$ with $\beta = \sum_{j=1}^{l} m_j \alpha_j$, $\sum_{j=1}^{l} m_j = M$. When $j \neq i$ we again obtain 0 using the Leibniz property (2.26). When $i = j$ we have

$$\widetilde{D}_i([e_{\alpha_i}, e_\beta]) = [e_{S_i(\lambda)}, e_\beta].$$

Using the Table 12.2 we find that for all simple algebras except C_l we have either

$$S_i(\lambda) = \lambda - \alpha_i$$

or

$$S_i(\lambda) = \lambda.$$

Consequently, no root of the form $S_i(\lambda)+\beta$, where $\beta=\sum_{j=1}^{l} m_j\alpha_j$, exists when $\sum_{j=1}^{l} m_j \geq 2$. Therefore the Lie bracket $[e_{S_i(\lambda)}, e_\beta]$ vanishes. In the special case of the algebras C_l we have

$$S_1(\lambda)=\lambda-2\alpha_1$$

and we must perform a more detailed analysis. Let us express e_β as a multiple Lie bracket of simple root vectors e_{α_j},

$$e_\beta=\left[e_{\alpha_{j_1}},\ldots,\left[e_{\alpha_{j_{M-1}}},e_{\alpha_{j_M}}\right]\ldots\right]$$

where $\beta=\sum_{K=1}^{M}\alpha_{j_K}$ and use the Jacobi identity sufficiently many times:

$$\begin{aligned}[e_{S_1(\lambda)},e_\beta]=&\left[[e_{S_1(\lambda)},e_{\alpha_{j_1}}],[\ldots,[e_{\alpha_{j_{M-1}}},e_{\alpha_{j_M}}]]\ldots\right]\\&+\cdots+\left[e_{\alpha_{j_1}},\ldots,\left[e_{\alpha_{j_{M-1}}},[e_{S_1(\lambda)},e_{\alpha_{j_M}}]\right]\ldots\right].\end{aligned}$$

Whenever $j_K\neq 1$ the corresponding commutator of $e_{\alpha_{j_K}}$ with $e_{S_1(\lambda)}$ vanishes. When $j_K=1$, we have

$$[e_{S_1(\lambda)},e_{\alpha_1}]=ce_{\lambda-\alpha_1}$$

for some integer constant c. The only positive root vector which does not commute with $e_{\lambda-\alpha_1}$ is e_{α_1}; therefore, any further commutator with any e_{α_j}, $j\neq 1$ gives zero. Finally, the commutator of $e_{\lambda-\alpha_1}$ with e_{α_1} can in our computation arise only through

$$\left[e_{S_1(\lambda)},[e_{\alpha_1},e_{\alpha_1}]\right]$$

which vanishes immediately. Therefore, in all cases we have shown that

$$\widetilde{D}_i([e_{\alpha_j},e_\beta])=0$$

when $\beta=\sum_{j=1}^{l} m_j\alpha_j$ and $\sum_{j=1}^{l} m_j=M$, concluding the induction step.

To sum up, we have just shown that for any simple complex Lie algebra $\mathfrak{g}$ the derivations $\widetilde{D}_i$ of the algebra $\mathrm{NR}\big(\mathfrak{b}(\mathfrak{g})\big)$ give zero whenever they act on e_β, $\beta\in\Delta^+\backslash\Delta^S$.

Let us assume from now on that $l>2$. Then we always have $S_i(\lambda)\notin\Delta^S$ for all $i=1,\ldots,l$ and consequently

$$\widetilde{D}_i\circ\widetilde{D}_j(e_{\alpha_k})=0 \tag{12.7}$$

for every $\alpha_k\in\Delta^S$. The previous calculation allows us to conclude that (12.7) must hold for any $\alpha\in\Delta^+$, i.e., we have

$$\widetilde{D}_i\circ\widetilde{D}_j=0$$

for any $i,j=1,\ldots,l$.

The derivations D_i commute among each other and act diagonally on $\widetilde{D}_j$,

$$[D_i,\widetilde{D}_j]\in\operatorname{span}\{\widetilde{D}_j\}. \tag{12.8}$$

To conclude, we have just seen that under the assumption that l is greater than 2, the $2l$ outer derivations $D_i,\widetilde{D}_i$ span a Lie subalgebra $\mathfrak{Out}\big(\mathrm{NR}\big(\mathfrak{b}(\mathfrak{g})\big)\big)$ of the algebra of all derivations $\mathfrak{Der}\big(\mathrm{NR}\big(\mathfrak{b}(\mathfrak{g})\big)\big)$. This algebra can be further decomposed into a semidirect sum of an l-dimensional Abelian ideal spanned by the nilpotent derivations $\widetilde{D}_i$ and an l-dimensional Abelian subalgebra spanned by D_i.

The cases $l=1$ and $l=2$ are discussed in the following section, above Theorem 12.2.

12.2. Solvable extensions of the Borel nilradicals $\mathrm{NR}(\mathfrak{b}(\mathfrak{g}))$

Let us now study the structure of any solvable Lie algebra with the nilradical $\mathrm{NR}(\mathfrak{b}(\mathfrak{g}))$, $l = \operatorname{rank}\mathfrak{g} > 2$.

From the fact that there are only l linearly nilindependent derivations D_i in $\mathfrak{Out}(\mathrm{NR}(\mathfrak{b}(\mathfrak{g})))$ we conclude that the maximal number of nonnilpotent basis elements in any solvable Lie algebra $\mathfrak{s}$ with the nilradical $\mathrm{NR}(\mathfrak{b}(\mathfrak{g}))$ is l. One algebra with this number of nonnilpotent basis elements is already known, namely the Borel subalgebra $\mathfrak{b}(\mathfrak{g})$ of the simple Lie algebra $\mathfrak{g}$. The following argument will show that it is the only one (up to isomorphisms).

12.2.1. Solvable extensions of the Borel nilradicals of maximal dimension.

Let us assume that we have a solvable Lie algebra $\mathfrak{s}$ with the nilradical $\mathfrak{n} = \mathrm{NR}(\mathfrak{b}(\mathfrak{g}))$ and $l = \operatorname{rank}\mathfrak{g}$ nonnilpotent basis elements f_i. They define l outer linearly nilindependent derivations $\widehat{D}^i$ such that $\widehat{D}^i = \mathrm{ad}(f_i)|_{\mathfrak{n}}$. Using the transformation (9.13) we may choose the basis vectors f_i so that

$$\widehat{D}^i = D_i + \sum_{j=1}^{l} \omega^i_j \widetilde{D}_j \tag{12.9}$$

where $D_i, \widetilde{D}_j$ are the derivations defined above. We proceed to show that we can put $\omega^i_j = 0$, $\forall i, j$. Since $\widehat{D}^i$ lie in the subalgebra $\mathfrak{Out}(\mathrm{NR}(\mathfrak{b}(\mathfrak{g})))$ of $\mathfrak{Der}(\mathrm{NR}(\mathfrak{b}(\mathfrak{g})))$ and at the same time $[\widehat{D}^i, \widehat{D}^j] \in \mathfrak{Inn}(\mathrm{NR}(\mathfrak{b}(\mathfrak{g})))$ must hold, we find that

$$[\widehat{D}^k, \widehat{D}^j] = 0. \tag{12.10}$$

This requirement together with (12.8) in turn implies

$$\omega^j_i [D_k, \widetilde{D}_i] + \omega^k_i [\widetilde{D}_i, D_j] = 0. \tag{12.11}$$

for every $i, j, k = 1, \dots, l$ such that $k \neq j$ (no summation over i). For any given i we can find $\tilde{\imath}$ such that $[\widetilde{D}_i, D_{\tilde{\imath}}] \neq 0$. Consequently, the value of $\omega^{\tilde{\imath}}_i$ together with the root system specifying the Lie brackets $[D_k, \widetilde{D}_i]$ completely determines all ω^j_i for $j \neq \tilde{\imath}$. Altogether, we still have one undetermined parameter $\omega^{\tilde{\imath}}_i$ for each $i = 1, \dots, l$. We proceed to eliminate these parameters through a suitable choice of automorphism in (9.14).

We have $(\widetilde{D}_i)^2 = 0$ and consequently $\tilde{\phi}_i(t_i) \equiv \exp(t_i \widetilde{D}_i) = \mathbf{1} + t_i \widetilde{D}_i$. Under the transformation (9.14) the derivations $D_j, \widetilde{D}_j, \widehat{D}^j$ change as follows

$$\begin{aligned} D_j &\to D_j + t_i[\widetilde{D}_i, D_j], \\ \widetilde{D}_j &\to \widetilde{D}_j, \\ \widehat{D}^j = D_j + \sum_{k=1}^{l} \omega^j_k \widetilde{D}_k &\to D_j + t_i[\widetilde{D}_i, D_j] + \sum_{k=1}^{l} \omega^j_k \widetilde{D}_k. \end{aligned} \tag{12.12}$$

Now we consider step by step each $i = 1, \dots, l$. We take $\widehat{D}^{\tilde{\imath}}$ which transforms nontrivially under the transformation (12.12) by definition of $\tilde{\imath}$, due to $[\widetilde{D}_i, D_{\tilde{\imath}}] \neq 0$. We use it to set $\omega^{\tilde{\imath}}_i = 0$ after the transformation. Equation (12.11) then implies that after the transformation all $\omega^j_i = 0$. The automorphisms $\tilde{\phi}_i(t_i)$ commute amongst each other, i.e., we can independently set to zero the constants ω^j_i belonging to different values of i without changing the others. Therefore we have found that our derivations $\widehat{D}^j$ can be brought to the form

$$\widehat{D}^j = D_j$$

through a conjugation by a suitable automorphism $\tilde{\Phi} = \tilde{\phi}_1(t_1) \circ \cdots \circ \tilde{\phi}_l(t_l)$ of $\mathrm{NR}\big(\mathfrak{b}(\mathfrak{g})\big)$.

As we have just seen, the action of the nonnilpotent basis vectors f_i on the nilradical $\mathrm{NR}\big(\mathfrak{b}(\mathfrak{g})\big)$ is the same as in the Borel algebra. In order to show the uniqueness of the maximal solvable extension of $\mathrm{NR}\big(\mathfrak{b}(\mathfrak{g})\big)$ it remains to be shown that we can always accomplish

$$[f_i, f_j] = 0.$$

The kernel of the adjoint representation $\mathrm{ad}\colon \mathfrak{n} \to \mathfrak{gl}(\mathfrak{n})$ of the nilradical $\mathfrak{n}$ is the center $\mathfrak{z}$ of the algebra $\mathfrak{n}$ which in our case is spanned by e_λ. Consequently we have

$$[f_i, f_j] = \gamma_{ij} e_\lambda, \quad \gamma_{ij} = -\gamma_{ji}$$

which is the inverse image of the relation $[\mathrm{ad}(f_i)|_{\mathfrak{n}}, \mathrm{ad}(f_j)|_{\mathfrak{n}}] = 0$ (cf. (12.10)). Let us write the highest root in terms of simple roots as

$$\lambda = \sum_{j=1}^{l} \lambda_j \alpha_j, \; \lambda_j \in \mathbb{N}.$$

The Jacobi identity (f_i, f_j, f_k) implies that

$$\lambda_i \gamma_{jk} + \lambda_j \gamma_{ki} + \lambda_k \gamma_{ij} = 0. \tag{12.13}$$

Let us perform a transformation

$$f_i \to f_i + \tau_i e_\lambda$$

which induces a change of the constants γ_{ij}

$$\gamma_{ij} \to \gamma_{ij} + \lambda_i \tau_j - \lambda_j \tau_i.$$

In this way we may set to zero for example all γ_{1j}, $j = 2, \ldots, l$ by selecting $\tau_1 = 1$ and choosing τ_j so that $\gamma_{1j} + \lambda_1 \tau_j - \lambda_j = 0$ (recall that all $\lambda_j \geq 1$). After such a transformation, (12.13) becomes

$$\lambda_1 \gamma_{jk} = 0, \quad j, k \neq 1$$

and implies that all γ_{jk} must now vanish. Therefore, we have found a basis $(e_\alpha, f_i \mid \alpha \in \Delta^+, i = 1, \ldots, l)$ in our solvable algebra such that

$$\begin{aligned} [f_i, e_\beta] &= m_i e_\beta, \quad \text{where } \beta = \sum_{j=1}^{l} m_j \alpha_j, \\ [f_i, f_j] &= 0. \end{aligned}$$

To sum up, we have found that for any complex simple Lie algebra $\mathfrak{g}$ such that $\operatorname{rank} \mathfrak{g} > 2$ the maximal solvable Lie algebra with the nilradical $\mathrm{NR}\big(\mathfrak{b}(\mathfrak{g})\big)$ is unique and isomorphic to the Borel subalgebra $\mathfrak{b}(\mathfrak{g})$ of $\mathfrak{g}$.

The same is true also when $\operatorname{rank} \mathfrak{g} = 1$ or $\operatorname{rank} \mathfrak{g} = 2$, i.e., $\mathfrak{g} = \mathfrak{sl}(2), \mathfrak{sl}(3), \mathfrak{so}(5)$ or G_2. In the case of $\mathfrak{g} = G_2$ the proof given above can be taken over without modifications, since $S_i(\lambda) \in \Delta^+ \backslash \Delta^S$ for all simple roots α_i. In the cases of $\mathfrak{g} = \mathfrak{sl}(2), \mathfrak{sl}(3), \mathfrak{so}(5)$ this is no longer true but the result follows from the investigation of Abelian nilradicals ($\mathfrak{g} = \mathfrak{sl}(2)$) in Chapter 10, Heisenberg nilradicals ($\mathfrak{g} = \mathfrak{sl}(3)$) in Section 11.3, and the nilradical $\mathfrak{n}_{4,1}$ of Section 13.1 ($\mathfrak{g} = \mathfrak{so}(5)$).

Thus, we have proven the following theorem:

Theorem 12.2. *Let* $\mathfrak{g}$ *be a complex simple Lie algebra,* $\mathfrak{b}(\mathfrak{g})$ *its Borel subalgebra and* $\mathfrak{n} = \mathrm{NR}\big(\mathfrak{b}(\mathfrak{g})\big)$ *the nilradical of* $\mathfrak{b}(\mathfrak{g})$. *The solvable Lie algebra with the nilradical* $\mathrm{NR}\big(\mathfrak{b}(\mathfrak{g})\big)$ *of the maximal dimension* $\dim \mathfrak{n} + \operatorname{rank} \mathfrak{g}$ *is unique and isomorphic to the Borel subalgebra* $\mathfrak{b}(\mathfrak{g})$ *of* $\mathfrak{g}$.

12.2.2. Solvable extensions of the split real form of the nilradical $\mathrm{NR}\big(\mathfrak{b}(\mathfrak{g})\big)$. As was already noted, the maximal solvable subalgebra of a real form of a complex Lie algebra is not necessarily unique. Let us consider the split real form $\mathfrak{g}$ and its maximal solvable subalgebra that consists of the maximally noncompact Cartan subalgebra (that remains diagonalizable over $\mathbb{R}$) and all positive root spaces. That means that it is spanned (over $\mathbb{R}$) by all h_α and e_α in the Weyl – Chevalley basis where $\alpha \in \Delta^+$. Let us call it the split Borel subalgebra and also denote it by $\mathfrak{b}(\mathfrak{g})$. The results of [67] also apply to real split Borel subalgebras.

Our Theorem 12.2 holds for the split real form of a complex simple Lie algebra $\mathfrak{g}$ with one exception: when $\mathfrak{g} = \mathfrak{sl}(3,\mathbb{R})$ there are several such maximal solvable algebras (see Section 11.2). The proof for the cases $\operatorname{rank} \mathfrak{g} > 2$ or $\mathfrak{g} = \mathrm{G}_2$ is the same as above, the proof for the case $\mathfrak{g} = \mathfrak{so}(3,2)$ (the split real form of $\mathfrak{so}(5,\mathbb{C})$) follows from Section 13.1 below.

12.2.3. Solvable extensions of the Borel nilradicals of less than maximal dimension. After investigating the case of the maximal dimension solvable extension in the previous Sections let us consider the cases where the number of nonnilpotent elements of $\mathfrak{s}$ is smaller than the rank of $\mathfrak{g}$, i.e., $q < l$.

In such a case, we have derivations

$$\widehat{D}^a = \sum_{j=1}^{l} \big(\sigma_j^a D_j + \omega_j^a \widetilde{D}_j\big), \quad a = 1, \ldots, q \tag{12.14}$$

representing the elements f_a in the adjoint representation of $\mathfrak{s}$ on $\mathfrak{n}$, $\widehat{D}^a = \mathrm{ad}(f_a)|_{\mathfrak{n}}$. The $q \times l$ matrix $\sigma = (\sigma_j^a)$ must have maximal rank, i.e., q, in view of the nilindependence of $\widehat{D}^a$. However we can no longer set σ_j^a equal to the Kronecker delta δ_j^a as was the case for $q = l$.

The condition (12.10), i.e., $[\widehat{D}^a, \widehat{D}^b] = 0$, remains and implies

$$\sum_{j,k=1}^{l} (\sigma_j^a \omega_k^b - \sigma_j^b \omega_k^a)\big(\lambda_j - \delta_{jk}(1 + s_j)\big)\widetilde{D}_k = 0$$

which after separation of coefficients of linearly independent derivations $\widetilde{D}_k$ gives a set of $\frac{1}{2}q(q-1)l$ conditions

$$\sum_{j=1}^{l} (\sigma_j^a \omega_k^b - \sigma_j^b \omega_k^a)\big(\lambda_j - \delta_{jk}(1 + s_j)\big) = 0, \quad k = 1, \ldots, l \tag{12.15}$$

for $2q{\cdot}l$ unknown constants σ_j^a, ω_k^b. When all $\omega_k^b = 0$, any choice of σ_j^a solves (12.15).

Let us now find out how the constants σ_j^a, ω_j^a transform under transformations Φ of the form (9.14), i.e., a change of basis in the nilradical. When Φ belongs to the connected component of $\mathrm{Aut}(\mathfrak{n})$ generated by exponentiation of $\mathfrak{Der}\big(\mathrm{NR}\big(\mathfrak{b}(\mathfrak{g})\big)\big)$, the transformation (9.14) will not change the constants σ_j^a, owing to the fact that $[D_i, D_j] = 0$, $[D_i, \widetilde{D}_j] \in \operatorname{span}\{\widetilde{D}_j\}$. We mention that certain permutations of the constants σ_j^a may become possible under parity transformations belonging to other components of Aut($\mathfrak{n}$). E.g., the Dynkin diagram and the root system of $\mathfrak{sl}(l+1,\mathbb{C})$ has the reflection symmetry $\alpha_j \leftrightarrow \alpha_{l+1-j}$ which gives rise to an automorphism of $\mathfrak{n}$ interchanging σ_j^a with σ_{l+1-j}^a. The constants ω_k^a transform as follows:

- under automorphisms $\exp(u_j D_j)$ the constants ω_k^a get scaled by a factor $\exp(u_j d_{jk})$ where $[D_j, \widetilde{D}_k] = d_{jk}\widetilde{D}_k$.
- under automorphisms $\tilde{\phi}_j(t_j) = \exp(t_j\widetilde{D}_j) = \mathbf{1} + t_j\widetilde{D}_j$ the constants ω_k^a are shifted to

$$\widehat{\omega}_k^a = \omega_k^a - \Big(\sum_{j=1}^{l} \sigma_j^a\lambda_j - \sigma_k^a(1+s_k)\Big)t_k \tag{12.16}$$

as follows from the evaluation of the commutator $\big[\widetilde{D}_j, \sum_{k=1}^{l}\sigma_k^a D_k\big]$.

Let us assume that $\omega_k^a \neq 0$ for certain a and k. We may add a suitable multiple of $\widehat{D}^a$ to all $\widehat{D}^b$, $b \neq a$ so that after the addition we have $\omega_k^b = 0$ for all $b \neq a$. Next, we check whether $\sum_{j=1}^{l}\sigma_j^a\lambda_j - \sigma_k^a(1+s_k) \neq 0$. If it holds, we can use the transformation (12.16) to set also $\omega_k^a = 0$. If not, we may attempt to modify $\widehat{D}^a$ by addition of any multiple of $\widehat{D}^b$, $b \neq a$. If

$$\sum_{j=1}^{l}\sigma_j^b\lambda_j - \sigma_k^b(1+s_k) \neq 0 \tag{12.17}$$

holds for at least one $b = 1,\ldots,q$, we can first change $\widehat{D}^a$ and next eliminate $\omega_k^a = 0$, i.e., now we have $\omega_k^b = 0$ for all b, including $b = a$. Since the automorphisms $\tilde{\phi}_k(t_k)$ for different k commute we may repeat the procedure for all values of k. Consequently, for any $k = 1,\ldots,l$ we may set all $\omega_k^a = 0$ whenever the condition (12.17) is satisfied for at least one $a \in \{1,\ldots,q\}$.

Let us find out for how many values of k we can always transform the parameters ω_k^a into $\omega_k^a = 0$. The matrix Q with entries $Q_k^j = (\lambda_j - (1+s_k)\delta_{jk})$ has the form

$$Q = \begin{pmatrix} \lambda_1 - (1+s_1) & \lambda_1 & \lambda_1 & \ldots & \lambda_1 \\ \lambda_2 & \lambda_2 - (1+s_2) & \lambda_2 & \ldots & \lambda_2 \\ \lambda_3 & \lambda_3 & \lambda_3 - (1+s_3) & \ldots & \lambda_3 \\ \vdots & \vdots & & \ddots & \vdots \\ \lambda_l & \lambda_l & \lambda_l & \ldots & \lambda_l - (1+s_l) \end{pmatrix}.$$

Subtracting the second column from the first, the third from the second, etc., and using $\lambda_k > 0$ and $s_i \geq 0$ we find that it has always the maximal rank l, i.e., Q is a regular matrix. The expressions $R_k^a = \big(\sum_{j=1}^{l}\sigma_j^a\lambda_j - \sigma_k^a(1+s_k)\big)$ can be written as elements of the matrix product $R = \sigma\cdot Q$. Whenever the k-th column $(R_k^a)_{a=1,\ldots,q}$ of the matrix R is nonvanishing we can eliminate all parameters ω_k^a from our derivations $\widehat{D}^a$. The number of indices k such that the column $(R_k^a)_{a=1,\ldots,q}$ does not vanish is bounded from below by the rank of matrix R, i.e., also by the rank of matrix σ, which is equal to q. Therefore, we have at most $l-q$ values of the index k such that some parameters ω_k^a are nonvanishing.

Now, after the simplifications introduced above, we revisit the commutativity condition (12.15). We observe that (12.15) can be written as a sum of terms, all of them containing the expression

$$\omega_k^d \sum_{j=1}^{l} \sigma_j^c \big(\lambda_j - \delta_{jk}(1+s_j)\big) \tag{12.18}$$

where $c = a$ and $d = b$ or vice versa. If the condition (12.17) is satisfied we have set $\omega_k^d = 0$; otherwise, $\sum_{j=1}^{l} \sigma_j^c \big(\lambda_j - \delta_{jk}(1+s_j)\big) = 0$. Either way, all terms of the form (12.18) vanish and consequently the commutativity condition (12.15) is always satisfied.

Finally, we consider the Lie brackets between elements f_a, f_b. Equation (9.7) takes the form

$$[f_a, f_b] = \gamma_{ab} e_\lambda, \quad a, b = 1, \ldots, q. \tag{12.19}$$

The constants γ_{ab} in (12.19) can be modified by transformations of the form

$$f_a \to f_a' = f_a + \tau_a e_\lambda.$$

Such a transformation does not change the derivations $\widehat{D}_a$, $a = 1, \ldots, q$ but may modify the Lie brackets $[f_a, f_b]$. Indeed, when there exists at least one $\widehat{D}^a$ such that $\kappa = \sum_{j=1}^{l} \lambda^j \sigma_j^a \neq 0$, we have

$$[f_a, e_\lambda] = \widehat{D}^a(e_\lambda) = \kappa e_\lambda \neq 0.$$

Now we may choose $\tau_a = 0$ and $\tau_b = -\gamma_{ab}/\kappa$, $a \neq b$ so that we obtain

$$[f_a', f_b'] = \gamma_{ab} e_\lambda - \frac{\gamma_{ab}}{\kappa} \kappa e_\lambda = 0,$$

i.e., we have set $\gamma_{ab} = 0$ for the given a and all $b = 1, \ldots, q$. The Jacobi identity (f_a, f_b, f_c)

$$\gamma_{ab} \widehat{D}_c(e_\lambda) + \gamma_{bc} \widehat{D}_a(e_\lambda) + \gamma_{ca} \widehat{D}_b(e_\lambda) = 0$$

reduces to

$$\gamma_{bc} \widehat{D}_a(e_\lambda) = \kappa \gamma_{bc} e_\lambda = 0, \quad \forall\, b, c = 2, \ldots, q$$

and implies that all constants $\gamma_{bc} = 0$ vanish. Therefore, the basis elements f_a in $\mathfrak{s}$ can be chosen so that they commute whenever the condition

$$\sum_{j=1}^{l} \lambda^j \sigma_j^a \neq 0$$

holds for at least one a. On the other hand, if $\sum_{j=1}^{l} \lambda^j \sigma_j^a = 0$ for all a, the Jacobi identity (f_a, f_b, f_c) is satisfied trivially.

Thus, we have proven the following theorem

Theorem 12.3. *Let $\mathfrak{g}$ be a complex simple Lie algebra or its split real form, $\mathfrak{b}(\mathfrak{g})$ its (split) Borel subalgebra. Let $\mathfrak{g}$ be different from $\mathfrak{sl}(3, \mathbb{R})$. Any solvable extension $\mathfrak{s}$ of the nilradical $\mathrm{NR}(\mathfrak{b}(\mathfrak{g}))$ by q nonnilpotent elements f_a, $a = 1, \ldots, q \leq$ rank $\mathfrak{g}$ is defined by q commuting derivations $\widehat{D}^a$ and a constant $q \times q$ antisymmetric matrix $\gamma = (\gamma_{ab})$. The derivations $\widehat{D}^a$ determine the Lie brackets*

$$[f_a, e_\alpha] = \widehat{D}^a(e_\alpha), \quad a = 1, \ldots, q, \ \alpha \in \Delta^+$$

and take the form

$$\widehat{D}^a = \mathrm{ad}(f_a)|_{\mathfrak{n}} = \sum_{j=1}^{l} \big(\sigma_j^a D_j + \omega_j^a \widetilde{D}_j\big), \quad a = 1, \ldots, q,$$

where the matrix $\sigma = (\sigma^a_j)$, $a = 1, \dots, q$, $j = 1, \dots, l$, *has the maximal possible rank* q. *For any given value of* k *all parameters* ω^a_k *are equal to zero when the condition*

$$\sum_{j=1}^{l} \sigma^a_j \lambda_j - \sigma^a_k (1 + s_k) \neq 0 \tag{12.20}$$

is satisfied for at least one $a \in \{1, \dots, q\}$. *The condition* (12.20) *is always satisfied for at least* q *values of the index* k, *i.e., there are at most* $l - q$ *values of* k *such that some of the parameters* ω^a_k *are nonvanishing.*

The matrix $\gamma = (\gamma_{ab})$ *defines the Lie brackets*

$$[f_a, f_b] = \gamma_{ab} e_\lambda, \quad a, b = 1, \dots, q.$$

When

$$\sum_{j=1}^{l} \lambda^j \sigma^a_j \neq 0$$

holds for at least one $a \in \{1, \dots, q\}$, *the constants* γ_{ab} *are all equal to* 0, *i.e.,*

$$[f_a, f_b] = 0.$$

Similarly as for Theorem 12.2, Theorem 12.3 was proven above under the assumption that rank $\mathfrak{g} > 2$. The derivation is the same when $\mathfrak{g} = \mathrm{G}_2$. When $\mathfrak{g} = \mathfrak{sl}(2, \mathbb{C})$ or $\mathfrak{g} = \mathfrak{sl}(2, \mathbb{R})$ the result is trivial. When $\mathfrak{g} = \mathfrak{sl}(3, \mathbb{C})$ the result follows from the consideration of Heisenberg nilradicals in Section 11. When $\mathfrak{g} = \mathfrak{sl}(3, \mathbb{R})$ the results of Theorem 12.3 do not hold as was already indicated in Section 12.2.2 (for more details, see Section 11.3). In the cases of $\mathfrak{g} = \mathfrak{so}(5, \mathbb{C})$ and $\mathfrak{g} = \mathfrak{so}(3, 2)$ it follows from the investigation of the nilradical $\mathfrak{n}_{4,1}$ in Section 13.1.

The conditions in Theorem 12.3 are sufficient, i.e., any set of constants σ^a_j, ω^a_j and γ_{ab} satisfying the properties listed in the theorem gives rise to a solvable extension of the nilradical $\mathrm{NR}(\mathfrak{b}(\mathfrak{g}))$. On the other hand, the description presented in Theorem 12.3 is not unique, i.e., different choices of σ^a_j, ω^a_j and γ_{ab} may lead to isomorphic algebras. As already noted, we may replace the derivations $\widehat{D}^a$ by any linearly independent combination of them thus changing all the parameters σ^a_j, ω^a_j and γ_{ab}. Also we may employ the scaling automorphisms to change the values of ω^a_j and γ_{ab}.

We remark that by virtue of indecomposability of the Borel nilradicals, all solvable Lie algebras described in Theorem 12.3 are indecomposable.

12.2.4. Solvable extensions of dimension $n_{\mathrm{NR}} + 1$. The results contained in Theorem 12.3 can be further refined in the particular case $q = 1$. Any single derivation $\widehat{D}^1$ of the form (12.14) with nonvanishing vector (σ^1_j) defines an $(n_{\mathrm{NR}} + 1)$-dimensional solvable extension of the nilradical $\mathfrak{n}$. Let us investigate what restrictions we may impose on the parameters ω^1_k through a choice of a suitable transformation (9.14).

In the transformation (12.16) we choose t_k so that $\widehat{\omega}^1_k = 0$ whenever the expression $\sum_{j=1}^{l} \sigma^1_j \lambda_j - \sigma^1_k (1 + s_k)$ is nonvanishing. Otherwise, the parameter ω^1_k remains nonvanishing after conjugation by any automorphism from $\exp(\mathfrak{Der}(\mathrm{NR}(\mathfrak{b}(\mathfrak{g}))))$.

We already know that there can be at most $l - 1$ nonvanishing parameters $\omega^1_k = 0$. On the other hand, for any choice of up to $l - 1$ indices k_u, $u = 1, \dots, l - 1$

we can find constants σ_j^1 such that

$$\sum_{j=1}^{l} \sigma_j^1 \lambda_j - \sigma_{k_u}^1 (1 + s_{k_u}) = 0, \quad u = 1, \dots, l-1$$

and consequently the parameters $\widehat{\omega}_{k_u}^1$ cannot be set to zero by conjugation by any automorphism from the connected part of the automorphism group $\mathrm{Aut}(\mathfrak{n})$. That means that there are up to $l-1$ nondiagonal nonremovable parameters in the derivation $\widehat{D}^1$ (whose positions may be changed by transformations belonging to other components of $\mathrm{Aut}(\mathfrak{n})$ if the root system of $\mathfrak{g}$ allows them).

The nonvanishing parameters ω_k^1 can be scaled to 1 over the field of complex numbers. Over the field of real numbers, the usual problems with square roots may arise, so in some cases we can only normalize ω_k^1 to $\omega_k^1 = \pm 1$.

In order to show this, let us consider the automorphisms $\hat{\phi}_j(v_j)$, $v_j \neq 0$ defined by

$$\hat{\phi}_j(v_j) e_\alpha = v_j^{m_j} e_\alpha, \quad \alpha = \sum_{j=1}^{l} m_j \alpha_j \in \Delta^+$$

which generalize the inner automorphisms $\phi_j(u_j) = \exp(u_j D_j)$ to include also scaling by negative numbers. Under the transformation $\hat{\phi}_j(v_j)$ the constant ω_k^1 gets scaled by the factor $v_j^{d_{jk}}$ where $d_{jk} = \lambda_j - \delta_{jk}(1 + s_k)$,

$$\omega_k^1 \to v_j^{\lambda_j - \delta_{jk}(1+s_k)} \omega_k^1.$$

Combining l such commuting automorphisms together,

$$\hat{\phi}(v) = \hat{\phi}_1(v_1) \circ \cdots \circ \hat{\phi}_l(v_l), \tag{12.21}$$

we transform

$$\omega_k^1 \to \frac{\prod_{j=1}^{l} v_j^{\lambda_j}}{v_k^{1+s_k}} \omega_k^1. \tag{12.22}$$

Let us consider what can be accomplished using (12.22). Let us first deal with all k such that $s_k = 0$ (and $\omega_k^1 \neq 0$ by assumption). We set $v_k = \omega_k^1$. We still have at our disposal at least one constant v_i such that $s_i \neq 0$. We can use it to set $\prod_{j=1}^{l} v_j^{\lambda_j} = 1$ over the field of complex numbers. Over the field of real numbers we can set $\prod_{j=1}^{l} v_j^{\lambda_j} = 1$ when λ_i is odd and $\prod_{j=1}^{l} v_j^{\lambda_j} = \pm 1$ when λ_i is even. Thus all nonvanishing constants ω_k^1 such that $s_k = 0$ were brought by a suitable automorphism (12.21) to 1 over the field of complex numbers and to $\epsilon = \pm 1$ over the field of real numbers (where the value of ϵ is the same for all k).

Assuming that the above described rescaling was already performed, we want to scale ω_i^1 with $s_i \neq 0$ to a convenient value. We recall that at most $l-1$ parameters ω_j^1 are nonvanishing. Therefore, if $\omega_i^1 \neq 0$ we have at least one k, $k \neq i$ such that $\omega_k^1 = 0$. We choose an automorphism $\hat{\phi}(v)$ such that $v_j = 1$ whenever $j \neq i, k$,

$$v_i^{1+s_i} = \omega_i^1$$

and v_k is chosen so that $\prod_{j=1}^{l} v_j^{\lambda_j} = 1$. Over the field of real numbers we may accomplish only $v_i^{1+s_i} = \pm\omega_i^1$ when s_i is odd and $\prod_{j=1}^{l} v_j^{\lambda_j} = \pm 1$ when λ_k is even. The automorphism $\hat{\phi}(v)$ constructed in this way does not affect the values

ω_j^1 set to 1 in the previous step (or scales them by a common minus factor when $\prod_{j=1}^{l} v_j^{\lambda_j} = -1$) and allows us to scale ω_i^1 with $s_i \neq 0$ to 1 (± 1, respectively). In the case of the algebra A_l the procedure is repeated once again for the second index $\hat{\imath}$ such that $s_{\hat{\imath}} \neq 0$.

To sum up, we have the following theorem.

Theorem 12.4. *Let $\mathfrak{g}$ and $\mathfrak{b}(\mathfrak{g})$ satisfy the conditions of Theorem 12.3. Any solvable extension of the nilradical $\mathrm{NR}(\mathfrak{b}(\mathfrak{g}))$ by one nonnilpotent element is up to isomorphism defined by a single derivation*

$$\widehat{D} = \mathrm{ad}(f_1)|_{\mathfrak{n}} = \sum_{j=1}^{l} (\sigma_j D_j + \omega_j \widetilde{D}_j) \tag{12.23}$$

chosen so that the first nonvanishing parameter σ_j is equal to one. The parameter ω_k vanishes whenever $\sum_{j=1}^{l} \sigma_j \lambda_j - \sigma_k(1+s_k) \neq 0$. At most $l-1$ parameters ω_k are nonvanishing. They are all equal to 1 over the field of complex numbers. Over the field of real numbers they are equal to ± 1 and all parameters ω_k with k such that $s_k = 0$ have the same sign.

12.3. Solvable Lie algebras with triangular nilradicals

12.3.1. The structure of the algebras of strictly upper triangular matrices and their derivations. Let us now apply the results introduced above in Sections 12.1, 12.2 to the case of the nilradical of the Borel subalgebra of the Lie algebra $\mathfrak{sl}(l+1)$ over the fields of complex and real numbers. The algebra $\mathfrak{sl}(l+1, \mathbb{R})$ is the split real form of $\mathfrak{sl}(l+1, \mathbb{C})$; therefore, the results of the previous sections apply in both cases. We shall perform some of the calculations explicitly to illustrate the general arguments of the previous subsections.

The Borel subalgebra of the simple matrix Lie algebra $\mathfrak{sl}(l+1, \mathbb{F})$ over the field $\mathbb{F}$ of complex or real numbers can be naturally identified with the algebra of upper triangular $(l+1) \times (l+1)$ matrices. A basis for this subalgebra can be chosen in the form

$$t_i, e_{ik} \in \mathbb{F}^{(l+1)\times(l+1)},$$

where

$$\begin{aligned} (t_i)_{aa} &= \delta_{ia} - \delta_{i+1,a}, \quad 1 \leq i \leq l, \qquad (t_i)_{ab} = 0, \quad a \neq b, \\ (e_{ik})_{ab} &= \delta_{ia}\delta_{kb}, \quad 1 \leq i < k \leq l+1. \end{aligned} \tag{12.24}$$

The vectors t_i span the Cartan subalgebra of $\mathfrak{sl}(l+1)$. Let ϕ_i be functionals on $\mathfrak{g}_0 = \mathrm{span}\{t_i \mid i = 1, \ldots, l\}$ defined as

$$\phi_i \colon \phi_i(t_j) = \delta_{ij} - \delta_{i,j-1}, \quad 1 \leq i, j \leq l$$

and

$$\phi_{l+1} = -\sum_{i=1}^{l} \phi_i.$$

The vectors e_{ik} are common eigenvectors of $\mathrm{ad}(t_j)$ and consequently they are root vectors. The root corresponding to the vector e_{ik} is

$$\alpha_{ik} = \phi_i - \phi_k.$$

In the chosen ordering of roots these roots α_{ik}, $1 \leq i < k \leq l+1$ form the set of positive roots Δ^+. Their negatives $\alpha_{ki} = -\alpha_{ik}$ correspond to root vectors which are obtained by matrix transposition from e_{ik}.

The nilradical $\mathfrak{t}(l+1)$ of the Borel subalgebra of the Lie algebra $\mathfrak{sl}(l+1)$ is spanned by the vectors e_{ik} and can be naturally identified with the algebra $\mathfrak{t}(l+1)$ of strictly upper triangular $(l+1) \times (l+1)$ matrices, henceforth called *triangular nilradical.* Its Lie brackets are

$$[e_{ik}, e_{ab}] = \delta_{ka} e_{ib} - \delta_{bi} e_{ak} \tag{12.25}$$

and its dimension is

$$n \equiv \dim\ \mathfrak{t}(l+1) = \tfrac{1}{2} l(l+1). \tag{12.26}$$

The simple roots, i.e., those positive roots which cannot be written as a sum of two positive roots, are

$$\alpha_i \equiv \alpha_{i,i+1} = \phi_i - \phi_{i+1}, \quad 1 \leq i \leq l. \tag{12.27}$$

The basis vectors t_i of the Cartan subalgebra were chosen so that they coincide with the vectors h_{α_i} of (2.34) up to a constant multiple, $h_{\alpha_i} = 2(l+1)t_i$. Any positive root α_{ik} is expressed in terms of the simple roots as

$$\alpha_{ik} = \sum_{j=i}^{k-1} \alpha_j. \tag{12.28}$$

The highest root is

$$\lambda = \alpha_{1,l+1} = \sum_{j=1}^{l} \alpha_j. \tag{12.29}$$

There are only two simple roots α such that $\lambda - \alpha$ is again a root, namely $\alpha = \alpha_1$ and $\alpha = \alpha_l$. Consequently, we have

$$S_{\alpha_1}(\lambda) = \sum_{j=2}^{l} \alpha_j = \alpha_{2,l+1}, \qquad S_{\alpha_l}(\lambda) = \sum_{j=1}^{l-1} \alpha_j = \alpha_{1,l},$$
$$S_{\alpha_i}(\lambda) = \lambda, \quad 2 \leq i \leq l-1.$$

From the knowledge of the root system of $\mathfrak{sl}(l+1)$ we can immediately construct a basis of outer derivations of the nilpotent algebra $\mathfrak{t}(l+1)$ using the results of [67] recalled earlier (cf. (12.6)). There are $2l$ linearly independent outer derivations of $\mathfrak{t}(l+1)$ which can be chosen as $D_i, \widetilde{D}_i$, $1 \leq i \leq l$:

- $D_i(e_{jk}) = e_{jk}$ when $j \leq i \leq k-1$, $D_i(e_{jk}) = 0$ otherwise,
- $\widetilde{D}_1(e_{12}) = e_{2,l+1}$, $\widetilde{D}_1(e_{ik}) = 0$ otherwise,
- $\widetilde{D}_l(e_{l,l+1}) = e_{1,l}$, $\widetilde{D}_l(e_{ik}) = 0$ otherwise,
- when $2 \leq i \leq l-1$ we have $\widetilde{D}_i(e_{j,j+1}) = \delta_{ij} e_{1,l+1}$ and $\widetilde{D}_i(e_{jk}) = 0$, $1 \leq j < k-1 \leq l$.

We observe that the derivations D_i are nonnilpotent whereas $\widetilde{D}_i$ are nilpotent operators. $\widetilde{D}_i$ together with the inner derivations span the nilradical of $\mathfrak{Der}(\mathfrak{t}(l+1))$. Let us denote by $\mathfrak{Out}(\mathfrak{t}(l+1))$ the subalgebra of $\mathfrak{Der}(\mathfrak{t}(l+1))$ spanned by the outer derivations $D_i, \widetilde{D}_i$.

The main results on solvable Lie algebras with triangular nilradicals were obtained in [123]. They were generalized to any Borel nilradical in [118] and were reproduced in Sections 12.1 and 12.2 above. For the sake of completeness we present the original result of [123] here as a proposition.

Proposition 12.5. *Every solvable Lie algebra* $\mathfrak{s}$ *with a triangular nilradical* $\mathfrak{t}(l+1)$ *can be transformed to a canonical basis* (e_{ik}, f_α), $\alpha = 1, \dots, f$, $1 \le i < k \le l+1$, *with commutation relations* (12.25) *and*

$$[f_\alpha, e_{ik}] = \sum_{p<q} d^\alpha_{ik,pq}\, e_{pq} \tag{12.30}$$

$$[f_\alpha, f_\beta] = \gamma_{\alpha\beta} e_{1(l+1)}, \quad 1 \le \alpha, \beta \le f,\ d^\alpha_{ik,pq}, \gamma_{\alpha\beta} \in \mathbb{F}. \tag{12.31}$$

The commutation relations (12.30) *can be rewritten as*

$$\begin{gathered} [f_\alpha, \vec{e}] = \vec{e} \cdot \widehat{D}_\alpha, \quad \widehat{D}_\alpha \in \mathbb{F}^{n\times n} \\ \vec{e} \equiv (e_{1(l+1)}, e_{1l}, e_{2(l+1)}, \dots, e_{12}, e_{23}, \dots, e_{l(l+1)}). \end{gathered} \tag{12.32}$$

We mention that the vector $\vec{e}$ *introduces an order in lines* (*columns*) *of the matrices* $\widehat{D}_\alpha$, *where each line* (*column*) *is represented by two numbers. The matrices* $\widehat{D}_\alpha = (d^\alpha_{ik,pq})$ *have the following canonical form.*

(1) *They are upper triangular.*
(2) *The only off-diagonal matrix elements that do not vanish identically and cannot be annulled by a redefinition of the elements* f_α *are*:

$$d^\alpha_{2(l+1),12}, \qquad d^\alpha_{1(l+1),j(j+1)} \quad (2 \le j \le l-1), \qquad d^\alpha_{1l,l(l+1)}. \tag{12.33}$$

(3) *The diagonal elements* $d^\alpha_{i(i+1),i(i+1)}$, $1 \le i \le l$ *are free. The remaining diagonal elements satisfy*

$$d^\alpha_{ik,ik} = \sum_{p=i}^{k-1} d^\alpha_{p(p+1),p(p+1)}, \quad i+1 < k. \tag{12.34}$$

The canonical forms of the matrices $\widehat{D}_\alpha$ *and the constants* $\gamma_{\alpha\beta}$ *satisfy the following conditions*:

(1) *The set of matrices* $\widehat{D}_\alpha$ *have the form specified above and are linearly nilindependent. For* $f \ge 2$ *they all commute, i.e.,*

$$\left[\widehat{D}_\alpha, \widehat{D}_\beta\right] = 0. \tag{12.35}$$

(2) *All constants* $\gamma_{\alpha\beta}$ *vanish unless we have* $d^\gamma_{1(l+1),1(l+1)} = 0$ *for* $\gamma = 1, \dots, f$ *simultaneously for all* γ.
(3) *The remaining off-diagonal elements* $d^\alpha_{ik,ab}$ *also vanish, unless the diagonal elements satisfy* $d^\beta_{ik,ik} = d^\beta_{ab,ab}$ *for* $\beta = 1, \dots, f$ *simultaneously for all* β.
(4) *The maximal number of nonnilpotent elements is* $f_{\max} = l$. *In this case the nonnilpotent elements always commute, i.e.,*

$$[f_\alpha, f_\beta] = 0 \tag{12.36}$$

and the matrices $\widehat{D}_\alpha$ *are explicitly given by the diagonal matrices*

$$a^\alpha_{ik} = \sum_{p=i}^{k-1} \delta_{\alpha p}, \quad 1 \le i < k \le l+1,\ 1 \le \alpha \le l. \tag{12.37}$$

(5) *For $f = 1$ the matrix $\widehat{D}_1$ has at most $l-1$ off-diagonal elements that can be normalized to $+1$ for $\mathbb{F} = \mathbb{C}$ and to ± 1 for $\mathbb{F} = \mathbb{R}$.*

For the sake of brevity we shall use the notation $\mathfrak{s}(l+1, f)$ for any solvable extension of the triangular nilradical $\mathfrak{t}(l+1)$ by f nonnilpotent elements in the following sections. The dimension of $\mathfrak{s}(l+1, f)$ satisfies

$$\dim \mathfrak{s}(l+1, f) = \frac{l(l+1)}{2} + f. \tag{12.38}$$

Proposition 12.5 is a specialization of Theorems 12.3 and 12.4 to the case of triangular nilradicals.

12.3.2. Illustration of the procedure for low dimensions. We shall now investigate in detail the lowest dimensional nontrivial case of solvable extensions described in Theorems 12.2, 12.3, 12.4. We shall proceed independently of the theorems, demonstrating several components of the proofs presented earlier in an explicit example.

In the case $l = 2$, the Lie algebra $\mathfrak{t}(3)$ is isomorphic to the Heisenberg algebra $\mathfrak{h}(1)$. Solvable Lie algebras with Heisenberg nilradicals were considered earlier in Chapter 11.

Let us consider the case $l = 3$. We shall use the basis of $\mathfrak{t}(4)$ ordered as $\mathcal{E} = (e_{14}, e_{13}, e_{24}, e_{12}, e_{23}, e_{34})$. In this basis $\mathcal{E}$, the six linearly independent outer derivations are represented by matrices

$$\begin{aligned}
D_1 &= \begin{pmatrix} 1&0&0&0&0&0\\ 0&1&0&0&0&0\\ 0&0&0&0&0&0\\ 0&0&0&1&0&0\\ 0&0&0&0&0&0\\ 0&0&0&0&0&0 \end{pmatrix}, &
\widetilde{D}_1 &= \begin{pmatrix} 0&0&0&0&0&0\\ 0&0&0&0&0&0\\ 0&0&0&1&0&0\\ 0&0&0&0&0&0\\ 0&0&0&0&0&0\\ 0&0&0&0&0&0 \end{pmatrix},\\
D_2 &= \begin{pmatrix} 1&0&0&0&0&0\\ 0&1&0&0&0&0\\ 0&0&1&0&0&0\\ 0&0&0&0&0&0\\ 0&0&0&0&1&0\\ 0&0&0&0&0&0 \end{pmatrix}, &
\widetilde{D}_2 &= \begin{pmatrix} 0&0&0&0&1&0\\ 0&0&0&0&0&0\\ 0&0&0&0&0&0\\ 0&0&0&0&0&0\\ 0&0&0&0&0&0\\ 0&0&0&0&0&0 \end{pmatrix},\\
D_3 &= \begin{pmatrix} 1&0&0&0&0&0\\ 0&0&0&0&0&0\\ 0&0&1&0&0&0\\ 0&0&0&0&0&0\\ 0&0&0&0&0&0\\ 0&0&0&0&0&1 \end{pmatrix}, &
\widetilde{D}_3 &= \begin{pmatrix} 0&0&0&0&0&0\\ 0&0&0&0&0&1\\ 0&0&0&0&0&0\\ 0&0&0&0&0&0\\ 0&0&0&0&0&0\\ 0&0&0&0&0&0 \end{pmatrix};
\end{aligned} \tag{12.39}$$

the matrices of derivations are defined by the formula

$$D(\vec{e}) = \vec{e} \cdot D, \quad \text{where } \vec{e} = (e_{14}, e_{13}, e_{24}, e_{12}, e_{23}, e_{34}).$$

Consequently, any collection of outer derivations $\widehat{D}_k$ can be brought by addition of suitable inner derivations to the form

$$(12.40)\qquad \widehat{D}_k = \begin{pmatrix} d_1^k+d_2^k+d_3^k & 0 & 0 & 0 & d_5^k & 0 \\ 0 & d_1^k+d_2^k & 0 & 0 & 0 & d_6^k \\ 0 & 0 & d_2^k+d_3^k & d_4^k & 0 & 0 \\ 0 & 0 & 0 & d_1^k & 0 & 0 \\ 0 & 0 & 0 & 0 & d_2^k & 0 \\ 0 & 0 & 0 & 0 & 0 & d_3^k \end{pmatrix} \in \mathfrak{Out}(\mathfrak{t}(4)).$$

Let us assume that a solvable extension $\mathfrak{s}$ of the nilradical $\mathfrak{t}(4)$,

$$\dim \mathfrak{s} = \dim \mathfrak{t}(4) + f,$$

is given and that the outer derivations $\widehat{D}_k$, $1 \leq k \leq f$ represent the action of nonnilpotent elements f_k on the nilradical $\mathfrak{t}(4)$ as in (9.11). The requirement of nilindependence of the matrices $\widehat{D}_k$ allows to conclude that

$$(12.41)\qquad f \leq 3.$$

Let us now look at the cases $f \geq 2$. The outer derivations of the form (12.40) form an Abelian subalgebra of $\mathfrak{Der}(\mathfrak{t}(4))$ by condition (12.10). The kernel of the adjoint representation of $\mathfrak{t}(4)$ is the center of $\mathfrak{t}(4)$, spanned by e_{14}. Therefore, the elements f_k must satisfy

$$(12.42)\qquad [f_j, f_k] = \gamma_{jk} e_{14}$$

for some constants $\gamma_{jk} = -\gamma_{kj}$.

Finally, let us consider the condition (9.10). It is nontrivial only for $f = 3$. Using (12.42) we find

$$(12.43)\qquad \gamma_{12}(d_1^3+d_2^3+d_3^3) + \gamma_{31}(d_1^2+d_2^2+d_3^2) + \gamma_{23}(d_1^1+d_2^1+d_3^1) = 0.$$

We may modify the constants γ_{jk} by adding $\beta^k e_{14}$ to f_k. Such a transformation does not alter the derivations $\widehat{D}_k$, because e_{14} is in the center of $\mathfrak{t}(4)$, but changes the constants γ_{jk} in (12.42).

$$(12.44)\qquad \tilde{\gamma}_{jk} = \gamma_{jk} + (d_1^j+d_2^j+d_3^j)\beta^k - (d_1^k+d_2^k+d_3^k)\beta^j.$$

Hence, by using (12.44) for $f = 2$ and (12.44), (12.43) for $f = 3$, we obtain

$$(12.45)\qquad [f_j, f_k] = \begin{cases} \gamma_{jk} e_{14} & \text{for } (d_1^1+d_2^1+d_3^1) = \cdots = (d_1^f+d_2^f+d_3^f) = 0, \\ 0 & \text{otherwise.} \end{cases}$$

To further simplify the structure of the solvable algebra $\mathfrak{s}$, let us perform a transformation of the type (9.14), i.e., conjugate our derivations $\widehat{D}_k$ by suitable automorphisms of $\mathfrak{t}(4)$. The connected part of the group of automorphisms is generated by derivations $\mathfrak{Der}(\mathfrak{t}(4))$ and can be enlarged to a bigger group by inclusion of several parity transformations. This allows negative values of group parameters g_1, g_2, g_3 below. We shall consider two particular 3-dimensional subgroups of the group of automorphisms, namely

$$G_1 = \begin{pmatrix} 1 & 0 & 0 & 0 & g_5 & 0 \\ 0 & 1 & 0 & 0 & 0 & g_6 \\ 0 & 0 & 1 & g_4 & 0 & 0 \\ 0 & 0 & 0 & 1 & 0 & 0 \\ 0 & 0 & 0 & 0 & 1 & 0 \\ 0 & 0 & 0 & 0 & 0 & 1 \end{pmatrix}, \quad G_2 = \begin{pmatrix} g_1g_2g_3 & 0 & 0 & 0 & 0 & 0 \\ 0 & g_1g_2 & 0 & 0 & 0 & 0 \\ 0 & 0 & g_2g_3 & 0 & 0 & 0 \\ 0 & 0 & 0 & g_1 & 0 & 0 \\ 0 & 0 & 0 & 0 & g_2 & 0 \\ 0 & 0 & 0 & 0 & 0 & g_3 \end{pmatrix} \tag{12.46}$$

where $g_1g_2g_3 \neq 0$.

Conjugation by G_1 simultaneously transforms all derivations $\widehat{D}_k$ to

$$G_1 \cdot \widehat{D}_k \cdot G_1^{-1} = \begin{pmatrix} d_1^k + d_2^k + d_3^k & 0 & 0 & 0 & \tilde{d}_5^k & 0 \\ 0 & d_1^k + d_2^k & 0 & 0 & 0 & \tilde{d}_6^k \\ 0 & 0 & d_2^k + d_3^k & \tilde{d}_4^k & 0 & 0 \\ 0 & 0 & 0 & d_1^k & 0 & 0 \\ 0 & 0 & 0 & 0 & d_2^k & 0 \\ 0 & 0 & 0 & 0 & 0 & d_3^k \end{pmatrix}, \tag{12.47}$$

where

$$\begin{aligned} \tilde{d}_4^k &= d_4^k + g_4(d_1^k - d_2^k - d_3^k), \\ \tilde{d}_5^k &= d_5^k - g_5(d_1^k + d_3^k), \\ \tilde{d}_6^k &= d_6 + g_6(d_3^k - d_1^k - d_2^k). \end{aligned} \tag{12.48}$$

Thus using a suitable choice of the group parameter g_4 we can eliminate d_4^k, $(1 \leq k \leq f)$ of the specific derivation $\widehat{D}_k$, provided that $d_1^k \neq d_2^k + d_3^k$. The parameters g_5 and g_6 are used in the same way. Therefore, at most 3 off-diagonal elements can be eliminated using this transformation.

Next we employ the second automorphism G_2. The matrices $\widehat{D}_k$ are transformed to

$$\begin{aligned} &G_2 \cdot \widehat{D}_k \cdot G_2^{-1} \\ &= \begin{pmatrix} d_1^k + d_2^k + d_3^k & 0 & 0 & 0 & g_1g_3d_5^k & 0 \\ 0 & d_1^k + d_2^k & 0 & 0 & 0 & (g_1g_2/g_3)d_6^k \\ 0 & 0 & d_2^k + d_3^k & (g_2g_3/g_1)d_4^k & 0 & 0 \\ 0 & 0 & 0 & d_1^k & 0 & 0 \\ 0 & 0 & 0 & 0 & d_2^k & 0 \\ 0 & 0 & 0 & 0 & 0 & d_3^k \end{pmatrix}, \end{aligned} \tag{12.49}$$

Therefore, we can normalize up to three nonvanishing off-diagonal elements. For $\mathbb{F} = \mathbb{C}$ they can be set equal to $+1$, for $\mathbb{F} = \mathbb{R}$ we must in some cases allow the possibility of normalizing to either $+1$ or to -1.

A further automorphism of $\mathfrak{t}(4)$ is given by

$$G_S = \begin{pmatrix} 1 & 0 & 0 & 0 & 0 & 0 \\ 0 & 0 & -1 & 0 & 0 & 0 \\ 0 & -1 & 0 & 0 & 0 & 0 \\ 0 & 0 & 0 & 0 & 0 & 1 \\ 0 & 0 & 0 & 0 & 1 & 0 \\ 0 & 0 & 0 & 1 & 0 & 0 \end{pmatrix}. \tag{12.50}$$

Its effect on the derivations $\widehat{D}_k$ is a simultaneous interchange of $d_1^k \leftrightarrow d_3^k$ and $d_4^k \leftrightarrow d_6^k$ accompanied by a change of sign of d_4^k, d_6^k as is seen from

$$G_S \cdot \widehat{D}_k \cdot G_S^{-1} = \begin{pmatrix} d_1^k + d_2^k + d_3^k & 0 & 0 & 0 & d_5^k & 0 \\ 0 & d_2^k + d_3^k & 0 & 0 & 0 & -d_4^k \\ 0 & 0 & d_1^k + d_2^k & -d_6^k & 0 & 0 \\ 0 & 0 & 0 & d_3^k & 0 & 0 \\ 0 & 0 & 0 & 0 & d_2^k & 0 \\ 0 & 0 & 0 & 0 & 0 & d_1^k \end{pmatrix}. \tag{12.51}$$

Let us now specialize to the individual cases of $\mathfrak{s}(4, f)$.

The Lie algebras $\mathfrak{s}(4,1)$. The derivation $\widehat{D} \equiv \widehat{D}_1$ has the "canonical" form given by (12.40). In agreement with Theorem 12.4, we can eliminate all off-diagonal elements using the transformation (12.47), unless the diagonal elements satisfy specific equations (e.g., $d_1^k = d_2^k + d_3^k$). At most, two of these equalities can hold, otherwise the derivation $\widehat{D}$ will be nilpotent. Thus, at most two off-diagonal entries remain. They can be normalized to $+1$, with one exception, namely, if we have $d_4 \neq 0$, $d_6 \neq 0$ for $\mathbb{F} = \mathbb{R}$. Then, we can transform to one of the two following inequivalent possibilities

$$d_4 = +1, \quad d_6 = \pm 1.$$

The final result is that for $\mathbb{F} = \mathbb{C}$, 8 inequivalent types of such matrices exist, one of them depending on 2 complex parameters, 3 depending on one complex parameter, 3 without parameters. For $\mathbb{F} = \mathbb{R}$ the parameters are restricted to real numbers and one nonparametric case splits into two nonisomorphic ones.

(1)

	e_{14}	e_{13}	e_{24}	e_{12}	e_{23}	e_{34}
f_1	$(1+a+b)e_{14}$	$(1+a)e_{13}$	$(1+b)e_{24}$	ae_{12}	e_{23}	be_{34}

generically
US $= [0]$
CS $= [7, 6]$
DS $= [7, 6, 3, 0]$

where $|b| \leq |a|$. When $|a| = |b|$ the algebras with the parameters (a, b) and (b, a) are isomorphic. For particular values of the parameters the dimension of characteristic ideals differ as follows. When $a + b = -1$ the upper central series is US $= [1]$. When $b = 0$, $a \neq 0$ we have CS $= [7, 5]$, DS $= [7, 5, 2, 0]$, when $a = b = 0$ we have CS $= [7, 4]$, DS $= [7, 4, 0]$.

(2)

	e_{14}	e_{13}	e_{24}	e_{12}	e_{23}	e_{34}
f_1	$(1+a)e_{14}$	e_{13}	ae_{24}	e_{12}	0	ae_{34}

generically
US $= [0]$
CS $= [7, 5]$
DS $= [7, 5, 1, 0]$

where $|a| \leq 1$. When $|a| = 1$ the algebras with the parameters a and $1/a$ are isomorphic. For particular values of the parameter a the dimension of characteristic ideals differ as follows. When $a = -1$ the upper central series is US $= [1]$. When $a = 0$ we have CS $= [7, 4, 3]$, DS $= [7, 4, 1, 0]$.

(3)

	e_{14}	e_{13}	e_{24}	e_{12}	e_{23}	e_{34}
f_1	$2(1+a)e_{14}$	$(2+a)e_{13}$	$(1+a)e_{24}$	$(1+a)e_{12}+e_{24}$	e_{23}	ae_{34}

generically
US = [0]
CS = [7, 6]
DS = [7, 6, 3, 0]

where $a \in \mathbb{F}$. For particular values of the parameters the dimension of characteristic ideals differ as follows. When $a = -1$ we have US = [1, 2], CS = [7, 5], DS = [7, 5, 2, 0]. When $a = 0$ we have CS = [7, 5], DS = [7, 5, 2, 0].

(4)

	e_{14}	e_{13}	e_{24}	e_{12}	e_{23}	e_{34}
f_1	$2e_{14}$	e_{13}	e_{24}	$e_{12}+e_{24}$	0	e_{34}

US = [0]
CS = [7, 5]
DS = [7, 5, 1, 0].

(5)

	e_{14}	e_{13}	e_{24}	e_{12}	e_{23}	e_{34}
f_1	e_{14}	$(1+a)e_{13}$	$(1-a)e_{24}$	ae_{12}	$e_{23}+e_{14}$	$-ae_{34}$

generically
US = [0]
CS = [7, 6]
DS = [7, 6, 3, 0]

where $a \in \mathbb{F}$. The algebras with the parameters a and $-a$ are isomorphic. When $a = 0$ we have CS = [7, 4], DS = [7, 4, 0].

(6)

	e_{14}	e_{13}	e_{24}	e_{12}	e_{23}	e_{34}
f_1	0	e_{13}	$-e_{24}$	e_{12}	e_{14}	$-e_{34}$

US = [1]
CS = [7, 5]
DS = [7, 5, 1, 0].

(7)

	e_{14}	e_{13}	e_{24}	e_{12}	e_{23}	e_{34}
f_1	$2e_{14}$	$3e_{13}$	e_{24}	$e_{12}+e_{24}$	$2e_{23}+e_{14}$	$-e_{34}$

US = [0]
CS = [7, 6]
DS = [7, 6, 3, 0].

(8)

	e_{14}	e_{13}	e_{24}	e_{12}	e_{23}	e_{34}
f_1	$2e_{14}$	e_{13}	e_{24}	$e_{12}+e_{24}$	0	$e_{34}+\epsilon e_{13}$

US = [0]
CS = [7, 5]
DS = [7, 5, 1, 0]

where $\epsilon = 1$ over $\mathbb{C}$, $\epsilon = \pm 1$ over $\mathbb{R}$.

The Lie algebras $\mathfrak{s}(4, 2)$. From Theorem 12.3 or, equivalently, from (12.45) and the previous results on the matrices $\widehat{D}_k$, we can determine all nonisomorphic structures of Lie algebras $\mathfrak{s}(4, 2)$ (of dimension 8). For $\mathbb{F} = \mathbb{C}$ or $\mathbb{R}$, 9 inequivalent types of such algebras exist, one depending on 2 parameters, 4 on one parameter, 4 without parameters.

(1)

	e_{14}	e_{13}	e_{24}	e_{12}	e_{23}	e_{34}	f_1
f_1	$(1+a)e_{14}$	e_{13}	$(1+a)e_{24}$	0	e_{23}	ae_{34}	0
f_2	$(1+b)e_{14}$	e_{13}	be_{24}	e_{12}	0	be_{34}	0

generically
US $= [0]$
CS $= [8, 6]$
DS $= [8, 6, 3, 0]$

where $a, b \in \mathbb{F}$ such that $(a, b) \neq (-1, -1)$. When $a = b = 0$ the lower central and derived series change to CS $= [8, 5]$, DS $= [8, 5, 2, 0]$.

(2)

	e_{14}	e_{13}	e_{24}	e_{12}	e_{23}	e_{34}	f_1
f_1	0	e_{13}	0	0	e_{23}	$-e_{34}$	0
f_2	0	e_{13}	$-e_{24}$	e_{12}	0	$-e_{34}$	ae_{14}

US $= [1]$
CS $= [8, 6]$
DS $= [8, 6, 3, 0]$

where $a \in \mathbb{F}$.

(3)

	e_{14}	e_{13}	e_{24}	e_{12}	e_{23}	e_{34}	f_1
f_1	e_{14}	e_{13}	e_{24}	0	$e_{23} + e_{14}$	0	0
f_2	0	e_{13}	$-e_{24}$	e_{12}	ae_{14}	$-e_{34}$	0

US $= [0]$
CS $= [8, 6]$
DS $= [8, 6, 3, 0]$

where $a \in \mathbb{F}$.

(4)

	e_{14}	e_{13}	e_{24}	e_{12}	e_{23}	e_{34}	f_1
f_1	e_{14}	e_{13}	e_{24}	0	e_{23}	0	0
f_2	0	e_{13}	$-e_{24}$	e_{12}	e_{14}	$-e_{34}$	0

US $= [0]$
CS $= [8, 6]$
DS $= [8, 6, 3, 0]$.

(5)

	e_{14}	e_{13}	e_{24}	e_{12}	e_{23}	e_{34}	f_1
f_1	0	e_{13}	0	e_{24}	e_{23}	$-e_{34}$	0
f_2	$2e_{14}$	e_{13}	e_{24}	$e_{12} + ae_{24}$	0	e_{34}	0

US $= [0]$
CS $= [8, 6]$
DS $= [8, 6, 3, 0]$

where $a \in \mathbb{F}$

(6)

	e_{14}	e_{13}	e_{24}	e_{12}	e_{23}	e_{34}	f_1
f_1	0	e_{13}	0	0	e_{23}	$-e_{34}$	0
f_2	$2e_{14}$	e_{13}	e_{24}	$e_{12} + e_{24}$	0	e_{34}	0

US $= [0]$
CS $= [8, 6]$
DS $= [8, 6, 3, 0]$.

(7)

	e_{14}	e_{13}	e_{24}	e_{12}	e_{23}	e_{34}	f_1
f_1	e_{14}	0	e_{24}	0	0	e_{34}	0
f_2	e_{14}	e_{13}	0	e_{12}	0	0	0

US $= [0]$
CS $= [8, 5]$
DS $= [8, 5, 1, 0]$.

The Lie algebra $\mathfrak{s}(4,3)$. We can choose a representative set of matrices $\widehat{D}_1$, $\widehat{D}_2, \widehat{D}_3$ by putting $d_j^k = \delta_{jk}$, where $j,k = 1,2,3$, in the general "canonical" form. For the matrix $\widehat{D}_1$, we use the transformation (12.47) to annul d_4^1, d_5^1 and d_6^1. The commutativity of the matrices $\widehat{D}_k$ imposes that also the matrices $\widehat{D}_2$ and $\widehat{D}_3$ must be diagonal.

Since $d_1^k + d_2^k + d_3^k$ are all different from zero, we can set

$$[f_j, f_k] = 0, \quad j,k = 1,2,3. \tag{12.52}$$

as in formula (12.45). Therefore, in agreement with Theorem 12.2 there exists only one Lie algebra $\mathfrak{s}(4,3)$ ($\mathbb{F} = \mathbb{C}$ or $\mathbb{R}$) and its dimension is 9. The structure of this algebra is given by the following table

	e_{14}	e_{13}	e_{24}	e_{12}	e_{23}	e_{34}	f_1	f_2
f_1	e_{14}	e_{13}	0	e_{12}	0	0	0	0
f_2	e_{14}	e_{13}	e_{24}	0	e_{23}	0	0	0
f_3	e_{14}	0	e_{24}	0	0	e_{34}	0	0

US = [0]
CS = [9, 6]
DS = [9, 6, 3, 0]

12.4. Casimir invariants of nilpotent and solvable triangular Lie algebras

At this stage no results are available on Casimir invariants of solvable Lie algebras with Borel nilradicals other than those associated with the Cartan series A_l of simple Lie algebras. We shall limit ourselves to this case, i.e., to solvable Lie algebras with triangular nilradicals $\mathfrak{t}(l+1)$ considered in Section 12.3. Let us first consider invariants of the nilpotent algebras $\mathfrak{t}(l+1)$ themselves.

12.4.1. Invariants of nilpotent triangular Lie algebras. We use the basis for the Borel subalgebra of $\mathfrak{sl}(l+1,\mathbb{F})$ given in (12.24). The basis for its nilradical $\mathfrak{t}(l+1)$ is given by the elements e_{ik}. The commutation relations of $\mathfrak{t}(l+1)$ are given by (12.25). We assume that $l \geq 3$ since $\mathfrak{t}(2)$ is trivial and $\mathfrak{t}(3)$ is isomorphic to the Heisenberg algebra $\mathfrak{h}(1)$ treated in Chapter 11. The dimension $l = 2$ is the only case for which there is an isomorphism between the triangular and the Heisenberg Lie algebras.

The differential operators $\widehat{E}_{ik}$ realizing the coadjoint representation of $\mathfrak{t}(l+1)$, are

$$\widehat{E}_{ik} = \sum_{b=k+1}^{l+1} e_{ib}\frac{\partial}{\partial e_{kb}} - \sum_{a=1}^{i-1} e_{ak}\frac{\partial}{\partial e_{ai}}. \tag{12.53}$$

Note that $\widehat{E}_{1(l+1)} \equiv 0$ in (12.53), since $e_{1(l+1)}$ commutes with all elements of $\mathfrak{t}(l+1)$.

We shall realize the coadjoint representation of $\mathfrak{t}(l+1)$ in the space of differentiable functions of $l(l+1)/2$ variables, i.e.,

$$I = I\big(e_{1(l+1)}, e_{1l}, e_{2(l+1)}, \ldots, e_{13}, \ldots, e_{(l-1)(l+1)}, e_{12}, e_{23}, \ldots, e_{l(l+1)}\big). \tag{12.54}$$

(The variables are ordered along lines of the matrix (e_{ix}) parallel to the main diagonal, starting from the top right corner). The function I is an invariant of the coadjoint representation of $\mathfrak{t}(l+1)$ if it satisfies the system of linear first order PDEs

$$\widehat{E}_{ik}(I) = 0, \quad 1 \leq i < k \leq l+1. \tag{12.55}$$

Let us now consider the strictly upper triangular $(l+1)\times(l+1)$ matrix-valued function $Q = Q(l+1)$ of the form

$$Q_{ik}(e_{1(l+1)}, e_{1l}, e_{2(l+1)}, \dots, e_{12}, \dots, e_{l(l+1)}) = \begin{cases} e_{ik} & \text{when } k-i \geq 1, \\ 0 & \text{otherwise.} \end{cases} \tag{12.56}$$

We define the determinant $Z_\mu = Z_\mu(l+1)$ constructed from the $\mu \times \mu$ upper right corner submatrix of the matrix Q, i.e.,

$$Z_\mu = \begin{vmatrix} e_{1(l-\mu+2)} & e_{1(l-\mu+3)} & \cdots & e_{1(l+1)} \\ e_{2(l-\mu+2)} & e_{2(l-\mu+3)} & \cdots & e_{2(l+1)} \\ \vdots & \vdots & & \vdots \\ e_{\mu(l-\mu+2)} & e_{\mu(l-\mu+3)} & \cdots & e_{\mu\,(l+1)} \end{vmatrix}, \quad 1 \leq \mu \leq \left[\frac{l+1}{2}\right] \tag{12.57}$$

where we used the notation $[x]$ for the integer part of a positive number. In particular,

$$p = \left[\frac{l+1}{2}\right] = \begin{cases} \dfrac{l+1}{2} & \text{when } l \text{ is odd,} \\ \dfrac{l}{2} & \text{when } l \text{ is even.} \end{cases} \tag{12.58}$$

Theorem 12.6. *The triangular Lie algebra* $\mathfrak{t}(l+1)$ *defined by equations* (12.24) *and* (12.25) *has exactly* $[(l+1)/2]$ *functionally independent invariants. A basis of invariants is given by*

$$I_\mu = Z_\mu, \quad \mu = 1, \dots, \left[\frac{l+1}{2}\right] \tag{12.59}$$

where Z_μ *is the determinant function given by* (12.57), *i.e., a polynomial of order* μ.

Proof. Let us first consider the case l even, i.e., $l = 2p$ for $p = 2, 3, \dots$

We begin by applying the set of $p(p+2)$ differential operators of (12.53), given by

$$\begin{matrix} \widehat{E}_{1(p+1)}, & \widehat{E}_{1(p+2)}, & \cdots & \widehat{E}_{1(l+1)}, \\ \widehat{E}_{2(p+1)}, & \widehat{E}_{2(p+2)}, & \cdots & \widehat{E}_{2(l+1)}, \\ \vdots & \vdots & & \vdots \\ \widehat{E}_{p(p+1)}, & \widehat{E}_{p(p+2)}, & \cdots & \widehat{E}_{p(l+1)}, \\ 0 & \widehat{E}_{(p+1)(p+2)}, & \cdots & \widehat{E}_{(p+1)(l+1)} \end{matrix} \tag{12.60}$$

on the functions (12.54). The action of all these operators eliminates the dependence on the $p(p+1)$ variables e_{ik} for

$$1 \leq i \leq p, \quad i+1 \leq k \leq p+1 \tag{12.61}$$

and

$$p+1 \leq i \leq l, \quad i+1 \leq k \leq l+1. \tag{12.62}$$

The p^2 remaining variables are

$$\begin{matrix} e_{1(p+2)}, & e_{1(p+3)}, & \cdots & e_{1(l+1)}, \\ e_{2(p+2)}, & e_{2(p+3)}, & \cdots & e_{2(l+1)}, \\ \vdots & \vdots & & \vdots \\ e_{p(p+2)}, & e_{p(p+3)}, & \cdots & e_{p(l+1)} \end{matrix} \tag{12.63}$$

and the $p(p-1)$ remaining differential operators $\widehat{E}_{ik}$ of (12.53) act on any function of these variables as

$$\widehat{E}_{ik} = \sum_{b=p+2}^{l+1} e_{ib} \frac{\partial}{\partial e_{kb}}, \quad 1 \leq i \leq p-1,\ i+1 \leq k \leq p, \tag{12.64}$$

$$\widehat{E}_{ik} = -\sum_{a=1}^{p} e_{ak} \frac{\partial}{\partial e_{ai}}, \quad p+2 \leq i \leq l,\ i+1 \leq k \leq l+1. \tag{12.65}$$

These differential operators are linearly independent. Therefore, the number of invariants for $\mathfrak{t}(2p+1)$ is p, i.e., the difference between the number of remaining variables and the number of remaining independent differential operators.

Now it suffices to verify that the remaining differential operators (12.64) and (12.65) annihilate determinants $Z_1, \ldots, Z_p$, i.e., $\widehat{E}_{ik}(Z_\alpha) = 0$ for $\alpha = 1, \ldots, p$.

Let us first consider the set of differential operators (12.64). A given differential operator $\widehat{E}_{ik}$ (i and k fixed) of (12.64) annihilates the determinant $Z_1, \ldots, Z_{k-1}$ trivially, since the variables e_{kb} ($p+2 \leq b \leq l+1$) do not figure in these determinants. Therefore, it is sufficient to check how $\widehat{E}_{ik}$ act on $Z_k, \ldots, Z_p$.

The determinant Z_β where $\beta \in \{k, k+1, \ldots, p\}$ can be expanded in terms of its kth row

$$Z_\beta = \sum_{b=2p+2-\beta}^{l+1} e_{kb}\, C_{kb}^{(\beta)} \tag{12.66}$$

where $C_{kb}^{(\beta)}$ is the cofactor of the $\beta \times \beta$ square matrix associated with the determinant Z_β. Hence, the differential operator $\widehat{E}_{ik}$ applied on these determinants gives

$$\widehat{E}_{ik}(Z_\beta) = \sum_{b=2p+2-\beta}^{l+1} e_{ib}\, C_{kb}^{(\beta)}. \tag{12.67}$$

The right-hand side of (12.67) vanishes, since it corresponds to the expansion of determinant in terms of the cofactors of a different row. This gives the determinant of a matrix with two identical rows, hence zero.

The procedure is very similar for the set of differential operators (12.65). An operator $\widehat{E}_{ik}$ of this set annihilates the determinants $Z_1, \ldots, Z_{l+1-i}$ since the operator acts only on variables not appearing in the determinants. Let us consider the action of $\widehat{E}_{ik}$ in (12.65) for the determinants Z_γ, where $\gamma \in \{l-i+2, l-i+3, \ldots, p\}$.

We can write the determinants Z_γ as

$$Z_\gamma = \sum_{a=1}^{\gamma} e_{ai}\, C_{ai}^{(\gamma)} \tag{12.68}$$

and the action of the differential operators $\widehat{E}_{ik}$ on these determinants is given by

$$\widehat{E}_{ik}(Z_\gamma) = -\sum_{a=1}^{\gamma} e_{ak}\, C_{ai}^{(\gamma)}. \tag{12.69}$$

Hence, we again obtain a determinant with two identical columns. More precisely, the action of differential operator $\widehat{E}_{ik}$ in (12.65) on determinants (12.68) is the following: the column e_{ai} in determinants Z_γ is replaced by the column $-e_{ak}$, for $1 \leq a \leq \gamma$. Therefore, by the property of determinants, this action annihilates Z_γ.

The proof for the case l odd is very similar to the even case and we omit it. $\square$

12.4.2. Invariants of the solvable triangular Lie algebras. In this section we shall consider the Casimir invariants of the Lie algebras $\mathfrak{s}(l+1,f)$ with f nonnilpotent elements added to the triangular nilradical $\mathfrak{t}(l+1)$.

Using Proposition 12.5 we can construct (as in Section 3.2) the differential operators realizing a basis for the coadjoint representation of the Lie algebras $\mathfrak{s}(l+1,f)$:

$$\widehat{E}_{ik} = \sum_{b=k+1}^{l+1} e_{ib}\frac{\partial}{\partial e_{kb}} - \sum_{a=1}^{i-1} e_{ak}\frac{\partial}{\partial e_{ai}} - \sum_{\alpha=1}^{f}(d_{ik}^{\alpha}e_{ik} + \Gamma_{ik}^{\alpha})\frac{\partial}{\partial f_{\alpha}} \tag{12.70}$$

$$\widehat{F}_{\alpha} = \sum_{i<k}(d_{ik}^{\alpha}e_{ik} + \Gamma_{ik}^{\alpha})\frac{\partial}{\partial e_{ik}} + \sum_{\beta=1}^{f}(\gamma_{\alpha\beta}e_{1(l+1)})\frac{\partial}{\partial f_{\beta}}. \tag{12.71}$$

We have introduced the notation

$$\begin{aligned}
d_{ik}^{\alpha} &\equiv d_{ik,ik}^{\alpha}, \\
\Gamma_{12}^{\alpha} &\equiv d_{2(l+1),12}^{\alpha}\, e_{2(l+1)}, \\
\Gamma_{j(j+1)}^{\alpha} &\equiv d_{1(l+1),j(j+1)}^{\alpha}\, e_{1(l+1)}, \qquad j = 2,3,\dots,l-1, \\
\Gamma_{l(l+1)}^{\alpha} &\equiv d_{1l,l(l+1)}^{\alpha}\, e_{1l}, \\
\Gamma_{lm}^{\alpha} &\equiv 0, \qquad m-l \geq 2.
\end{aligned} \tag{12.72}$$

In the generic case the differential operators (12.71) will not contain the second summation since $\gamma_{\alpha\beta} = 0$ unless $d_{1(l+1)}^{\gamma} = 0$ for $\gamma = 1,\dots,f$.

Equation (3.2) for the invariants in our case amounts to the system of equations

$$\widehat{E}_{ik}I(e_{1(l+1)}, e_{1l}, \dots, e_{l(l+1)}, f_1, \dots, f_f) = 0, \quad 1 \leq i < k \leq l+1, \tag{12.73}$$

$$\widehat{F}_{\alpha}I(e_{1(l+1)}, e_{1l}, \dots, e_{l(l+1)}, f_1, \dots, f_f) = 0, \quad \alpha = 1,\dots,f. \tag{12.74}$$

It is useful to construct linear combinations of these operators with variable coefficients so as to obtain a subset of equations involving only derivatives with respect to $f_1,\dots,f_f$. These linear combinations do not represent elements of the Lie algebra $\mathfrak{s}(l+1,f)$ but this is permitted when we treat equations (12.73) and (12.74) as a system of linear homogeneous partial differential equations.

Let us associate a differential operator $\widehat{Z}_{\mu}$ with each invariant Z_{μ} of the nilpotent Lie algebra $\mathfrak{t}(l+1)$ (see (12.57)). For each Z_{μ} we take a sum of μ determinants of the form (12.57) and in each of them we replace one column of scalars by a column of operators $\widehat{E}_{ik}$. For example, we have

$$\widehat{Z}_1 = \widehat{E}_{1(l+1)}, \quad \widehat{Z}_2 = \begin{vmatrix} \widehat{E}_{1(l)} & e_{1(l+1)} \\ \widehat{E}_{2(l)} & e_{2(l+1)} \end{vmatrix} + \begin{vmatrix} e_{1(l)} & \widehat{E}_{1(l+1)} \\ e_{2(l)} & \widehat{E}_{2(l+1)} \end{vmatrix} \tag{12.75}$$

and in general, we have the formula

$$\widehat{Z}_{\mu} = \sum_{j=1}^{\mu} \begin{vmatrix} e_{1(l+1-\mu+1)} & e_{1(l+1-\mu+2)} & \cdots & \widehat{E}_{1(l+1-\mu+j)} & \cdots & e_{1(l+1)} \\ e_{2(l+1-\mu+1)} & e_{2(l+1-\mu+2)} & \cdots & \widehat{E}_{2(l+1-\mu+j)} & \cdots & e_{2(l+1)} \\ \vdots & \vdots & & \vdots & & \vdots \\ e_{\mu(l+1-\mu+1)} & e_{\mu(l+1-\mu+2)} & \cdots & \widehat{E}_{\mu(l+1-\mu+j)} & \cdots & e_{\mu\,(l+1)} \end{vmatrix}, \quad 1 \leq \mu \leq \left[\frac{l+1}{2}\right]. \tag{12.76}$$

It is a straightforward calculation to prove that we have

$$\widehat{Z}_\mu = \sum_{\alpha=1}^{f} \phi_\alpha(e_{ik}) \frac{\partial}{\partial f_\alpha} \tag{12.77}$$

for some functions $\phi_\alpha(e_{1(l+1)}, e_{1l}, \ldots, e_{l(l+1)})$, i.e., that all derivatives with respect to e_{ik} drop out. When the structure matrices $\widehat{D}_\alpha$ are diagonal we obtain the formula

$$\widehat{Z}_j = -Z_j \sum_{\alpha=1}^{f} \sum_{\mu=1}^{j} \left(d^{\alpha}_{\mu(l-\mu+2)} \frac{\partial}{\partial f_\alpha} \right). \tag{12.78}$$

Remark. For nondiagonal matrices $\widehat{D}_\alpha$, this formula is generic for even l. However, for odd l, off-diagonal terms will appear.

We can construct $[(l+1)/2]$ such operators; at most f of them are linearly independent.

12.4.3. Casimir invariants of solvable extensions of $\mathfrak{t}(4)$. To illustrate the procedure we first consider the lowest dimensional nontrivial case, namely solvable extensions of $\mathfrak{t}(4)$, i.e., the Borel nilradical of $\mathfrak{sl}(4, \mathbb{F})$. Hence the solvable extensions have $f = 1, 2$ or 3 nonnilpotent elements. For each algebra $\mathfrak{s}(4, f)$ we will state results concerning the form and the number of invariants.

For each lemma, the strategy that we will adopt to prove it is the following. We separate the proof in two parts:

(A) We find the invariants depending only on the variables e_{ab}, $1 \le a < b \le 4$.
(B) We find the invariants which depend on both variables e_{ik} and f_α, $\alpha = 1, \ldots, f$.

In each of these cases, we will apply the differential operators $\widehat{E}_{ik}$ and $\widehat{F}_\alpha$ of (12.70) and (12.71) on the functions $I = I\big((e_{ab}, f_\alpha)_{a<b, 1\le\alpha\le f}\big)$. However in the case A, since we assume that the functions I only depend on the variables e_{ab}, the differential operators $\widehat{E}_{ik}$ will be the same as the operators of the nilpotent Lie algebra $\mathfrak{t}(4)$ (the f_α derivatives do not act on I). Therefore, by using the results of Theorem 12.6, we will only have to apply the differential operators $\widehat{F}_\alpha$ on functions of the type

$$I = I(Z_1, Z_2) \tag{12.79}$$

where $Z_1 = e_{14}$ and $Z_2 = e_{13}\, e_{24} - e_{23}\, e_{14}$.

In the case B, we will begin by imposing

$$\widehat{Z}_j I(e_{ab}, f_\alpha) = 0, \quad j = 1, 2 \tag{12.80}$$

such that the dependence on the f_α variables is preserved in I. Then we will apply all the differential operators (12.70) and (12.71) of the coadjoint representation of $\mathfrak{s}(4, f)$.

The case $f = 1$. The matrix $\widehat{D}_1$ of the Lie algebras $\mathfrak{s}(4,1)$ has the form (12.40)

$$\widehat{D}_1 = \begin{pmatrix} d_1 + d_2 + d_3 & 0 & 0 & 0 & d_5 & 0 \\ 0 & d_1 + d_2 & 0 & 0 & 0 & d_6 \\ 0 & 0 & d_2 + d_3 & d_4 & 0 & 0 \\ 0 & 0 & 0 & d_1 & 0 & 0 \\ 0 & 0 & 0 & 0 & d_2 & 0 \\ 0 & 0 & 0 & 0 & 0 & d_3 \end{pmatrix}, \tag{12.81}$$

where at most 2 of the 3 off-diagonal elements d_i, $i = 4, 5, 6$ do not vanish.

Lemma 12.7. *A solvable triangular Lie algebra of the type* $\mathfrak{s}(4,1)$ *has either* 3 *invariants, or* 1 *invariant.*

(1) *Three invariants exist if and only if the conditions*

$$d_1 + d_3 = d_2 = d_5 = 0 \tag{12.82}$$

are satisfied. In this case the algebra can be characterized by $d_1 = -d_3 = 1$, $d_2 = 0$, $d_6 = d_5 = d_4 = 0$ *in the matrix* (12.81). *A basis for the invariants is:*

$$I_1 = Z_1, \tag{12.83}$$

$$I_2 = Z_2, \tag{12.84}$$

$$I_3 = (e_{12}\, e_{24} + e_{13}\, e_{34}) + e_{14}\, f_1. \tag{12.85}$$

Otherwise there exists precisely one invariant. Two types of Lie algebras occur.

(2) $(d_1 + d_3, d_2) \neq (0, 0)$ *and* $d_5 = 0$ *in matrix* (12.81). *The invariant is*

$$I = \frac{(Z_2)^{d_1+d_2+d_3}}{(Z_1)^{d_1+2d_2+d_3}}. \tag{12.86}$$

(3) $d_1 + d_3 = 0$, $d_5 = 1$, d_2 *is a free parameter in matrix* (12.81) *and the invariant is*

$$I = d_2 \frac{Z_2}{(Z_1)^2} + \ln Z_1. \tag{12.87}$$

Proof. (A) We impose that the differential operator $\widehat{F}_1$ of (12.71) should annihilate the functions of type $I = I(Z_1, Z_2)$, i.e.,

$$\begin{aligned}
\widehat{F}_1(I) = \Bigg[&(d_1\, e_{12} + d_6\, e_{24}) \frac{\partial}{\partial e_{12}} + (d_2\, e_{23} + d_5\, e_{14}) \frac{\partial}{\partial e_{23}} \\
&+ (d_3\, e_{34} + d_4\, e_{13}) \frac{\partial}{\partial e_{34}} + (d_2 + d_3)\, e_{13} \frac{\partial}{\partial e_{13}} \\
&+ (d_1 + d_2)\, e_{24} \frac{\partial}{\partial e_{24}} + (d_1 + d_2 + d_3)\, e_{14} \frac{\partial}{\partial e_{14}} \Bigg] I \\
= (d_1 + d_2 + d_3)&\, Z_1 \frac{\partial I}{\partial Z_1} + \left[(d_1 + 2d_2 + d_3)\, Z_2 - d_5\, (Z_1)^2\right] \frac{\partial I}{\partial Z_2} = 0.
\end{aligned} \tag{12.88}$$

We first note that if we have $d_1 + d_3 = d_2 = d_5 = 0$, i.e., conditions (12.82) then both Z_1 and Z_2 are invariants. Also, the matrix $\widehat{D}_1$ can be diagonalized with no loss of generality and set equal to

$$\widehat{D}_1 = \operatorname{diag}(0, 1, -1, 1, 0, -1). \tag{12.89}$$

In all other cases (12.88) implies that just one invariant of this type exists. We obtain it using the method of characteristics.

Two cases arise:

(i) $d_5 = 0$. The invariant is then given by (12.86), with $(d_1 + d_3, d_2) \neq (0, 0)$.

(ii) $d_5 \neq 0$. From Section 12.3.2 we know that in this case we can normalize d_5 to 1 and we necessarily have $d_1 + d_3 = 0$. Hence, we obtain the invariant (12.87), where d_2 is a free parameter.

(B) In this case we impose $\widehat{Z}_j(I) = 0$ $(j = 1, 2)$ for functions of the type $I = I(e_{14}, e_{13}, e_{24}, e_{12}, e_{23}, e_{34}, f_1)$ and the differential operators $\widehat{Z}_j$ are given by

$$\widehat{Z}_1 \equiv \widehat{E}_{14} = -(d_1 + d_2 + d_3)\, Z_1 \frac{\partial}{\partial f_1}, \tag{12.90}$$

$$\begin{aligned} \widehat{Z}_2 &\equiv e_{13}\, \widehat{E}_{24} - e_{23}\, \widehat{E}_{14} + e_{24}\, \widehat{E}_{13} - e_{14}\, \widehat{E}_{23} \\ &= \left[-(d_1 + 2d_2 + d_3)\, Z_2 + d_5\, (Z_1)^2\right] \frac{\partial}{\partial f_1}. \end{aligned} \tag{12.91}$$

Hence the required dependence on f_1 will survive only if we have $d_1 + d_3 = d_2 = d_5 = 0$. This coincides with (12.82), the condition for Z_1 and Z_2 to be invariant. Furthermore, we can normalize d_1 to 1 and cancel d_6 and d_4 by transformation (12.47).

We now apply all the differential operators of the coadjoint representation of $\mathfrak{s}(4,1)$ and the final result is that we obtain two invariants (12.83) and (12.84) independent of f_1 and one invariant (12.85) depending on f_1. □

The case $f = 2$. The Lie algebras $\mathfrak{s}(4,2)$ are given by the following matrices of derivations (12.40):

$$\begin{aligned} \widehat{D}_1 &= \begin{pmatrix} d_1 + d_2 + d_3 & 0 & 0 & 0 & 0 & 0 \\ 0 & d_1 + d_2 & 0 & 0 & 0 & 0 \\ 0 & 0 & d_2 + d_3 & 0 & 0 & 0 \\ 0 & 0 & 0 & d_1 & 0 & 0 \\ 0 & 0 & 0 & 0 & d_2 & 0 \\ 0 & 0 & 0 & 0 & 0 & d_3 \end{pmatrix}, \\ \widehat{D}_2 &= \begin{pmatrix} \tilde{d}_1 + \tilde{d}_2 + \tilde{d}_3 & 0 & 0 & 0 & \tilde{d}_5 & 0 \\ 0 & \tilde{d}_1 + \tilde{d}_2 & 0 & 0 & 0 & \tilde{d}_6 \\ 0 & 0 & \tilde{d}_2 + \tilde{d}_3 & \tilde{d}_4 & 0 & 0 \\ 0 & 0 & 0 & \tilde{d}_1 & 0 & 0 \\ 0 & 0 & 0 & 0 & \tilde{d}_2 & 0 \\ 0 & 0 & 0 & 0 & 0 & \tilde{d}_3 \end{pmatrix} \end{aligned} \tag{12.92}$$

where we have at most one off-diagonal element in $\widehat{D}_2$. Furthermore, the coefficient γ_{12} in (12.31) is in the generic case zero (i.e., the two nonnilpotent elements f_1, f_2 commute). However, for the particular case $d_1 + d_2 + d_3 = 0 = \tilde{d}_1 + \tilde{d}_2 + \tilde{d}_3$, we can have $\gamma_{12} \neq 0$ in (12.31).

Lemma 12.8. *A solvable triangular Lie algebra of the type* $\mathfrak{s}(4,2)$ *has either 2 invariants or none. Two invariants exist if the conditions*

$$\tilde{d}_2\,(d_1 + d_3) - d_2\,(\tilde{d}_1 + \tilde{d}_3) = 0, \tag{12.93}$$

$$(d_1 + d_2 + d_3)\, \tilde{d}_5 = 0 \tag{12.94}$$

are satisfied simultaneously. They lead to the following algebras and invariants.

(1) $d_1 = -d_3 = \tilde{d}_2 = \tilde{d}_5 = 1$ *and* $d_2 = \tilde{d}_1 = \tilde{d}_3 = \tilde{d}_6 = \tilde{d}_4 = \gamma_{12} = 0$ *in matrices* (12.92) *and a basis for the invariants is*:

$$\begin{aligned} I_1 &= \frac{Z_2}{(Z_1)^2} + \ln Z_1, \\ I_2 &= \frac{e_{12}\, e_{24} + e_{13}\, e_{34}}{e_{14}} + f_1. \end{aligned} \tag{12.95}$$

(2) $d_1 = -d_3 = \tilde{d}_2 = 1$, $d_2 = \tilde{d}_1 = \tilde{d}_6 = \tilde{d}_5 = \tilde{d}_4 = \gamma_{12} = 0$ *and* $\tilde{d}_3$ *a free parameter in matrices* (12.92). *A basis for the invariants is:*

$$I_1 = \frac{Z_2^{1+\tilde{d}_3}}{Z_1^{2+\tilde{d}_3}}, \qquad I_2 = \frac{e_{12}\, e_{24} + e_{13}\, e_{34}}{e_{14}} + f_1. \tag{12.96}$$

(3) $d_1 = \tilde{d}_3 = 1$ *and* $d_2 = d_3 = \tilde{d}_1 = \tilde{d}_2 = \tilde{d}_6 = \tilde{d}_5 = \tilde{d}_4 = \gamma_{12} = 0$ *in matrices* (12.92). *A basis for the invariants is:*

$$I_1 = \frac{Z_2}{Z_1}, \qquad I_2 = \frac{e_{12}e_{24} + e_{13}e_{34}}{e_{14}} + (f_1 - f_2). \tag{12.97}$$

(4) $d_1 = -d_3 = \tilde{d}_2 = -\tilde{d}_3 = 1$ *and* $d_2 = \tilde{d}_1 = \tilde{d}_6 = \tilde{d}_5 = \tilde{d}_4 = 0$ *in matrices* (12.92) *and the invariants are:*

$$I_1 = Z_1,$$
$$I_2 = e_{12}e_{24} + e_{13}e_{34} + Z_1 f_1 + \gamma_{12}(Z_1)^2 \ln Z_2.$$

Otherwise, there is no invariant.

PROOF. (A) We first apply differential operators $\widehat{F}_1$ and $\widehat{F}_2$ on functions of type $I = I(Z_1, Z_2)$. We obtain a system of two linear partial differential equations given by

$$\begin{pmatrix} \widehat{F}_1(I) \\ \widehat{F}_2(I) \end{pmatrix} = \begin{pmatrix} (d_1 + d_2 + d_3)\, Z_1 & (d_1 + 2d_2 + d_3)\, Z_2 \\ (\tilde{d}_1 + \tilde{d}_2 + \tilde{d}_3)\, Z_1 & (\tilde{d}_1 + 2\tilde{d}_2 + \tilde{d}_3)\, Z_2 - \tilde{d}_5\, (Z_1)^2 \end{pmatrix} \begin{pmatrix} \dfrac{\partial I}{\partial Z_1} \\ \dfrac{\partial I}{\partial Z_2} \end{pmatrix} = 0. \tag{12.98}$$

The rank of the 2×2 matrix in (12.98) cannot be zero, since then matrices $\widehat{D}_1$ and $\widehat{D}_2$ would not be linearly nilindependent. Also, if the rank is 2 there is no invariant that depends only on Z_1 and Z_2. However, solutions exist if the rank of the matrix is 1 for all values of Z_1 and Z_2. This gives conditions (12.93) and (12.94).

Let us now assume that the condition (12.93) is satisfied. We consider the diagonal and the nondiagonal cases separately.

(i) $\tilde{d}_5 = 0$. In this case, we obtain the invariant (12.86) for $(d_1 + d_3, d_2) \neq (0, 0)$.

(ii) $\tilde{d}_5 \neq 0$, $(d_1 + d_2 + d_3) = 0$. Since $\tilde{d}_5$ is nonzero in $\widehat{D}_2$, we necessarily have $\tilde{d}_1 + \tilde{d}_3 = 0$, i.e., $\tilde{d}_1 + \tilde{d}_3 = 0$ which gives the condition $d_2\, \tilde{d}_2 = 0$ by (12.93). Two cases are possible under these condition.
One case gives the invariant (12.95) for $d_1 = -d_3 = \tilde{d}_2 = \tilde{d}_5 = 1$ and $d_2 = \tilde{d}_1 = \tilde{d}_3 = \tilde{d}_6 = \tilde{d}_4 = 0$.
In the other case, we simply obtain the invariant $I = Z_1$ for the Lie algebra characterized by $d_2 = \tilde{d}_1 = -\tilde{d}_3 = \tilde{d}_5 = 1$, $\tilde{d}_2 = \tilde{d}_6 = \tilde{d}_4 = 0$ and $d_3 = -(d_1 + 1)$ (with d_1 a free parameter).

Remark. The subcase $d_2 = 0 = \tilde{d}_2$ in (ii) gives two nildependent matrices $\widehat{D}_1, \widehat{D}_2$ and is therefore not considered.

(B) In this case, we begin by applying the differential operators $\widehat{Z}_1, \widehat{Z}_2$ on functions of type $I = I(e_{14}, e_{13}, e_{24}, e_{12}, e_{23}, e_{34}, f_1, f_2)$, i.e.,

$$(12.99)\quad \begin{pmatrix} \widehat{Z}_1(I) \\ \widehat{Z}_2(I) \end{pmatrix} = \begin{pmatrix} -(d_1+d_2+d_3)\,Z_1 & -(\tilde{d}_1+\tilde{d}_2+\tilde{d}_3)\,Z_1 \\ (d_1+2d_2+d_3)\,Z_2 & (\tilde{d}_1+2\tilde{d}_2+\tilde{d}_3)\,Z_2 - \tilde{d}_5\,(Z_1)^2 \end{pmatrix} \begin{pmatrix} \dfrac{\partial I}{\partial f_1} \\ \dfrac{\partial I}{\partial f_2} \end{pmatrix} = 0.$$

The dependence on f_1 and f_2 can exist only if the determinant of the 2×2 matrix in (12.99) is zero. This again imposes the conditions (12.93) and (12.94).

Let us assume that the condition (12.93) is satisfied. We separate the problem into three distinct cases.

(i) $\big((d_1+d_2+d_3), (\tilde{d}_1+\tilde{d}_2+\tilde{d}_3)\big) \neq (0,0)$, $\tilde{d}_5 \neq 0$. The condition $\tilde{d}_5 \neq 0$ implies two consequences. First we have from (12.94) that $(d_1+d_2+d_3) = 0$ and therefore $(\tilde{d}_1+\tilde{d}_2+\tilde{d}_3) \neq 0$. Second, we necessarily have $\tilde{d}_1+\tilde{d}_3 = 0$ which implies from (12.94) that $\tilde{d}_2\, d_2 = 0$.
In this case, the invariants are (12.95) and the Lie algebra $\mathfrak{s}(4,2)$ satisfies $d_1 = -d_3 = \tilde{d}_2 = \tilde{d}_5 = 1$, $d_2 = \tilde{d}_1 = \tilde{d}_3 = \tilde{d}_6 = \tilde{d}_4 = \gamma_{12} = 0$.

(ii) $\big((d_1+d_2+d_3), (\tilde{d}_1+\tilde{d}_2+\tilde{d}_3)\big) \neq (0,0)$, $\tilde{d}_5 = 0$. In this case, two triangular solvable Lie algebras are associated with the invariants (12.96) and (12.97). One Lie algebra is characterized by the parameters $d_1 = -d_3 = \tilde{d}_2 = 1$, $d_2 = \tilde{d}_1 = \tilde{d}_6 = \tilde{d}_5 = \tilde{d}_4 = \gamma_{12} = 0$ and $\tilde{d}_3$ a free parameter. The other Lie algebra is characterized by $d_1 = \tilde{d}_3 = 1$ and $d_2 = d_3 = \tilde{d}_1 = \tilde{d}_2 = \tilde{d}_6 = \tilde{d}_5 = \tilde{d}_4 = \gamma_{12} = 0$.

(iii) $\big((d_1+d_2+d_3), (\tilde{d}_1+\tilde{d}_2+\tilde{d}_3)\big) = (0,0)$. In this case, we see that conditions (12.93) and (12.94) are automatically respected. Also, we can have a nonzero γ_{12} in (12.71).
Since $(d_1+d_2+d_3) = 0 = (\tilde{d}_1+\tilde{d}_2+\tilde{d}_3)$, we can substitute d_3 by $-(d_1+d_2)$ and $\tilde{d}_3$ by $-(\tilde{d}_1+\tilde{d}_2)$ in the matrices (12.92). However, by imposing the commutativity (12.35) and the nilindependence of the matrices $\widehat{D}_1$ and $\widehat{D}_2$, we obtain $d_1 = -d_3 = \tilde{d}_2 = -\tilde{d}_3 = 1$ and $d_2 = \tilde{d}_1 = \tilde{d}_6 = \tilde{d}_5 = \tilde{d}_4 = 0$. Hence, we obtain the two invariants (12.98). □

The Lie algebra $\mathfrak{s}(4,3)$. The matrices of derivations $\widehat{D}_\alpha$ are diagonal and given by

$$(12.100)\quad \begin{gathered} \widehat{D}_1 = \operatorname{diag}(1,1,0,1,0,0), \quad \widehat{D}_2 = \operatorname{diag}(1,1,1,0,1,0), \\ \widehat{D}_3 = \operatorname{diag}(1,0,1,0,0,1). \end{gathered}$$

Furthermore, the nonnilpotent elements commute, i.e., $\gamma_{\alpha\,\beta} = 0$, $\alpha, \beta = 1,2,3$ (see equations (12.31) and (12.36)).

Lemma 12.9. *The triangular solvable Lie algebra* $\mathfrak{s}(4,3)$ *has precisely* 1 *invariant given by*

$$(12.101)\quad I = \frac{e_{12}\,e_{24} + e_{13}\,e_{34}}{e_{14}} + (f_1 - f_3).$$

PROOF. (A) In this case, it is easy to demonstrate that after we have applied the differential operator $\widehat{F}_1$ on functions of type $I = I(Z_1, Z_2)$, we obtain the quotient of Z_2 over Z_1. However, when we apply operator $\widehat{F}_2$ on functions $\tilde{I}(I) = \tilde{I}(Z_2/Z_1)$, we obtain

$$\begin{aligned} 0 = \widehat{F}_2\tilde{I} &= \left(e_{23}\frac{\partial}{\partial e_{23}} + e_{13}\frac{\partial}{\partial e_{13}} + e_{24}\frac{\partial}{\partial e_{24}} + e_{14}\frac{\partial}{\partial e_{14}}\right)\tilde{I} \\ &= I\frac{\partial \tilde{I}}{\partial I}. \end{aligned} \tag{12.102}$$

Therefore, there is no invariant in this class.

(B) We first impose that the differential operators $\widehat{Z}_1$ and $\widehat{Z}_2$ annihilate the functions of type $I = I(e_{14}, e_{13}, e_{24}, e_{12}, e_{23}, e_{34}, f_1, f_2, f_3)$, where

$$\widehat{Z}_1 = -Z_1\left(\frac{\partial}{\partial f_1} + \frac{\partial}{\partial f_2} + \frac{\partial}{\partial f_3}\right), \tag{12.103}$$

$$\widehat{Z}_2 = -Z_2\left(\frac{\partial}{\partial f_1} + 2\frac{\partial}{\partial f_2} + \frac{\partial}{\partial f_3}\right). \tag{12.104}$$

Since the Lie algebra $\mathfrak{s}(4,3)$ has no parameters, these conditions are not on the parameters of the algebra (as before) but on the f_α dependence of the invariant. Hence, the new functions on which we will apply all the differential operators of the coadjoint representation of $\mathfrak{s}(4,3)$ are of the type

$$\tilde{I} = \tilde{I}(e_{12}, e_{23}, e_{34}, e_{13}, e_{24}, e_{14}, f_1 - f_3).$$

Next, we obtain the invariant (12.101) by imposing that the operators of the coadjoint representation of $\mathfrak{s}(4,3)$ annihilate $\tilde{I}$. □

12.4.4. General results.

Proposition 12.10. *The triangular solvable Lie algebra $\mathfrak{s}(l+1,l)$ has precisely $[l/2]$ functionally independent invariants. A basis is given by*

$$I_\mu = \frac{(-1)^{\mu+1}}{Z_\mu}\left(\sum_{\rho=1}^{l+1-2\mu} W_\rho^{(\mu)}\right) + (f_\mu - f_{l+1-\mu}) \tag{12.105}$$

for $\mu = 1, \ldots, [l/2]$. The function Z_μ is the determinant given by (12.57) *and $W_\rho^{(\mu)}$ is also a determinant function given by the determinant of the $(\mu+1)\times(\mu+1)$ matrix:*

$$W_\rho^{(\mu)} = \begin{vmatrix} e_{1(\rho+\mu)} & e_{1(l+1-\mu+1)} & e_{1(l+1-\mu+2)} & \cdots & e_{1(l+1)} \\ e_{2(\rho+\mu)} & e_{2(l+1-\mu+1)} & e_{2(l+1-\mu+2)} & \cdots & e_{2(l+1)} \\ \vdots & \vdots & \vdots & & \vdots \\ e_{\mu(\rho+\mu)} & e_{\mu(l+1-\mu+1)} & e_{\mu(l+1-\mu+2)} & \cdots & e_{\mu(l+1)} \\ 0 & e_{(\rho+\mu)(l+1-\mu+1)} & e_{(\rho+\mu)(l+1-\mu+2)} & \cdots & e_{(\rho+\mu)(l+1)} \end{vmatrix}. \tag{12.106}$$

Proposition 12.11. *A diagonal solvable Lie algebra of the type $\mathfrak{s}(l+1,1)$ has either $[(l+3)/2]$ or $[(l-1)/2]$ functionally independent invariants.*

(1) *$[(l+3)/2]$ invariants exist if and only if the conditions*

$$d_{i(i+1)} + d_{(l+1-i)(l+1-i+1)} = 0, \quad i = 1, \ldots, \left[\frac{l+1}{2}\right] \tag{12.107}$$

are satisfied. A basis is given by $[(l+1)/2]$ *invariants independent of* f_1 *and one invariant depending on* f_1*:*

$$I_\mu = Z_\mu, \quad \mu = 1, \dots, \left[\frac{l+1}{2}\right], \tag{12.108}$$

$$I_{[(l+1)/2]+1} = \sum_{\mu=1}^{[l/2]} \sum_{\rho=1}^{l+1-2\mu} \frac{(-1)^{\mu+1}}{Z_\mu} d_{\mu(\mu+1)} W_\rho^{(\mu)} + f_1 \tag{12.109}$$

where the function Z_μ *and* $W_\rho^{(\mu)}$ *are determinant functions given by the equations*(12.57) *and* (12.106)*, respectively.*

(2) *Otherwise there exist precisely* $[(l-1)/2]$ *invariants, all independent of* f_1*. A basis is given by*

$$I_\mu = \frac{(Z_{\mu+1})^\alpha}{(Z_1)^\beta}, \quad \mu = 1, \dots, \left[\frac{l-1}{2}\right] \tag{12.110}$$

with

$$\frac{\alpha}{\beta} = \frac{d_{1(l+1)}}{\sum_{k=1}^{\mu+1} d_{k(l+1+1-k)}} \tag{12.111}$$

where the function Z_μ *is the determinant function given by* (12.57)*.*

By *diagonal* solvable Lie algebra of the type $\mathfrak{s}(l+1,1)$ in Proposition 12.11, we mean that the matrix $\widehat{D}_1$ of (12.30) is diagonal.

Propositions 12.10 and 12.11 contain two types of information on the invariants: They give the form of the invariant functions and the number of functionally independent invariants. It is an easy calculation to prove that the functions I_μ of Proposition 12.10 and Proposition 12.11 are annihilated by the coadjoint representation (12.70), (12.71) of the Lie algebras $\mathfrak{s}(l+1,l)$ and $\mathfrak{s}(l+1,1)$, respectively. However, it is much more difficult to establish the number of functionally independent invariants in Proposition 12.10 and Proposition 12.11. The difficulty is to prove that no further invariants exists. One way of doing that is to calculate the rank of the antisymmetric matrix $C = C(\mathfrak{s}(l+1,l))$ and $C = C(\mathfrak{s}(l+1,1))$ of the commutation relations for the corresponding Lie algebra. The number of invariants is then given by the difference between the dimension of the solvable Lie algebra and the rank of the matrix C (see (3.4)).

For the Lie algebra $\mathfrak{s}(l+1,l)$ of dimension $\frac{1}{2}l(l+3)$, C is the antisymmetric matrix given by the elements

$$C = ([e_{ik}, e_{ab}], [e_{ik}, f_\alpha]), \quad 1 \le i < k \le l+1, \ 1 \le a < b \le l+1, \ \alpha = 1, \dots, l \tag{12.112}$$

and for the Lie algebra $\mathfrak{s}(l+1,1)$ of dimension $\frac{1}{2}(l^2+l+2)$, the matrix C is given by the elements

$$C = ([e_{ik}, e_{ab}], [e_{ik}, f_1]), \quad 1 \le i < k \le l+1, \ 1 \le a < b \le l+1. \tag{12.113}$$

For example, the antisymmetric matrix C of the 7-dimensional Lie "diagonal" algebra $\mathfrak{s}(4,1)$ is given by

$$C = \begin{pmatrix} 0 & 0 & 0 & 0 & 0 & 0 & -d_{123}e_{14} \\ 0 & 0 & 0 & 0 & 0 & e_{14} & -d_{12}e_{13} \\ 0 & 0 & 0 & -e_{14} & 0 & 0 & -d_{23}e_{24} \\ 0 & 0 & e_{14} & 0 & e_{13} & 0 & -d_1e_{12} \\ 0 & 0 & 0 & -e_{13} & 0 & e_{24} & -d_2e_{23} \\ 0 & -e_{14} & 0 & 0 & -e_{24} & 0 & -d_3e_{34} \\ d_{123}e_{14} & d_{12}e_{13} & d_{23}e_{24} & d_1e_{12} & d_2e_{23} & d_3e_{34} & 0 \end{pmatrix}. \tag{12.114}$$

where we define $d_{23} = d_2 + d_3$, $d_{12} = d_1 + d_2$ and $d_{123} = d_1 + d_2 + d_3$. Hence, it is easy to calculate that

$$\operatorname{rank}(C) = \begin{cases} 4 & \text{for } d_1 + d_3 = d_2 = 0, \\ 6 & \text{otherwise} \end{cases} \tag{12.115}$$

giving, respectively, three and one invariants (in accordance with Proposition 12.11 and Lemma 12.7).

12.4.5. Conclusions. The problem of finding all invariants of the coadjoint representation of the triangular nilpotent algebras $\mathfrak{t}(l+1)$ is solved completely by Theorem 12.6. A basis for the invariants consists of polynomials and provides Casimir operators in the enveloping algebra of $\mathfrak{t}(l+1)$.

The situation with the solvable triangular Lie algebras $\mathfrak{s}(l+1,f)$ is more complicated. We have provided guidelines for calculating the invariants for all values of l, but presented comprehensive results only for $l = 3$. The results for $l = 3$ show that all invariants are polynomial only in special cases. In general, rational, irrational and logarithmic type invariants must be allowed in any basis of invariants.

We have also presented results for the generalized Casimir invariants of $\mathfrak{s}(l+1,l)$ and $\mathfrak{s}(l+1,1)$ for arbitrary values of l. Propositions 12.10 and 12.11 on the Casimir invariants for general l and $f = l$ or $f = 1$ were presented as conjectures in [124]. Later these statements were proved to be correct by V. Boyko et al. [10].

The generalized Casimir invariants of $\mathfrak{s}(l+1,f)$ algebras with $1 \leq f \leq l$ were calculated in closed form by V. Boyko et al. in [11] for all values of f whenever $\mathfrak{s}(l+1,f)$ is a subalgebra of the Borel subalgebra of $\mathfrak{sl}(l+1,\mathbb{F})$.

CHAPTER 13

Solvable Lie Algebras with Filiform and Quasifiliform Nilradicals

In this chapter we shall consider nilradicals with maximal or almost maximal degree of nilpotency.

For any nilpotent Lie algebra we have $\dim \mathfrak{n}^2 \leq \dim \mathfrak{n} - 2 = n-2$. The reason is that if the contrary held we would necessarily have $[\mathfrak{n}, \mathfrak{n}^2] = \mathfrak{n}^2$, i.e., $\mathfrak{n}^k = \mathfrak{n}^2$ for all $k \geq 2$ and this would violate the nilpotency. Consequently, the degree of nilpotency of any given nilpotent Lie algebra can be at most $K = n-1$. A nilpotent Lie algebra of maximal degree of nilpotency $K = n-1$ is called *filiform* [131]. It immediately follows that for any filiform algebra we have $\dim(\mathfrak{n}/\mathfrak{n}^2) = 2$ and $\dim(\mathfrak{n}^k/\mathfrak{n}^{k+1}) = 1$ for $k = 2, \ldots, n-1$. Properties of such nilpotent algebras were already investigated in great detail, see, e.g., [45, 52, 53, 131].

To make the structure of a given filiform Lie algebra more transparent one should choose a suitable basis. Following M. Vergne [131] we shall use an *adapted basis* $\mathcal{E} = (e_1, \ldots, e_n)$ which exists for any filiform Lie algebra. It satisfies

$$\begin{aligned} \mathfrak{n}^k &= \operatorname{span}\{e_1, \ldots, e_{n-k}\}, \quad && k \geq 2, \\ [e_k, e_n] &= e_{k-1}, && k = 2, \ldots, n-1, \\ [e_1, e_j] &= 0, && j = 2, \ldots, n, \\ [e_j, e_{n-j+1}] &= (-1)^j \alpha e_1, && j = 2, \ldots, n-1, \\ [e_j, e_k] &= 0 \bmod \mathfrak{n}^{2n-j-k+1}, && 3 \leq j < k \leq n-1,\ n-1 < j+k \end{aligned} \tag{13.1}$$

with $\alpha = 0$ for n odd and $\alpha = 0$ or $\alpha = 1$ for n even.

To any nilpotent Lie algebra we can associate the *graded* Lie algebra $\mathfrak{gr}(\mathfrak{n})$ of $\mathfrak{n}$

$$\mathfrak{gr}(\mathfrak{n}) = \sum_{k=1}^{K} \mathfrak{n}^k/\mathfrak{n}^{k+1} \simeq \sum_{k=1}^{K} \mathfrak{m}_k$$

(cf. (9.19)) with the new Lie bracket

$$[x, y]_{\mathfrak{gr}} = [x, y] \bmod \mathfrak{n}^{k+j+1}, \quad \forall x \in \mathfrak{m}_k \simeq \mathfrak{n}^k/\mathfrak{n}^{k+1}, /y \in \mathfrak{m}_j \simeq \mathfrak{n}^j/\mathfrak{n}^{j+1}.$$

As was shown in [131], n-dimensional filiform algebras can be divided into two classes depending on the structure of their graded algebras. One class has

$$\mathfrak{gr}(\mathfrak{n}) = \mathfrak{n}_{n,1}$$

where

$$\mathfrak{n}_{n,1} = \operatorname{span}\{e_1, \ldots, e_n\}, \qquad [e_k, e_n] = e_{k-1}, \quad k = 2, \ldots, n-1. \tag{13.2}$$

$\mathfrak{n}_{n,1}$ is called the *model filiform algebra*.

The second class is present only when n is even and has

$$\mathfrak{gr}(\mathfrak{n}) = \mathfrak{Q}_n$$

where

$$(13.3)\qquad \begin{gathered}\mathfrak{Q}_n = \operatorname{span}\{e_1, \ldots, e_n\},\\ [e_k, e_n] = e_{k-1}, \quad [e_k, e_{n-k+1}] = (-1)^k e_1, \qquad k = 2, \ldots, n-1.\end{gathered}$$

$\mathfrak{Q}_n$ is called the *special filiform algebra*.

In (13.2) and (13.3) all commutators not listed explicitly vanish. A notation resembling the one in [3, 116] was used here, with a minor modification: the index in $\mathfrak{Q}_n$ is equal to the dimension of $\mathfrak{Q}_n$, in [3] it was half of it.

These two classes correspond to the two different values of the parameter α in the adapted basis (13.1).

In Sections 13.1 and 13.2 we will present in detail all solvable extensions of the model filiform algebra $\mathfrak{n}_{n,1}$ of (13.2) and of the algebra $\mathfrak{n}_{n,2}$ with the Lie brackets defined in (13.21) below. These results were originally obtained in [116] and [117], respectively. In Section 13.3 we will review results obtained for other filiform algebras. In Section 13.4 based on [115] we shall study an example of a nilradical which is not filiform but contains a model filiform subalgebra of codimension 2.

13.1. Classification of solvable Lie algebras with the model filiform nilradical $\mathfrak{n}_{n,1}$

13.1.1. Nilpotent algebra $\mathfrak{n}_{n,1}$. The Lie algebra $\mathfrak{n}_{n,1}$ is defined by the Lie brackets (13.2). We shall consider the dimension $n \geq 4$ ($\mathfrak{n}_{2,1}$ does not exist and $\mathfrak{n}_{3,1}$ is the three-dimensional Heisenberg algebra that was considered in Chapter 11). The dimensions of the subalgebras in the characteristic series are

$$(13.4)\quad \mathrm{US} = [1, 2, \ldots, n-2, n], \quad \mathrm{CS} = [n, n-2, n-3, \ldots, 1, 0], \quad \mathrm{DS} = [n, n-2, 0].$$

Its maximal Abelian ideal $\mathfrak{a}$ can be identified with the centralizer of the highest center $\mathfrak{z}_{n-2} = \operatorname{span}\{e_1, \ldots, e_{n-2}\}$, i.e., $\mathfrak{a} = \operatorname{span}\{e_1, \ldots, e_{n-1}\}$. Hence $\mathfrak{a}$ is uniquely defined.

In order to find all nonnilpotent derivations D of $\mathfrak{n}$ we assign to D its matrix

$$D(e_a) = e_b D^b_a$$

and evaluate the condition (2.26) for basis elements $x = e_i$, $y = e_j$. We find

- $i = 1 < j < n$: $D^n_1 e_{j-1} = 0$,
- $1 < i < j < n$: $D^n_i e_{j-1} - D^n_j e_{i-1} = 0$,
- $i = 1$, $j = n$: $\sum_{k=1}^{n-1} D^{k+1}_1 e_k = 0$,
- $1 < i < j = n$: $(D^{i-1}_{i-1} - D^i_i - D^n_n)e_{i-1} + \sum_{k=1, k\neq i-1}^{n-1} (D^k_{i-1} - D^{k+1}_i)e_k = 0$.

From the first and second equations we immediately get

$$D^n_i = 0, \quad 1 \leq i < n,$$

from the third

$$D^k_1 = 0, \quad 1 < k.$$

In the last one the coefficients of linearly independent elements in the sum must be zero, therefore considering

$$(D^k_{i-1} - D^{k+1}_i) = 0, \quad k \neq i-1$$

we obtain by induction

$$D_i^j = 0, \quad D_j^i = D_{j-1}^{i-1} = D_{j-i+1}^1, \qquad i < j. \tag{13.5}$$

The remaining recursion relations

$$D_{i-1}^{i-1} - D_i^i - D_n^n = 0$$

can most easily be solved from the "lower right corner" up. Denoting

$$D_n^n = \alpha, \quad D_{n-1}^{n-1} = \beta$$

we find

$$D_i^i = (n-i-1)\alpha + \beta, \ 1 \le i \le n-1. \tag{13.6}$$

Thus we have solved the derivation property (2.26) for all basis elements of $\mathfrak{n}$. Finding the inner derivations is elementary and we write

Lemma 13.1. *The algebra of derivations of the nilpotent algebra $\mathfrak{n}_{n,1}$ is expressible in the standard basis $(e_1, \dots, e_n)$ of $\mathfrak{n}_{n,1}$ as the algebra of upper triangular matrices D whose elements satisfy*

$$\begin{aligned} D_i^i &= (n-i+1)D_n^n + D_{n-1}^{n-1}, & 1 \le i \le n-1, \\ D_j^i &= D_{j-i+1}^1, & i < j. \end{aligned}$$

Its subalgebra of inner derivations is

$$\operatorname{span}\{\operatorname{ad}_{e_2}, \dots, \operatorname{ad}_{e_n}\},$$

where

$$(\operatorname{ad}_{e_j})_b^a = \delta_b^n \delta_{j-1}^a, \quad \forall j \le n-1, \qquad (\operatorname{ad}_{e_n})_b^a = -\delta_{b+1}^a.$$

13.1.2. Construction of solvable Lie algebras with the nilradical $\mathfrak{n}_{n,1}$. As was explained in Section 9.2, to find all solvable Lie algebras with the nilradical $\mathfrak{n}_{n,1}$ we must find all nonequivalent nilindependent sets $\{D_1, \dots, D_f\}$ of derivations $\mathfrak{n}_{n,1}$. The equivalence is generated by the following transformations:

(1) We may add any inner derivation to D_k.

(2) We may perform a change of basis in $\mathfrak{n}$ such that the Lie brackets are not changed. Since any such change must inter alia preserve the lower central series and the ideal $\mathfrak{a}$, the matrix of such transformations must be upper triangular. The preservation of Lie brackets then imposes certain further relations. We may decompose any such transformation into a composition of scaling

$$e_n \to \tilde{e}_n = \omega e_n, \qquad e_k \to \tilde{e}_k = \tau \omega^{n-k-1} e_k, \quad k \le n-1, \tag{13.7}$$

where $\omega, \tau \in \mathbb{F}\backslash\{0\}$, and the transformation

$$\begin{aligned} e_k \to \tilde{e}_k &= e_k + \sum_{j=1}^{k-1} u_{k-j} e_j, & k \le n-1, \ u_1, \dots, u_{n-2} \in \mathbb{F}, \\ e_n \to \tilde{e}_n &= e_n + \sum_{j=1}^{n} v_j e_j, & v_1, \dots, v_{n-1} \in \mathbb{F}. \end{aligned} \tag{13.8}$$

(To prove that (13.8) gives a general form of such a transformation it is sufficient to consider $[\tilde{e}_k, e_n] = [\tilde{e}_k, \tilde{e}_n] = \tilde{e}_{k-1}$ and to use induction on k.) The scaling (13.7) acts on D_k as

$$D_k \to S D_k S^{-1}$$

where $S = \mathrm{diag}(\tau\omega^{n-2}, \tau\omega^{n-3}, \dots, \tau, \omega)$, the transformation (13.8) as

$$D_k \to U D_k U^{-1}$$

where

$$U = \begin{pmatrix} 1 & u_1 & u_2 & \dots & u_{n-2} & v_1 \\ 0 & 1 & u_1 & \dots & u_{n-3} & v_2 \\ & \ddots & \ddots & \ddots & \ddots & \\ & & 0 & 1 & u_1 & v_{n-1} \\ & & & 0 & 1 & v_{n-1} \\ 0 & & \dots & & 0 & 1 \end{pmatrix}.$$

(3) We can change the basis in the space $\mathrm{span}\{D_1, \dots, D_p\}$.

By adding inner derivations we can transform all D_k into the form

$$D_k = \begin{pmatrix} d_1^k & 0 & a_3^k & a_4^k & \dots & a_{n-1}^k & 0 \\ 0 & d_2^k & 0 & a_3^k & \dots & a_{n-2}^k & 0 \\ & \ddots & \ddots & \ddots & & & \\ & & 0 & d_{n-3}^k & 0 & a_3^k & 0 \\ & & & 0 & d_{n-2}^k & 0 & 0 \\ & & & & 0 & \beta^k & a_n^k \\ 0 & & \dots & & & 0 & \alpha^k \end{pmatrix}, \tag{13.9}$$

where

$$d_1^k = (n-2)\alpha^k + \beta^k, \dots, d_j^k = (n-1-j)\alpha^k + \beta^k, \dots, d_{n-2}^k = \alpha^k + \beta^k.$$

We see that the number of nilindependent elements f can be at most 2 since a set of three or more derivations of the form (13.9) cannot be linearly nilindependent.

Case 1: $f = 1$. The entire structure of the associated solvable Lie algebra is encoded in the matrix $D \equiv D_1$. The Lie brackets of the nonnilpotent element f_1 with nilpotent elements are given by

$$[f_1, e_k] = D(e_k) = D_k^l e_l.$$

We shall divide our investigation into subcases determined by values of the parameters α, β, at least one of which must be nonzero.

(1) $\alpha \neq 0$. We rescale D to put $\alpha = 1$. Then by a change of basis (13.8) in $\mathfrak{n}$

$$\begin{aligned} \tilde{e}_k &= e_k - \frac{1}{l-1} a_l e_{k-l+1}, \quad && l \leq k \leq n-1, \\ \tilde{e}_k &= e_k, && 1 \leq k \leq l-1 \end{aligned}$$

we set equal to zero first a_3, then a_4, etc., up to a_{n-1}. From now on we assume that $a_k = 0, k \leq n-1$. If $\beta \neq 1$ $(= \alpha)$ then a further change of basis

$$\tilde{e}_n = e_n - \frac{a_n}{\beta - 1} e_{n-1} \tag{13.10}$$

transforms a_n into zero and the matrix D is diagonal

$$D = \mathrm{diag}(n-2+\beta, n-3+\beta, \dots, \beta, 1). \tag{13.11}$$

If $\beta = 1$ then a_n cannot be removed. The only remaining transformation is scaling (13.7) which allows us to scale any nonzero a_n to 1. Therefore we find in addition to (13.11) with $\beta = 1$ another possibility, namely

$$D = \begin{pmatrix} n-1 & 0 & \dots & 0 & 0 \\ 0 & n-2 & \dots & 0 & 0 \\ & & \ddots & & \\ 0 & 0 & \dots & 1 & 1 \\ 0 & 0 & \dots & 0 & 1 \end{pmatrix}. \tag{13.12}$$

(2) $\alpha = 0$. We rescale D to put $\beta = 1$. We use (13.8) to change e_n

$$\tilde{e}_n = e_n - a_n e_{n-1}$$

and transform a_n into $a_n = 0$. If D is diagonal, it cannot be further simplified. Let us assume that D is not diagonal. For $\alpha = 0$ the matrix D is invariant with respect to transformations (13.8) preserving $a_n = 0$, i.e., the parameters a_k cannot be removed. The only transformation we still have at our disposal is the scaling (13.7) which allows us to scale one chosen nonzero a_k to 1 over the field $\mathbb{C}$. Over $\mathbb{R}$ one value a_k can be scaled to 1 if k is even, or to ± 1 if k is odd.

Case 2: $f = 2$. By taking linear combinations of D_1, D_2 we obtain $\alpha^1 = 1$, $\beta^1 = 0$, $\alpha^2 = 0$, $\beta^2 = 1$. Further by a change of basis in $\mathfrak{n}$ (13.8) we take D_1 to its canonical form

$$D_1 = \operatorname{diag}(n-2, n-3, \dots, 2, 1, 0, 1)$$

found for $f = 1$. In order to define a solvable Lie algebra $\mathfrak{g}$ with nilradical $\mathfrak{n}$, the two derivations D_1, D_2 must commute to an inner derivation

$$[D_1, D_2] \in \operatorname{span}\{\operatorname{ad}_{e_2}, \dots, \operatorname{ad}_{e_n}\}.$$

Computing the commutator for the above given forms of D_1, D_2 (note that D_1 is diagonal) we immediately find that

$$a_k^2 = 0, \quad 3 \leq k \leq n$$

must hold.

Therefore there is a single canonical form of the derivations D_1, D_2

$$D_1 = \operatorname{diag}(n-2, n-3, \dots, 2, 1, 0, 1), \quad D_2 = \operatorname{diag}(1, 1, \dots, 1, 1, 0). \tag{13.13}$$

The corresponding solvable Lie algebra is now almost specified, the Lie brackets of nonnilpotent elements f_1, f_2 being

$$\begin{aligned} &[f_1, e_k] = (n-1-k)e_k, \quad k < n, \qquad [f_1, e_n] = e_n, \\ &[f_2, e_k] = e_k, \qquad\qquad\quad\;\; k < n, \qquad [f_2, e_n] = 0. \end{aligned}$$

It remains to fix the Lie bracket between f_1, f_2. Since D_1 and D_2 are commuting matrices representing f_1, f_2 in the adjoint representation of $\mathfrak{g}$ restricted to the ideal $\mathfrak{n}$, the Lie bracket of f_1, f_2 must be in the kernel of the representation map, i.e., in the center of $\mathfrak{n}$

$$[f_1, f_2] = \gamma e_1. \tag{13.14}$$

The transformation

$$f_1 \to \tilde{f}_1 = f_1 + \gamma e_1$$

takes γ in (13.14) into $\gamma = 0$ while leaving all other Lie brackets invariant. We conclude that in the case $f = 2$ the solvable Lie algebra with the nilradical $\mathfrak{n}_{n,1}$ is unique.

13.1.3. Standard forms of solvable Lie algebras with the nilradical $\mathfrak{n}_{n,1}$. The results obtained above can be summed up as theorems. We give them for algebras over the field $\mathbb{F} = \mathbb{R}$ and $\mathbb{C}$. We specify $\mathbb{F} = \mathbb{R}$, or $\mathbb{C}$ only when the two cases differ. In all cases we also give the dimensions of the subalgebras in the characteristic series. The nilradical in all cases is $\mathfrak{n}_{n,1}$ with the Lie brackets (13.2).

Theorem 13.2. *Any solvable, indecomposable, nonnilpotent Lie algebra $\mathfrak{s}$ with the nilradical $\mathfrak{n}_{n,1}$ will have dimension $\dim \mathfrak{s} = n+1$, or $\dim \mathfrak{s} = n+2$.*

Three types of solvable Lie algebras of dimension $\dim \mathfrak{s} = n+1$ exist for any $n \geq 4$. They are represented by the following:

(1) *The derivation D_1 diagonal,*

$$[f_1, e_k] = \big((n-k-1)\alpha + \beta\big)e_k, \quad k \leq n-1, \qquad [f_1, e_n] = \alpha e_n.$$

The mutually nonisomorphic algebras of this type are

$\mathfrak{s}_{n+1,1}(\beta)$: $\alpha = 1$, $\beta \in \mathbb{F}\backslash\{0, n-2\}$,

$$\mathrm{US} = [0], \quad \mathrm{CS} = [n+1, n], \quad \mathrm{DS} = [n+1, n, n-2, 0],$$

$\mathfrak{s}_{n+1,2}$: $\alpha = 1$, $\beta = 0$,

$$\mathrm{US} = [0], \quad \mathrm{CS} = [n+1, n-1, n-1], \quad \mathrm{DS} = [n+1, n-1, n-3, 0],$$

$\mathfrak{s}_{n+1,3}$: $\alpha = 1$, $\beta = 2-n$,

$$\mathrm{US} = [1], \quad \mathrm{CS} = [n+1, n], \quad \mathrm{DS} = [n+1, n, n-2, 0],$$

$\mathfrak{s}_{n+1,4}$: $\alpha = 0$, $\beta = 1$,

$$\mathrm{US} = [0], \quad \mathrm{CS} = [n+1, n-1], \quad \mathrm{DS} = [n+1, n-1, 0].$$

(2) *The derivation D_1 nondiagonal, its diagonal determined by $\alpha = \beta = 1$. We have*

$\mathfrak{s}_{n+1,5}$: $[f_1, e_k] = (n-k)e_k$, $k \leq n-1$, $[f_1, e_n] = e_n + e_{n-1}$,

$$\mathrm{US} = [0], \quad \mathrm{CS} = [n+1, n], \quad \mathrm{DS} = [n+1, n, n-2, 0].$$

(3) *The derivation D_1 nondiagonal, its diagonal determined by $\alpha = 0$, $\beta = 1$.*

$\mathfrak{s}_{n+1,6}(a_3, \ldots, a_{n-1})$: $[f_1, e_k] = e_k + \sum_{l=1}^{k-2} a_{k-l+1} e_l$, $k \leq n-1$, $[f_1, e_n] = 0$, $a_j \in \mathbb{F}$, *at least one a_j satisfies $a_j \neq 0$.*

Over $\mathbb{C}$, the first nonzero a_j satisfies $a_j = 1$.

Over $\mathbb{R}$, the first nonzero a_j for even j satisfies $a_j = 1$. If all $a_j = 0$ for j even, then the first nonzero a_j (j odd) satisfies $a_j = \pm 1$. We have

$$\mathrm{US} = [0], \quad \mathrm{CS} = [n+1, n-1], \quad \mathrm{DS} = [n+1, n-1, 0].$$

Precisely one class of solvable Lie algebras $\mathfrak{s}_{n+2}$ of $\dim \mathfrak{s} = n+2$ with the nilradical $\mathfrak{n}_{n,1}$ exists. It is represented by a basis $(e_1, \ldots, e_n, f_1, f_2)$ and the Lie brackets involving f_1 and f_2 are

$$\begin{aligned}
&[f_1, e_k] = (n-1-k)e_k, && 1 \leq k \leq n-1, && [f_1, e_n] = e_n, \\
&[f_2, e_k] = e_k, && 1 \leq k \leq n-1, && [f_2, e_n] = 0, \\
&[f_1, f_2] = 0.
\end{aligned}$$

For this algebra we have

$$\mathrm{US} = [0], \quad \mathrm{CS} = [n+2, n], \quad \mathrm{DS} = [n+2, n, n-2, 0].$$

The results presented in Theorem 13.2 also imply the classification of algebras with the nilradicals $\mathfrak{n}_{4,1}$ and $\mathfrak{n}_{5,5}$ in Sections 18.4, 19.9 and 19.13 of Part 4.

13.1.4. The generalized Casimir invariants of $\mathfrak{n}_{n,1}$ and of its solvable extensions. The differential operators corresponding to the basis elements of $\mathfrak{n}_{n,1}$ are

$$\widehat{E}_1 = 0, \qquad \widehat{E}_k = e_{k-1}\frac{\partial}{\partial e_n}, \quad 1 < k < n, \qquad \widehat{E}_n = -\sum_{k=2}^{n-1} e_{k-1}\frac{\partial}{\partial e_k}. \tag{13.15}$$

The form of $\widehat{E}_k$ with $1 < k < n$ implies that the invariants do not depend on e_n. Solving the equation $\widehat{E}_n\ I(e_1, e_2, \ldots, e_{n-1}) = 0$ by the method of characteristics, we obtain the following result

Theorem 13.3. *The nilpotent Lie algebra $\mathfrak{n}_{n,1}$ has $n-2$ functionally independent invariants. They can be chosen to be the following polynomials*

$$\begin{aligned} &\xi_0 = e_1, \\ &\xi_k = \frac{(-1)^k k}{(k+1)!} e_2^{k+1} + \sum_{j=0}^{k-1} (-1)^j \frac{e_2^j\, e_{k+2-j}\, e_1^{k-j}}{j!}, \quad 1 \le k \le n-3. \end{aligned} \tag{13.16}$$

Let us now consider the $(n+1)$-dimensional solvable Lie algebras of Theorem 13.2. The operators $\widehat{E}_i$ representing $\mathfrak{n}_{n,1}$ will each contain an additional term involving a derivative with respect to f_1. However, from the form of these operators we see that the invariants cannot depend on f_1. Moreover, they can only depend on the invariants (13.16) of $\mathfrak{n}_{n,1}$. To find the invariants of the algebras $\mathfrak{s}_{n+1,k}$ we must represent the element $f_1 \in \mathfrak{s}_{n+1,k}$ by the appropriate "truncated" differential operator $\widehat{F}_T$ (by "truncated" we mean that we keep only the part acting on $e_1, \ldots, e_{n-1}$). Next, we solve the equation

$$\widehat{F}_{1T} I(\xi_0, \ldots, \xi_{n-3}) = 0. \tag{13.17}$$

For the algebras $\mathfrak{s}_{n+1,1}, \ldots, \mathfrak{s}_{n+1,5}$ of Theorem 13.2, we have

$$\widehat{F}_{1T} = \sum_{k=1}^{n-1} ((n-1-k)\alpha + \beta) e_k \frac{\partial}{\partial e_k} \tag{13.18}$$

with α and β as in Theorem 13.2. For $\mathfrak{s}_{n+1,6}(a_3, \ldots, a_{n-1})$ we have

$$\widehat{F}_{1T} = e_1\frac{\partial}{\partial e_1} + e_2\frac{\partial}{\partial e_2} + \sum_{l=1}^{n-3}\left(e_{l+2} + \sum_{j=1}^{l} a_{l+3-j}e_j\right)\frac{\partial}{\partial e_{l+2}}. \tag{13.19}$$

Solving (13.17) in each case we obtain the results, listed below in Theorem 13.4.

Finally, let us consider the $(n+2)$-dimensional algebra $\mathfrak{s}_{n+2}$. The invariants can again depend only on $\xi_0, \ldots, \xi_{n-3}$. We have two additional truncated differential operators, namely

$$\widehat{F}_{1T} = \sum_{k=1}^{n-1} (n-1-k) e_k \frac{\partial}{\partial e_k}, \quad \widehat{F}_{2T} = \sum_{k=1}^{n-1} e_k \frac{\partial}{\partial e_k}. \tag{13.20}$$

Imposing two equations of the form (13.17) for $\widehat{F}_{1T}, \widehat{F}_{2T}$ we obtain the invariants listed below.

We collect our results into

Theorem 13.4. *The algebras $\mathfrak{s}_{n+1,1}(\beta), \ldots, \mathfrak{s}_{n+1,5}$ have $n-3$ invariants each. Their form is*

(1) $\mathfrak{s}_{n+1,1}(\beta)$, $\mathfrak{s}_{n+1,2}$ *and* $\mathfrak{s}_{n+1,5}$

$$\chi_k = \frac{\xi_k^{n-2+\beta}}{\xi_0^{(k+1)(n-3+\beta)}}, \qquad 1 \le k \le n-3.$$

For $\mathfrak{s}_{n+1,2}$ *and* $\mathfrak{s}_{n+1,5}$ *we have* $\beta = 0$ *and* $\beta = 1$, *respectively.*

(2) $\mathfrak{s}_{n+1,3}$

$$\chi_1 = \xi_0, \qquad \chi_k = \frac{\xi_k^2}{\xi_1^{k+1}}, \quad 2 \le k \le n-3.$$

(3) $\mathfrak{s}_{n+1,4}$

$$\chi_k = \frac{\xi_k}{\xi_0^{k+1}}, \quad 1 \le k \le n-3.$$

(4) $\mathfrak{s}_{n+1,6}(a_3, \dots, a_{n-1})$

$$\chi_k = \sum_{m=0}^{[(k+1)/2]} (-1)^m \frac{(\ln \xi_0)^m}{m!} \Bigg(\sum_{i_1+\cdots+i_m=k-2m+1} a_{i_1+3}a_{i_2+3}\cdots a_{i_m+3} + \sum_{j+i_1+\cdots+i_m=k-2m-1} \frac{\xi_{j+1}}{\xi_0^{j+2}} a_{i_1+3}a_{i_2+3}\cdots a_{i_m+3} \Bigg),$$
$$1 \le k \le n-3.$$

The summation indices take the values $0 \le j, i_1, \dots, i_m \le k+1$.

The Lie algebra $\mathfrak{s}_{n+2}$ *has* $n-4$ *functionally independent invariants that can be chosen to be*

$$\chi_k = \frac{\xi_{k+1}^2}{\xi_1^{k+2}}, \quad 1 \le k \le n-4.$$

We see that for the algebra $\mathfrak{s}_{n+1,6}(a_3, \dots, a_{n-1})$ the invariants involve powers of the logarithm $\ln \xi_0$. In all other cases we obtain sets of ratios of powers of ξ_k.

13.2. Classification of solvable Lie algebras with the nilradical $\mathfrak{n}_{n,2}$

13.2.1. Nilpotent algebra $\mathfrak{n}_{n,2}$ and its structure. The filiform Lie algebra $\mathfrak{n} = \mathfrak{n}_{n,2}$ exists for $n \ge 5$ and is defined by the Lie brackets

$$\begin{aligned}
&[e_j, e_k] = 0, && 1 \le j,k \le n-2,\\
&[e_1, e_{n-1}] = [e_2, e_{n-1}] = 0, &&\\
&[e_k, e_{n-1}] = e_{k-2}, && 3 \le k \le n-2,\\
&[e_1, e_n] = 0, &&\\
&[e_k, e_n] = e_{k-1}, && 2 \le k \le n-1.
\end{aligned} \tag{13.21}$$

We shall first consider $n \ge 6$. The final result is the same for $n = 5$ but there is a small difference in the proof, explained below (13.28). The dimensions of the subalgebras in the characteristic series are

$$\mathrm{US} = [1, 2, \dots, n-2, n], \quad \mathrm{CS} = [n, n-2, n-3, \dots, 1, 0], \quad \mathrm{DS} = [n, n-2, 0]. \tag{13.22}$$

Its maximal Abelian ideal $\mathfrak{a}$ coincides with the derived algebra $\mathfrak{n}^{(1)} = \mathfrak{n}^2$, i.e., $\mathfrak{a} = \operatorname{span}\{e_1, \dots, e_{n-2}\}$.

In order to find all nonnilpotent derivations of $\mathfrak{n}$ we first consider the structure of automorphisms of $\mathfrak{n}_{n,2}$. There exists a flag of ideals which is invariant under any automorphism

$$\mathfrak{n} \supset \operatorname{cent}_{\mathfrak{n}}(\mathfrak{n}^{n-2}) \supset \mathfrak{n}^2 \supset \mathfrak{n}^3 \supset \cdots \supset \mathfrak{n}^{n-1} \tag{13.23}$$

where each element in the flag has codimension one in the previous one. (We recall that $\operatorname{cent}_{\mathfrak{n}}(\mathfrak{n}^{n-2})$ is the centralizer of $\mathfrak{n}^{n-2}$ in $\mathfrak{n}$.) In any basis respecting the flag, e.g., the one used to write down the Lie brackets (13.21), any automorphism will be represented by a triangular matrix.

Furthermore, the whole algebra $\mathfrak{n}$ is generated via multiple commutators of the elements e_{n-1} and e_n, e.g., $e_{n-3} = [[e_{n-1}, e_n], e_n]$. That means that due to the definition of an automorphism (2.25) the knowledge of

$$\Phi(e_{n-1}) = \sum_{k=1}^{n-1} \phi_k e_k, \qquad \Phi(e_n) = \sum_{k=1}^{n} \psi_k e_k \tag{13.24}$$

in principle amounts to full knowledge of Φ. It remains to establish which choices of ϕ_k, $1 \le k \le n-1$ and ψ_k, $1 \le k \le n$ are consistent with the definition (2.25) of an automorphism.

Because of the triangular structure respecting (13.23) we immediately have

$$\begin{aligned} [\Phi(e_j), \Phi(e_k)] &= 0, & 1 \le j, k \le n-2, \\ [\Phi(e_1), \Phi(e_{n-1})] &= [\Phi(e_2), \Phi(e_{n-1})] = 0, & \\ [\Phi(e_1), \Phi(e_n)] &= 0. & \end{aligned}$$

The relation

$$[\Phi(e_k), \Phi(e_n)] = \Phi(e_{k-1}), \quad 2 \le k \le n-1 \tag{13.25}$$

can be viewed as a definition of $\Phi(e_{k-1})$, $2 \le k \le n-1$ in accordance with (13.24). Consequently it only remains to check when

$$[\Phi(e_k), \Phi(e_{n-1})] = \Phi(e_{k-2}), \quad 3 \le k \le n-2$$

or, equivalently,

$$[\Phi(e_{n-1}), \Phi(e_k)] = -\big[\Phi(e_n), [\Phi(e_n), \Phi(e_k)]\big], \quad 1 \le k \le n-2 \tag{13.26}$$

(the change in the index range is just for convenience, the added two relations are satisfied trivially). Since any automorphism Φ restricted to $\mathfrak{n}^2 = \operatorname{span}\{e_1, \dots, e_{n-2}\}$ is also a regular (invertible) map, we can write (13.26) as a relation between restrictions to $\mathfrak{n}^2$ of two operators

$$\operatorname{ad}|_{\mathfrak{n}^2}(\Phi(e_{n-1})) = -\Big(\operatorname{ad}|_{\mathfrak{n}^2}(\Phi(e_n))\Big)^2. \tag{13.27}$$

Writing down the matrices of the operators we find

$$\operatorname{ad}|_{\mathfrak{n}^2}(\Phi(e_{n-1})) = \begin{pmatrix} 0 & 0 & -\phi_{n-1} & 0 & 0 & \dots & 0 \\ & 0 & 0 & -\phi_{n-1} & 0 & \dots & 0 \\ & & 0 & 0 & -\phi_{n-1} & \dots & 0 \\ & & & \ddots & \ddots & \ddots & \\ & & & & 0 & 0 & -\phi_{n-1} \\ & & & & & 0 & 0 \\ & & & & & & 0 \end{pmatrix}$$

and

$$\mathrm{ad}|_{\mathfrak{n}^2}(\Phi(e_n)) = \begin{pmatrix} 0 & -\psi_n & -\psi_{n-1} & 0 & 0 & \dots & 0 \\ & 0 & -\psi_n & -\psi_{n-1} & 0 & \dots & 0 \\ & & 0 & -\psi_n & -\psi_{n-1} & \dots & 0 \\ & & & \ddots & \ddots & \ddots & \\ & & & & 0 & -\psi_n & -\psi_{n-1} \\ & & & & & 0 & -\psi_n \\ & & & & & & 0 \end{pmatrix}.$$

Consequently, the condition (13.27) gives us two constraints on ϕ_k, ψ_k, namely

$$\phi_{n-1} = \psi_n^2, \quad \psi_{n-1} = 0. \tag{13.28}$$

Notice that here the case $n = 5$ differs — the matrices above become 3×3 and only the first condition remains. Accordingly, the dimension of the group of automorphisms and of the algebra of derivations is by one higher than in the generic case. Even so, the final conclusion about the number and structure of the solvable Lie algebras with the nilradical $\mathfrak{n}_{5,2}$ also fits into the general pattern shown below.

To sum up, we have found that automorphisms Φ of $\mathfrak{n}$ are uniquely determined by a set of $2n - 3$ parameters

$$\phi_k, \psi_k, \psi_n \quad 1 \leq k \leq n-2 \tag{13.29}$$

where $\psi_n \neq 0$ and ϕ_k, ψ_k are arbitrary. Such an automorphism acts on the basis elements e_k in the following way

$$\begin{aligned} \Phi(e_k) &= \sum_{j=1}^{k-1} \beta_k^j \, e_j + (\psi_n)^{n-k+1} \, e_k, \qquad 1 \leq k \leq n-2, \\ \Phi(e_{n-1}) &= \sum_{j=1}^{n-2} \phi_j \, e_j + \psi_n^2 \, e_{n-1}, \\ \Phi(e_n) &= \sum_{j=1}^{n-2} \psi_j \, e_j + \psi_n \, e_n \end{aligned} \tag{13.30}$$

where the coefficient of the e_k term in $\Phi(e_k)$ was found from (2.25) and β_k^j denote some rather complicated functions of the parameters ϕ_k, ψ_k, ψ_n which can be deduced from (2.25) but which we shall not need in the following.

The derivations of $\mathfrak{n}$ are now easily found by considering automorphisms infinitesimally close to identity, i.e., differentiating one-parameter subgroups in $\mathrm{Aut}(\mathfrak{n})$. We find that the algebra of derivations is $(2n-3)$-dimensional. An arbitrary derivation D depends on $2n - 3$ parameters c_k, d_k, d_n, $1 \leq k \leq n-2$ and has the form

(13.31)
$$\begin{aligned} D(e_k) &= \sum_{j=1}^{k-1} b_k^j\, e_j + (n-k+1)\, d_n\, e_k, \qquad 1 \le k \le n-2, \\ D(e_{n-1}) &= \sum_{j=1}^{n-2} c_j\, e_j + 2\, d_n\, e_{n-1}, \\ D(e_n) &= \sum_{j=1}^{n-2} d_j\, e_j + d_n\, e_n \end{aligned}$$

where b_k^j denote some particular linear functions of the $2n-3$ parameters c_k, d_k, d_n. The derivation (13.31) is nonnilpotent if and only if d_n is nonvanishing.

13.2.2. Construction of solvable Lie algebras with the nilradical $\mathfrak{n}_{n,2}$. As was explained in Section 9.2, to find all solvable Lie algebras with nilradical $\mathfrak{n}_{n,2}$ we must find all nonequivalent nilindependent sets $\{D_1, \dots, D_f\}$ of derivations $\mathfrak{n}_{n,2}$.

Looking at (13.31) we immediately recognize that we can have at most one nilindependent derivation (such that $d_n \neq 0$). If there were more of them, say D and $\widetilde{D}$ then obviously by taking a linear combination $\tilde{d}_n D - d_n \widetilde{D}$ we obtain a nilpotent operator (namely one represented by a strictly upper triangular matrix). Therefore any solvable but not nilpotent Lie algebra with the nilradical $\mathfrak{n}_{n,2}$ must be $(n+1)$-dimensional. The question which remains is how many such algebras are nonisomorphic.

By proper choice of the multiple of D and adding suitable inner derivations we can transform D into the form

(13.32)
$$\begin{aligned} D(e_k) &= \sum_{j=1}^{k-1} b_{jk}\, e_j + (n-k+1)\, e_k, \qquad 1 \le k \le n-2, \\ D(e_{n-1}) &= \sum_{j=1}^{n-3} c_j\, e_j + 2\, e_{n-1}, \\ D(e_n) &= e_n. \end{aligned}$$

Notice that there are $n-1$ nontrivial inner derivations $\mathrm{ad}(e_k)$, $2 \le k \le n$ and one choice of scaling, so we are able to remove n parameters

$$d_j \to 0, \quad j = 1, \dots, n-2, \qquad d_n \to 1, \qquad c_{n-2} \to 0$$

in any nonnilpotent outer derivation (13.31). There are still $n-3$ parameters left in (13.32).

Next we perform a change of basis in $\mathfrak{n}$ such that the Lie brackets (13.21) are preserved, i.e., conjugate the derivation D by a suitable automorphism Φ

$$D \to \widetilde{D} = \Phi^{-1} \circ D \circ \Phi. \tag{13.33}$$

Our aim is to diagonalize the action of D, if possible. We find it convenient to perform this in $n-3$ steps, setting one parameter c_k equal to $c_k = 0$ in each step. Thus our Φ will be expressed as

$$\Phi = \Phi_{n-3} \circ \Phi_{n-2} \circ \cdots \circ \Phi_1 \tag{13.34}$$

where the automorphisms Φ_k are constructed as follows.

Let us assume that for a given $k \leq n-3$ we have already set $c_j = 0$ for all $k < j \leq n-2$ (assuming of course the form (13.32) for D). We construct an automorphism Φ_k defined by

$$\Phi_k(e_{n-1}) = \alpha_k e_k + e_{n-1}, \quad \Phi_k(e_n) = e_n$$

where α_k is to be determined. We have

$$\begin{aligned} D(\Phi_k(e_{n-1})) &= D(e_{n-1}) + \alpha_k D(e_k) \\ &= 2e_{n-1} + c_k e_k + (n-k+1)\alpha_k e_k + \sum_{j=1}^{k-1}(\cdots)e_j \\ &= 2(e_{n-1} + \tfrac{1}{2}(c_k + (n-k+1)\alpha_k)e_k) + \sum_{j=1}^{k-1}(\cdots)e_j. \end{aligned}$$

We find that

$$D(\Phi_k(e_{n-1})) = 2\Phi_k(e_{n-1}) + \sum_{j=1}^{k-1}(\cdots)\, e_j$$

precisely when

$$\alpha_k = \frac{c_k}{k-n+1}$$

(we recall that $k \leq n-3$). By this choice of α_k we set c_k to 0 and proceed to the next step, namely elimination of c_{k-1}.

To conclude, we are able to eliminate all c_k's using suitably chosen automorphisms Φ_k in (13.34). Indeed we have found that up to addition of inner derivations, conjugation by automorphisms and rescaling, there exists just one nilindependent set of outer derivations, consisting of a unique element D

$$D(e_k) = (n-k+1)\, e_k, \quad 1 \leq k \leq n. \tag{13.35}$$

Consequently, we have the following theorem.

Theorem 13.5. *For the given nilradical $\mathfrak{n}_{n,2}$, $n \geq 5$, there exists precisely one solvable nonnilpotent Lie algebra $\mathfrak{s}_{n+1}$ with the nilradical $\mathfrak{n}_{n,2}$. It has dimension* $\dim \mathfrak{s}_{n+1} = n+1$ *and its Lie brackets are as follows*

$$\begin{aligned} \mathfrak{s}_{n+1} &= \operatorname{span}\{e_1, \ldots, e_n, f_1\}, \\ [e_j, e_k] &= 0, && 1 \leq j, k \leq n-2, \\ [e_1, e_{n-1}] &= [e_2, e_{n-1}] = 0, \\ [e_k, e_{n-1}] &= e_{k-2}, && 3 \leq k \leq n-2, \\ [e_1, e_n] &= 0, \\ [e_k, e_n] &= e_{k-1}, && 2 \leq k \leq n-1, \\ [e_k, f_1] &= (n-k+1)e_k, && 1 \leq k \leq n. \end{aligned} \tag{13.36}$$

The dimensions of the characteristic series are

$$\mathrm{US} = [0], \quad \mathrm{CS} = [n+1, n], \quad \mathrm{DS} = [n+1, n, n-2, 0].$$

When $n = 5$ the nilradical $\mathfrak{n}_{n,2}$ becomes the nilpotent algebra $\mathfrak{n}_{5,6}$ of Section 18.1 in Part 4. Theorem 13.5 thus shows that there is a unique solvable extension of $\mathfrak{n}_{5,6}$ in Chapter 19.10, denoted as $\mathfrak{s}_{6,197}$.

13.2.3. Generalized Casimir invariants of $\mathfrak{n}_{n,2}$ and of its solvable extension. The differential operators corresponding to the basis elements of $\mathfrak{n}_{n,2}$ are

$$\begin{gathered} \widehat{E}_1 = 0, \qquad \widehat{E}_2 = e_1 \frac{\partial}{\partial e_n}, \\ \widehat{E}_k = e_{k-2} \frac{\partial}{\partial e_{n-1}} + e_{k-1} \frac{\partial}{\partial e_n}, \quad 3 \le k \le n-2, \\ \widehat{E}_{n-1} = -\sum_{k=3}^{n-2} e_{k-2} \frac{\partial}{\partial e_k} + e_{n-2} \frac{\partial}{\partial e_n}, \qquad \widehat{E}_n = -\sum_{k=2}^{n-1} e_{k-1} \frac{\partial}{\partial e_k}. \end{gathered} \tag{13.37}$$

The form of $\widehat{E}_k$, $1 \le k \le n$ implies that the invariants do not depend on e_{n-1}, e_n. Using (3.5) we find that the nilpotent Lie algebra $\mathfrak{n}_{n,2}$ has $n-4$ functionally independent invariants but it is rather complicated to directly solve the remaining two partial differential equations defining the invariants, namely

$$\widehat{E}_{n-1}\, I(e_1, e_2, \ldots, e_{n-2}) = 0, \quad \widehat{E}_n\, I(e_1, e_2, \ldots, e_{n-2}) = 0.$$

Let us instead employ the method of moving frames, as described in Section 3.3. First, we construct the flows of the vector fields $\widehat{E}_{n-1}, \widehat{E}_n$ acting on the space spanned by $e_1, \ldots, e_{n-2}$ only. We find

$$e_k(\Psi^{\alpha_{n-1}}_{\widehat{E}_{n-1}} p) = \sum_{j=0}^{[(k-1)/2]} \frac{(-1)^j}{j!} \alpha_{n-1}^j e_{k-2j}(p), \quad 1 \le k \le n-2 \tag{13.38}$$

(where $[\;]$ denotes the integer part and α_a are group parameters) and

$$e_k(\Psi^{\alpha_n}_{\widehat{E}_n} p) = \sum_{j=0}^{k-1} \frac{(-1)^j}{j!} \alpha_n^j e_{k-j}(p), \quad 1 \le k \le n-2. \tag{13.39}$$

Combining these two expressions together we find

$$\Psi(\vec{\alpha})_k(e_1, \ldots, e_n) = \sum_{l=0}^{k-1} \sum_{m=0}^{[(k-l-1)/2]} \frac{(-1)^{l+m}}{l!\, m!} \alpha_n^l \alpha_{n-1}^m e_{k-l-2m}, \quad 1 \le k \le n-2. \tag{13.40}$$

We see that these $n-2$ functions involve only the group parameters α_{n-1}, α_n. We can easily determine them out of two equations defining our section Σ. We choose them to be

$$0 = \Psi(\vec{\alpha})_2 = e_2 - \alpha_n e_1, \quad 0 = \Psi(\vec{\alpha})_3 = e_3 - \alpha_n e_2 + \frac{\alpha_n^2}{2} e_1 - \alpha_{n-1} e_1. \tag{13.41}$$

We find

$$\alpha_{n-1} = \frac{1}{e_1^2}\left(e_1 e_3 - \frac{1}{2} e_2^2\right), \quad \alpha_n = \frac{e_2}{e_1}. \tag{13.42}$$

Substituting these back into the remaining equations (13.40) and multiplying all of them by e_1^{k-2} (in order to get polynomial expressions) we find the invariants

$$\begin{aligned} \xi_0 &= e_1, \\ \xi_j &= e_1^{j+1} \tilde{e}_{j+3} \\ &= e_1^{j+1} \sum_{l=0}^{j+2} \sum_{m=0}^{[(j-l)/2+1]} \frac{(-1)^{l+m}}{l!\, m!} \alpha_n^l \alpha_{n-1}^m e_{j+3-l-2m}, \quad 1 \le j \le n-5. \end{aligned} \tag{13.43}$$

where the substitution (13.42) is assumed. Performing it explicitly we arrive at the following theorem.

Theorem 13.6. *The nilpotent Lie algebra $\mathfrak{n}_{n,2}$ has $n-4$ functionally independent Casimir invariants. They can be chosen to be the following polynomials*

$$\xi_0 = e_1,$$

$$\xi_j = \sum_{l=0}^{j+2} \sum_{m=0}^{\left[\frac{j-l}{2}+1\right]} \sum_{q=0}^{m} \frac{(-1)^{l+m+q}}{2^q\, l!(m-q)!q!}\, e_1^{j+1-l-m-q}\, e_2^{l+2q}\, e_3^{m-q}\, e_{j+3-l-2m} \tag{13.44}$$

where $1 \le j \le n-5$. All other Casimir invariants are functions of $\xi_0, \dots, \xi_{n-5}$.

We notice that ξ_j for $j \ge 1$ is a homogeneous polynomial of degree $j+2$. For the reader's convenience we list a few lowest order invariants explicitly (note that the dimension n of $\mathfrak{n}_{n,2}$ does not enter directly into the formulas, it just specifies where the list terminates)

$$\begin{aligned}
\xi_1 &= e_1^2 e_4 - e_1 e_2 e_3 + \tfrac{1}{3} e_2^3,\\
\xi_2 &= e_1^3 e_5 - e_1^2 e_2 e_4 - \tfrac{1}{2} e_1^2 e_3^2 + e_1 e_2^2 e_3 - \tfrac{1}{4} e_2^4,\\
\xi_3 &= e_1^4 e_6 - e_2 e_1^3 e_5 - e_1^3 e_3 e_4 + e_2^2 e_1^2 e_4 + e_1^2 e_2 e_3^2 - e_1 e_2^3 e_3 + \tfrac{1}{5} e_2^5,\\
\xi_4 &= e_1^5 e_7 - e_1^4 e_2 e_6 + e_1^3 e_2^2 e_5 - e_1^4 e_3 e_5 + e_1^3 e_2 e_3 e_4 - \tfrac{2}{3} e_1^2 e_2^3 e_4\\
&\qquad + \tfrac{2}{3} e_1 e_2^4 e_3 + \tfrac{1}{3} e_1^3 e_3^3 - e_1^2 e_2^2 e_3^2 - \tfrac{1}{9} e_2^6.
\end{aligned} \tag{13.45}$$

Let us now consider the $(n+1)$-dimensional solvable Lie algebra $\mathfrak{s}_{n+1}$ of Theorem 13.5. The operators $\widehat{E}_i$ representing elements in the nilradical $\mathfrak{n}_{n,2}$ will each contain an additional term involving a derivative with respect to f_1 and there is one additional operator, namely

$$\widehat{F}_1 = -\sum_{k=1}^{n} (n-k+1) e_k \frac{\partial}{\partial e_k}. \tag{13.46}$$

However, from the form of these operators, namely $\widehat{E}_1$, we see that the invariants cannot depend on f_1. Moreover, they can only depend on the invariants (13.44) of $\mathfrak{n}_{n,2}$. To find the invariants of the algebra $\mathfrak{s}_{n+1}$ we represent the nonnilpotent element $f_1 \in \mathfrak{s}_{n+1}$ by the appropriate "truncated" differential operator acting only on $(e_1, \dots, e_{n-2})$

$$\widehat{F}_{1T} = -\sum_{k=1}^{n-2} (n-k+1) e_k \frac{\partial}{\partial e_k}. \tag{13.47}$$

We must then solve the equation

$$\widehat{F}_{1T} I(\xi_0, \dots, \xi_{n-5}) = 0. \tag{13.48}$$

Let us again employ the method of moving frames, finding the flow of the vector field F_{1T}

$$e_k\big(\Psi_{\widehat{F}_{1T}}^{\alpha_{n+1}} p\big) = \exp(-(n-k+1)\alpha_{n+1})\, e_k(p), \quad 1 \le k \le n-2. \tag{13.49}$$

The full action of the group S_{n+1} on the space with coordinates $e_1, \ldots, e_{n-2}$ gives

$$\begin{aligned} &\Psi(\vec{\alpha})_k(e_1, \ldots, e_n) \\ &= \exp(-(n-k+1)\alpha_{n+1}) \sum_{l=0}^{k-1} \sum_{m=0}^{[(k-l-1)/2]} \frac{(-1)^{l+m}}{l!\, m!} \alpha_n^l \alpha_{n-1}^m e_{k-l-2m}, \\ &\qquad 1 \leq k \leq n-2. \end{aligned} \tag{13.50}$$

We choose our section Σ in the truncated space to be $\{(1, 0, 0, e_4, \ldots, e_{n-2})\}$, i.e., we have one more equation in addition to (13.41)

$$1 = \Psi(\vec{\alpha})_1 = \exp(-n\alpha_{n+1})e_1. \tag{13.51}$$

Solving it we find

$$\exp(-\alpha_{n+1}) = \left(\frac{1}{e_1}\right)^{1/n} \tag{13.52}$$

and substituting it together with (13.42) back into (13.50) we find invariants which can be succinctly expressed in the form

$$\tilde{\xi}_k = \frac{\xi_{k-3}}{e_1^{(jn+2n-j-2)/n}}, \quad 4 \leq k \leq n-2. \tag{13.53}$$

By taking suitable powers (in order to express invariants as ratios of polynomials) we arrive at a theorem.

Theorem 13.7. *The $(n+1)$-dimensional solvable Lie algebra $\mathfrak{s}_{n+1}$ of Theorem 13.5 has $n-5$ functionally independent invariants. They can be chosen in the form*

$$\xi_j = \frac{\xi_j^n}{\xi_0^{(n-1)(j+2)}}, \quad 1 \leq j \leq n-5, \tag{13.54}$$

i.e., they are rational in ξ_k and consequently in e_k.

13.3. Solvable Lie algebras with other filiform nilradicals

Other classes of filiform nilradicals whose solvable extensions have been classified are the special filiform algebra $\mathfrak{Q}_n$ [3] and $\mathbb{N}$-graded filiform algebras [19]. Let us briefly sum up their results.

The results of R. Campoamor-Stursberg et al. for the special filiform algebra $\mathfrak{Q}_n$ resemble the results for the standard filiform algebra $\mathfrak{n}_{n,1}$. Namely:

- The solvable extensions of $\mathfrak{Q}_n$ have at most two nonnilpotent elements.
- The extension with two nonnilpotent elements is unique and has the structure

$$\begin{gathered} [e_k, e_n] = e_{k-1}, \quad [e_k, e_{n-k+1}] = (-1)^k e_1, \qquad k = 2, \ldots, n-1, \\ [f_1, e_k] = (n-k)e_k, \quad k = 1, \ldots, n-1, \qquad [f_1, e_n] = e_n, \\ [f_2, e_1] = 2e_1, \qquad [f_2, e_k] = e_k, \quad k = 2, \ldots, n-1, \qquad [f_2, e_n] = e_{n-1}. \end{gathered} \tag{13.55}$$

- There are several extensions with one nonnilpotent element, some involving arbitrary parameters.

The detailed analysis is performed in the same way as for the nilradical $\mathfrak{n}_{n,1}$ and we shall not present it here.

We remark that the algebras $\mathfrak{n}_{n,1}$ and $\mathfrak{Q}_n$ are the only filiform algebras allowing solvable extensions by two nonnilpotent elements [53, 114]. Any other filiform algebra which is not characteristically nilpotent, has solvable extension(s) by one nonnilpotent element only.

The nilradical $\mathfrak{n}_{n,2}$ can be viewed as a prototype of such algebras. In [19] it was included by R. Campoamor-Stursberg into a larger class of filiform nilradicals, namely $\mathbb{N}$-*graded filiform nilradicals.* By definition these are nilpotent Lie algebras which possess a grading by vector subspaces labelled by natural numbers $1, \dots, \dim \mathfrak{n} - 1$

$$
\begin{aligned}
&\mathfrak{n} = \dot{+}\{\mathfrak{n}_i \mid 1 \le i\}, \qquad \dim \mathfrak{n}_1 = 2, \qquad \dim \mathfrak{n}_i = 1, \quad 2 \le i \le n-1, \\
&\mathfrak{n}_i \equiv 0, \quad n \le i, \qquad [\mathfrak{n}_i, \mathfrak{n}_j] \subseteq \mathfrak{n}_{i+j}, \quad 1 \le i, j, \le n-1.
\end{aligned} \tag{13.56}
$$

(The algebras $\mathfrak{n}_{n,1}$ also belong to this class.) The detailed analysis which was performed in [19] shows that all these algebras with the exception of $\mathfrak{n}_{n,1}$ have the same pattern of their solvable extensions. Namely, for each of the $\mathbb{N}$-graded filiform nilradicals a single, i.e., unique, solvable extension exists and its basis can be chosen so that the nonnilpotent element f_1 leaves the grading subspaces invariant and acts diagonally on them.

13.4. Example of an almost filiform nilradical

In this section we shall consider a series of nilradicals $\mathfrak{n}_{n,3}$ which is not filiform but is nilpotent of degree $n-3$ and contains the model filiform algebra $\mathfrak{n}_{n-2,1}$ as its subalgebra. We shall see that its structure is so restrictive that solvable extensions of this nilradical are closely related to solvable extensions of the nilradical $\mathfrak{n}_{n-2,1}$. Solvable extensions of the nilradical $\mathfrak{n}_{n,3}$ were first classified in [115]; our calculation here is based on this article.

The nilradical is

$$\mathfrak{n}_{n,3} = \operatorname{span}\{x_1, \dots, x_n\}, \quad n \ge 5$$

with the nonvanishing Lie brackets

$$
\begin{aligned}
[x_2, x_n] &= x_1, \\
[x_3, x_{n-1}] &= x_1, \\
[x_k, x_{n-1}] &= x_{k-1}, \qquad 4 \le k \le n-2, \\
[x_{n-1}, x_n] &= x_2.
\end{aligned} \tag{13.57}
$$

During the computation we shall assume that the dimension satisfies $n \ge 7$. Dimensions $n = 5$ and $n = 6$ will be treated separately below.

The n-dimensional nilpotent Lie algebra $\mathfrak{n}_{n,3}$ has an $(n-2)$-dimensional maximal Abelian ideal spanned by $\{x_1, \dots, x_{n-2}\}$ and one-dimensional center $C(\mathfrak{n}_{n,3}) = \operatorname{span}\{x_1\}$. The immersion of $\mathfrak{n}_{n-2,1}$ into $\mathfrak{n}_{n,3}$ is denoted by $\tilde{\mathfrak{n}}_{n-2,1}$ and is spanned by $x_1, x_3, \dots, x_{n-1}$. Similarly, $\mathfrak{n}_{n,3}$ also contains a subalgebra $\tilde{\mathfrak{n}}_{6,3}$ spanned by $x_1, x_2, x_3, x_4, x_{n-1}, x_n$. Neither $\tilde{\mathfrak{n}}_{n-2,1}$, nor $\tilde{\mathfrak{n}}_{6,3}$ are ideals.

In general, the knowledge of solvable extensions of a subalgebra of the given nilradical does not provide much help in the classification of all solvable extensions of the nilradical. That is because the outer derivations of the nilradical need not leave the subalgebra invariant. Indeed, the subalgebra is not invariant even with respect to inner derivations. However, in the particular case of the nilradical $\mathfrak{n}_{n,3}$ the classification can be reduced to the cases of $\mathfrak{n}_{n-2,1}$ treated above in Section 13.1 complemented by a reduction to the case of the nilradical $\mathfrak{n}_{6,3}$.

In the following we shall first determine the general form of an automorphism and a derivation of $\mathfrak{n}_{n,3}$. Next, we use this knowledge in the construction of all solvable extensions of the nilradical $\mathfrak{n}_{n,3}$. Finally, we deduce generalized Casimir invariants of both $\mathfrak{n}_{n,3}$ and its solvable extensions.

13.4.1. Automorphisms and derivations of the nilradical $\mathfrak{n}_{n,3}$. The nilpotent algebra $\mathfrak{n} = \mathfrak{n}_{n,3}$ has the following complete flag of ideals

$$
\begin{aligned}
(13.58)\quad & 0 \subset \mathfrak{n}^{n-3} \subset \mathfrak{n}^{n-4} \subset \mathfrak{z}_2 \subset \mathfrak{z}_3 \cap \mathfrak{n}^2 \subset \dots \\
& \dots \subset \mathfrak{z}_{n-5} \cap \mathfrak{n}^2 \subset \mathfrak{n}^2 \subset \operatorname{cent}_{\mathfrak{n}}(\mathfrak{z}_2) \subset \operatorname{cent}_{\mathfrak{n}}(\mathfrak{n}^{n-4}) \subset \mathfrak{n}.
\end{aligned}
$$

The flag (13.58) is invariant with respect to any automorphism of the Lie algebra $\mathfrak{n}_{n,3}$, i.e., in the basis respecting the flag any automorphism will be represented by an upper triangular matrix. The same triangular form holds also for derivations.

Therefore, we find it convenient to change the basis (x_k) of $\mathfrak{n}$ defined in (13.57) to a seemingly less natural (i.e., Lie brackets appear more cumbersome) basis (e_k) whose essential advantage over the original one is that it respects the flag (13.58), i.e., the kth subspace in the flag is $\operatorname{span}\{e_1, \dots, e_k\}$ for all k. Namely, we take

$$
\begin{aligned}
& e_1 = x_1, \quad e_2 = x_3, \quad e_3 = x_2, \quad e_4 = x_4, \dots, \\
& e_{n-2} = x_{n-2}, \quad e_{n-1} = x_n, \quad e_n = x_{n-1}.
\end{aligned} \tag{13.59}
$$

The nonvanishing Lie brackets now become

$$
\begin{aligned}
[e_2, e_n] &= e_1, \\
[e_3, e_{n-1}] &= e_1, \\
[e_4, e_n] &= e_2, \\
[e_k, e_n] &= e_{k-1}, \qquad 5 \le k \le n-2, \\
[e_{n-1}, e_n] &= -e_3.
\end{aligned} \tag{13.60}
$$

The important subalgebras isomorphic to $\mathfrak{n}_{n-2,1}$ and $\mathfrak{n}_{6,3}$ are now expressed as

$$
\tilde{\mathfrak{n}}_{n-2,1} = \operatorname{span}\{e_1, e_2, e_4, \dots, e_{n-2}, e_n\}, \quad \tilde{\mathfrak{n}}_{6,3} = \operatorname{span}\{e_1, e_2, e_3, e_4, e_{n-1}, e_n\},
$$

respectively. The ideals in the derived, lower central and upper central series are

$$
\begin{gathered}
\mathfrak{n}^2 = \mathfrak{n}^{(1)} = \operatorname{span}\{e_1, \dots, e_{n-3}\}, \qquad \mathfrak{n}^{(2)} = 0, \\
\mathfrak{n}^k = \operatorname{span}\{e_1, e_2, e_4, \dots, e_{n-k-1}\}, \quad 3 \le k \le n-5, \\
\mathfrak{n}^{n-4} = \operatorname{span}\{e_1, e_2\}, \qquad \mathfrak{n}^{n-3} = \operatorname{span}\{e_1\}, \qquad \mathfrak{n}^{n-2} = 0, \\
\mathfrak{z}_1 = \mathfrak{n}^{n-3}, \qquad \mathfrak{z}_2 = \operatorname{span}\{e_1, e_2, e_3\}, \\
\mathfrak{z}_k = \operatorname{span}\{e_1, \dots, e_{k+1}, e_{n-1}\}, \quad 3 \le k \le n-4, \qquad \mathfrak{z}_{n-3} = \mathfrak{n}.
\end{gathered}
$$

In order to find the structure of an arbitrary automorphism of $\mathfrak{n}_{n,3}$ we consider its matrix in the basis (13.59)

$$
\Phi(e_k) = \sum_{j=1}^{n} e_j \Phi^j_k. \tag{13.61}
$$

As mentioned above, such a matrix must necessarily be upper triangular because the flag (13.58) is preserved. It is also obvious that the knowledge of its last three columns, i.e., of its action on the generators $\Phi(e_{n-2}), \Phi(e_{n-1})$ and $\Phi(e_n)$, is sufficient for the knowledge of the whole matrix Φ. A natural question is the following: Under which conditions do the relations

$$\begin{aligned} \Phi(e_{n-2}) &= \alpha e_{n-2} + \sum_{k=1}^{n-3} \phi_k e_k, \\ \Phi(e_{n-1}) &= \beta e_{n-1} + \gamma e_{n-2} + \sum_{k=1}^{n-3} \psi_k e_k, \\ \Phi(e_n) &= \kappa e_n + \lambda e_{n-1} + \mu e_{n-2} + \sum_{k=1}^{n-3} \rho_k e_k \end{aligned} \tag{13.62}$$

give rise to an automorphism of $\mathfrak{n}_{n,3}$?

We must have $\alpha\beta\kappa \neq 0$ to have an invertible map. The preservation of $\mathfrak{z}_3$ implies $\gamma = 0$, $\psi_k = 0$, $k = 5, \dots, n-3$. The remaining conditions are found as follows

- $0 = \Phi([e_{n-2}, e_{n-1}])$ implies $\phi_3 = 0$,
- $0 = \Phi([[e_{n-1}, e_n], e_n])$ leads to $\psi_4 = (\lambda/\kappa)\beta$,
- $0 = \Phi([[e_{n-1}, e_n], e_{n-1}]) + \Phi((-\operatorname{ad}_{e_n})^{n-4} e_{n-2})$ leads to $\alpha = \beta^2\kappa^{5-n}$.

All other Lie brackets are either used to define $\Phi(e_k)$, $1 \leq k \leq n-3$ or are preserved trivially. Therefore, we conclude that any automorphism Φ of $\mathfrak{n}_{n,3}$ is defined in terms of $2n$ parameters which have been denoted by $\beta, \kappa, \lambda, \psi_1, \psi_2, \psi_3, \phi_1, \phi_2, \phi_4, \dots, \phi_{n-3}, \rho_1, \dots, \rho_{n-3}$. It acts on the generators of the Lie algebra $\mathfrak{n}_{n,3}$ in the following way:

$$\begin{aligned} \Phi(e_{n-2}) &= \beta^2\kappa^{5-n} e_{n-2} + \sum_{k=4}^{n-3} \phi_k e_k + \phi_2 e_2 + \phi_1 e_1, \\ \Phi(e_{n-1}) &= \beta e_{n-1} + \frac{\lambda}{\kappa}\beta e_4 + \sum_{k=1}^{3} \psi_k e_k, \\ \Phi(e_n) &= \kappa e_n + \lambda e_{n-1} + \mu e_{n-2} + \sum_{k=1}^{n-3} \rho_k e_k. \end{aligned} \tag{13.63}$$

Taking automorphisms infinitesimally close to unity, i.e., constructing the Lie algebra of the group of automorphisms, we find the algebra of derivations $\mathfrak{Der}(\mathfrak{n}_{n,3})$. It consists of all linear maps D which act on the generators e_{n-2}, e_{n-1}, e_n as follows:

$$\begin{aligned} D(e_{n-2}) &= (2c_{n-1} + (5-n)d_n)e_{n-2} + \sum_{k=4}^{n-3} b_k e_k + b_2 e_2 + b_1 e_1, \\ D(e_{n-1}) &= c_{n-1}e_{n-1} + d_{n-1}e_4 + \sum_{k=1}^{3} c_k e_k, \\ D(e_n) &= \sum_{k=1}^{n} d_k e_k; \end{aligned} \tag{13.64}$$

the action of D on the remaining basis elements $e_1, \dots, e_{n-3}$ is uniquely determined using multiple brackets and the definition of a derivation (2.26). The $2n$-dimensional algebra of derivations $\mathfrak{Der}(\mathfrak{n}_{n,3})$ contains an $(n-1)$-dimensional ideal of inner derivations $\mathfrak{Inn}(\mathfrak{n}_{n,3})$ having the form

$$D(e_{n-2}) = -c_3 e_{n-3},\quad D(e_{n-1}) = c_3 e_3 + c_1 e_1,\quad D(e_n) = \sum_{k=1}^{n-3} d_k e_k. \tag{13.65}$$

Indeed, such a derivation D can be expressed as

$$D = \operatorname{ad}\left(d_1 e_2 + c_1 e_3 + d_2 e_4 + \sum_{k=4}^{n-3} d_k e_{k+1} - d_3 e_{n-1} + c_3 e_n\right). \tag{13.66}$$

Since e_1 spans the kernel of ad, i.e., the center of $\mathfrak{n}_{n,3}$, derivations of the form (13.65) exhaust all inner derivations.

13.4.2. Construction of solvable Lie algebras with the nilradical $\mathfrak{n}_{n,3}$. Let us start by considering one additional basis element $f_1 \equiv f$, i.e., one derivation D. The elements of $\mathfrak{Der}(\mathfrak{n}_{n,3})/\mathfrak{Inn}(\mathfrak{n}_{n,3})$ can be uniquely represented by outer derivations of the form

$$\begin{aligned} D(e_{n-2}) &= (2c_{n-1} + (5-n)d_n)e_{n-2} + \sum_{k=4}^{n-4} b_k e_k + b_2 e_2 + b_1 e_1, \\ D(e_{n-1}) &= c_{n-1} e_{n-1} + d_{n-1} e_4 + c_3 e_3 + c_2 e_2, \\ D(e_n) &= d_n e_n + d_{n-1} e_{n-1} + d_{n-2} e_{n-2}. \end{aligned} \tag{13.67}$$

Above, a suitable inner derivation (13.65) was added to an arbitrary derivation (13.64), eliminating $n-1$ parameters. We mention that the form (13.67) of the representative of the coset $[D]$ is not invariant under conjugation by an automorphism

$$D \to D_\Phi = \Phi^{-1} \circ D \circ \Phi$$

so that we may be forced to use a representative $\Phi(D)'$ of the coset $[\Phi(D)]$ different from $\Phi(D)$. Such a change of representative amounts to an addition of an inner derivation and is understood in all simplifications below whenever we employ an automorphism. Due to the triangular shape of D we see that the sought-after Abelian subalgebras are at most two-dimensional since any higher-dimensional subalgebra in $\mathfrak{Der}(\mathfrak{n}_{n,3})/\mathfrak{Inn}(\mathfrak{n}_{n,3})$ will necessarily involve nonvanishing nilpotent elements.

Next, we find all possible canonical forms of the coset (13.67) up to conjugation by automorphisms and rescaling. In order to reduce the problem to the one already investigated in Section 13.1 we realize that the derivation of the form (13.67) leaves

$$\tilde{\mathfrak{n}}_{n-2,1} = \operatorname{span}\{e_1, e_2, e_4, \ldots, e_{n-2}, e_n\}$$

invariant if and only if $d_{n-1} = 0$. We conjugate a given derivation D by the automorphism defined by

$$\Phi(e_{n-2}) = e_{n-2}, \quad \Phi(e_{n-1}) = e_{n-1} + \frac{d_{n-1}}{d_n - c_{n-1}} e_4, \quad \Phi(e_n) = e_n + \frac{d_{n-1}}{d_n - c_{n-1}} e_{n-1}$$

whenever possible, i.e., when $d_n \neq c_{n-1}$. Now we have $\hat{d}_{n-1} = 0$, i.e., $D_\Phi \equiv \hat{D}$ leaves $\tilde{\mathfrak{n}}_{n-2,1}$ invariant. The case when none of the conjugate derivations D_Φ leaves $\tilde{\mathfrak{n}}_{n-2,1}$ invariant, which necessarily means that $d_n = c_{n-1}$, $d_{n-1} \neq 0$, will be dealt with later on, on page 195.

Provided we set $d_{n-1} = 0$, the outer derivation (13.67) restricted to $\tilde{\mathfrak{n}}_{n-2,1}$ has the same structure as in (13.9) in Section 13.1. Consequently, we may consider all solvable extensions of $\tilde{\mathfrak{n}}_{n-2,1}$ and then extend these to solvable extensions of $\mathfrak{n}_{n,3}$, i.e., determine the parameters c_{n-1}, c_3, c_2. In this way we obtain all solvable extensions of $\mathfrak{n}_{n,3}$ except the case $d_n = c_{n-1}$, $d_{n-1} \neq 0$.

The value of the parameter c_{n-1} is fixed by the structure of the solvable extension of $\tilde{\mathfrak{n}}_{n-2,1}$. Namely, in relation to parameters α, β introduced in Theorem 13.2 we have

$$c_{n-1} = \tfrac{1}{2}(\beta + (n-5)\alpha), \quad d_n = \alpha.$$

When $c_{n-1} \neq 0$ any derivation D can be brought to D_Φ with $c_2 = 0$ using an automorphism Φ specified by

$$\Phi(e_{n-2}) = e_{n-2}, \quad \Phi(e_{n-1}) = e_{n-1} - \frac{c_2}{c_{n-1}} e_2, \quad \Phi(e_n) = e_n.$$

When $c_{n-1} = 0$ we cannot eliminate nonvanishing c_2 by any automorphism but we scale it to 1 by rescaling of e_k's provided such scaling remains available by the structure of the solvable extension of the subalgebra $\tilde{\mathfrak{n}}_{n-2,1}$. It turns out that for $c_{n-1} = 0$ two nonconjugate extensions of a derivation of $\tilde{\mathfrak{n}}_{n-2,1}$ exist, namely those determined by $c_2 = 0, 1$.

A similar consideration can be applied also to the parameter c_3. When $d_n \neq 0$ any derivation D can be brought to D_Φ with $c_3 = 0$ using the automorphism Φ specified by

$$\Phi(e_{n-2}) = e_{n-2}, \quad \Phi(e_{n-1}) = e_{n-1} - \frac{c_3}{d_n} e_3, \quad \Phi(e_n) = e_n.$$

When $d_n = 0$ we cannot eliminate nonvanishing c_3 by any automorphism. Whether or not c_3 can be rescaled depends on the residual automorphisms still available—if the diagonal part of automorphisms is completely fixed by the structure of the solvable extension of the subalgebra $\tilde{\mathfrak{n}}_{n-2,1}$ nothing can be done, otherwise we can scale c_3 to 1 using an automorphism

$$\Phi(e_{n-2}) = e_{n-2}, \quad \Phi(e_{n-1}) = e_{n-1}, \quad \Phi(e_n) = \frac{1}{c_3} e_n.$$

To sum up, the extension to a derivation of the nilradical $\mathfrak{n}_{n,3}$ is unique up to conjugation when $d_n \neq 0$ and $c_{n-1} \neq 0$; otherwise, several nonequivalent extensions exist.

Now we employ Theorem 13.2 and extend its results to $\mathfrak{n}_{n,3}$. However there is a small caveat in this procedure. If we work over the field $\mathbb{R}$ the group of automorphisms of $\mathfrak{n}_{n-2,1}$ used in the derivation of Theorem 13.2 is slightly larger than the one we have available for the subalgebra $\tilde{\mathfrak{n}}_{n-2,1}$, i.e., inherited from automorphisms of $\mathfrak{n}_{n,3}$. In other words, the available automorphisms form a group only locally isomorphic to the group of automorphisms of $\mathfrak{n}_{n-2,1}$. More specifically, the sign of $\alpha = \beta^2 \kappa^{5-n}$ in (13.63) is restricted, i.e., for given n we have $\operatorname{sgn}\alpha = (\operatorname{sgn}\kappa)^{n-5}$. As a consequence, for our purposes we must for n even consider $[\tilde{f}, \tilde{e}_m] = \tilde{e}_m \pm \tilde{e}_{m-1}$ in $\tilde{\mathfrak{s}}_{m+1,5}$ $(m = n-2)$.

The corresponding solvable extensions of the nilradical $\mathfrak{n}_{n,3}$ are summarized in Theorem 13.8 below.

Coming back to the case $d_n = c_{n-1}$, $d_{n-1} \neq 0$, we first rescale D to get $d_n = c_{n-1} = 1$ and by scaling of e_k's we set $d_{n-1} = 1$. Using the automorphism

$$\Phi(e_{n-2}) = e_{n-2}, \quad \Phi(e_{n-1}) = e_{n-1}, \quad \Phi(e_n) = e_n + \frac{d_{n-2}}{n-6} e_{n-2}$$

we get rid of d_{n-2}; this is possible since $n \neq 6$. We get D which preserves the subalgebra $\tilde{\mathfrak{n}}_{6,3}$. Therefore, it is enough to consider its solvable extensions (with $d_n = c_{n-1} = 1$) and then extend these to solvable algebras with the nilradical $\mathfrak{n}_{n,3}$. It turns out that such an extension is unique up to conjugation, i.e., fully determined by $d_n = c_{n-1} = 1, d_{n-1} = 1, d_{n-2} = 0$, the remaining parameters in (13.67) vanish.

Finally, the two-dimensional Abelian subalgebras $\mathfrak{a}$ in $\mathfrak{Der}(\mathfrak{n}_{n,3})/\mathfrak{Inn}(\mathfrak{n}_{n,3})$ are easily obtained using the results of the previous analysis. Such subalgebras must contain two linearly independent elements D_1', D_2', whose diagonal parameters can be chosen to have the values $c_{n-1} = 1$, $d_n = -1$ and $c_{n-1} = 1$, $d_n = 0$, respectively. Due to the chosen values for D_1 we can always go over to $\widetilde{D}_1 = (D_1')_\Phi, \widetilde{D}_2 = (D_2')_\Phi$ where $\widetilde{D}_1$ was diagonalized by a suitable automorphism Φ. The restriction $[\widetilde{D}_1, \widetilde{D}_2] \in \mathfrak{Inn}(\mathfrak{n}_{n,3})$ now restricts $\widetilde{D}_2$ to be also diagonal. Therefore, all elements of $\mathfrak{a}$ act diagonally on $\mathfrak{n}_{n,3}$ in the chosen basis and can be expressed, e.g., in the basis defined by D_1 ($c_{n-1} = 0$, $d_n = 1$) and D_2 ($c_{n-1} = 1$, $d_n = 0$). The corresponding nonnilpotent elements f_1, f_2 in $\mathfrak{s}$ in general satisfy

$$[f_1, f_2] = \alpha e_1 \in C(\mathfrak{n})$$

but a simple redefinition $f_1 \to f_1 + (\alpha/2)e_1$ gives an isomorphic solvable algebra $\mathfrak{s}$ with $[f_1, f_2] = 0$.

To sum up, we have the following theorem

Theorem 13.8. *Let $\mathbb{F}$ be the field of real or complex numbers and n be an integer number greater or equal to 7. Any solvable Lie algebra $\mathfrak{s}$ over the field $\mathbb{F}$ with the nilradical $\mathfrak{n}_{n,3}$ has dimension $\dim \mathfrak{s} = n + 1$ or $\dim \mathfrak{s} = n + 2$.*

Five types of solvable Lie algebras of dimension $\dim \mathfrak{s} = n + 1$ with the nilradical $\mathfrak{n}_{n,3}$ exist. They are represented by the following:

(1) *The Lie brackets with f_1 acting diagonally on $\mathfrak{n}_{n,3}$*

$$[f_1, e_1] = (\alpha + 2\beta)e_1, \qquad [f_1, e_2] = 2\beta e_2, \qquad [f_1, e_3] = (\alpha + \beta)e_3,$$
$$[f_1, e_k] = ((3 - k)\alpha + 2\beta)e_k, \quad 4 \leq k \leq n - 2,$$
$$[f_1, e_{n-1}] = \beta e_{n-1}, \qquad [f_1, e_n] = \alpha e_n.$$

The classes of mutually nonisomorphic algebras of this type are

$\mathfrak{s}_{n+1,1}(\beta)$: $\alpha = 1$, $\beta \in \mathbb{F}\backslash\{0, -\frac{1}{2}, (n-5)/2\}$,

$$\mathrm{US} = [0], \quad \mathrm{CS} = [n+1, n], \quad \mathrm{DS} = [n+1, n, n-3, 0],$$

$\mathfrak{s}_{n+1,2}$: $\alpha = 1$, $\beta = (n-5)/2$,

$$\mathrm{US} = [0], \quad \mathrm{CS} = [n+1, n-1], \quad \mathrm{DS} = [n+1, n-1, n-4, 0],$$

$\mathfrak{s}_{n+1,3}$: $\alpha = 1$, $\beta = 0$,

$$\mathrm{US} = [0], \quad \mathrm{CS} = [n+1, n-1], \quad \mathrm{DS} = [n+1, n-1, n-4, 0],$$

$\mathfrak{s}_{n+1,4}$: $\alpha = 1$, $\beta = -\frac{1}{2}$,

$$\mathrm{US} = [1], \quad \mathrm{CS} = [n+1, n], \quad \mathrm{DS} = [n+1, n, n-3, 0],$$

$\mathfrak{s}_{n+1,5}$: $\alpha = 0$, $\beta = 1$,

$$\mathrm{US} = [0], \quad \mathrm{CS} = [n+1, n-1], \quad \mathrm{DS} = [n+1, n-1, 1, 0].$$

(2) $\mathfrak{s}_{n+1,6}(\epsilon)$:

$$[f_1, e_1] = (n-3)e_1, \qquad [f_1, e_2] = (n-4)e_2,$$
$$[f_1, e_3] = \left(1 + \frac{n-4}{2}\right)e_3, \qquad [f_1, e_k] = (n-1-k)e_k, \quad 4 \leq k \leq n-2,$$
$$[f_1, e_{n-1}] = \frac{n-4}{2}e_{n-1}, \qquad [f_1, e_n] = e_n + \epsilon e_{n-2},$$

where $\epsilon = 1$ over $\mathbb{C}$, whereas over $\mathbb{R}$ $\epsilon = 1$ for n odd, $\epsilon = \pm 1$ for n even.

$$\mathrm{US} = [0], \quad \mathrm{CS} = [n+1, n], \quad \mathrm{DS} = [n+1, n, n-3, 0].$$

(3) $\mathfrak{s}_{n+1,7}$:

$$[f_1, e_1] = e_1, \qquad [f_1, e_2] = 0, \qquad [f_1, e_3] = e_3 - e_1,$$
$$[f_1, e_k] = (3-k)e_k, \quad 4 \leq k \leq n-2,$$
$$[f_1, e_{n-1}] = e_2, \qquad [f_1, e_n] = e_n,$$
$$\mathrm{US} = [0], \quad \mathrm{CS} = [n+1, n-1], \quad \mathrm{DS} = [n+1, n-1, n-4, 0].$$

(4) $\mathfrak{s}_{n+1,8}(a_2, a_3, \ldots, a_{n-3})$:

$$[f_1, e_1] = e_1, \qquad [f, e_2] = e_2, \qquad [f_1, e_3] = \tfrac{1}{2}e_3,$$
$$[f_1, e_k] = e_k + \sum_{l=4}^{k-2} a_{k-l+1}e_l + a_{k-2}e_2 + a_{k-1}e_1, \qquad 4 \leq k \leq n-2,$$
$$[f_1, e_{n-1}] = \tfrac{1}{2}e_{n-1} + a_2 e_3, \qquad [f_1, e_n] = 0,$$

$a_j \in \mathbb{F}$, at least one a_j satisfies $a_j \neq 0$ and:

- *when $\mathbb{F} = \mathbb{C}$ the first nonzero a_j satisfies $a_j = 1$.*
- *when $\mathbb{F} = \mathbb{R}$ the first nonzero a_j for even j satisfies $a_j = 1$. If all $a_j = 0$ for j even, then the first nonzero a_j (j odd) satisfies $a_j = \pm 1$.*

$$\mathrm{US} = [0], \quad \mathrm{CS} = [n+1, n-1], \quad \mathrm{DS} = [n+1, n-1, 1, 0].$$

(5) $\mathfrak{s}_{n+1,9}$:

$$[f_1, e_1] = 3e_1, \qquad [f_1, e_2] = 2e_2, \qquad [f_1, e_3] = 2e_3 - e_2,$$
$$[f_1, e_k] = (5-k)e_k, \quad 4 \leq k \leq n-2,$$
$$[f_1, e_{n-1}] = e_{n-1} + e_4, \qquad [f_1, e_n] = e_n + e_{n-1},$$
$$\mathrm{US} = [0], \quad \mathrm{CS} = [n+1, n], \quad \mathrm{DS} = [n+1, n, n-3, 0].$$

Exactly one solvable Lie algebra $\mathfrak{s}_{n+2}$ of $\dim \mathfrak{s} = n+2$ with the nilradical $\mathfrak{n}_{n,3}$ exists. It is presented in a basis $(e_1, \ldots, e_n, f_1, f_2)$ where the Lie brackets involving

f_1 and f_2 are

$$\begin{array}{lll} [f_1, e_1] = e_1, & [f_2, e_1] = 2e_1, & \\ [f_1, e_2] = 0, & [f_2, e_2] = 2e_2, & \\ [f_1, e_3] = e_3, & [f_2, e_3] = e_3, & \\ [f_1, e_k] = (3-k)e_k, & [f_2, e_k] = 2e_k, & 4 \leq k \leq n-2, \\ [f_1, e_{n-1}] = 0, & [f_2, e_{n-1}] = e_{n-1}, & \\ [f_1, e_n] = e_n, & [f_2, e_n] = 0, & [f_1, f_2] = 0. \end{array}$$

For this algebra we have

$$\mathrm{US} = [0], \quad \mathrm{CS} = [n+2, n], \quad \mathrm{DS} = [n+2, n, n-3, 0].$$

We note that the class $\mathfrak{s}_{n+1,8}(a_2, a_3, \ldots, a_{n-3})$ encompasses both extensions of $\tilde{\mathfrak{s}}_{m+1,7}(a_3, \ldots, a_{m-1})$ and an extension of $\tilde{\mathfrak{s}}_{m+1,4}$ with $c_3 \neq 0$ in (13.67). The parameter brought to ± 1 was selected in the most convenient form for presentation and consequently is equivalent, but slightly different from a direct extension of $\tilde{\mathfrak{s}}_{m+1,7}(a_3, \ldots, a_{m-1})$ to the nilradical $\mathfrak{n}_{n,3}$; for that choice the nonequivalent values of parameters would be more cumbersome to write down.

Next, we investigate the classification of solvable extensions of $\mathfrak{n}_{n,2}$ in low dimensions $n = 6, 5$. Results in these dimensions differ somewhat from the general ones presented in Theorem 13.8.

13.4.3. Dimension $n = 6$. When $n = 6$ the results are as follows: all the algebras presented in Theorem 13.8 exist (with $e_{n-2} \equiv e_4$) but they do not exhaust all the possibilities. The reason for this is that in this particular dimension we have $[f_1, e_{n-2}] = (2c_5 - d_6)e_{n-2} + \cdots$. Therefore, if $d_6 = c_5$ then also $[f_1, e_{n-2}] = d_6 e_{n-2} + \cdots$. That implies that if we have $d_6 = c_5 \to 1$, $d_5 \neq 0$, $d_4 \neq 0$ in the derivation (13.67) then we cannot set d_5, nor d_4 equal to zero by any choice of automorphism Φ and we are left with only one scaling available — preferably used to set $d_5 \to 1$.

That means that for the 6-dimensional nilradical $\mathfrak{n}_{6,3}$ we have solvable extensions $\mathfrak{s}_{7,1}(\beta), \mathfrak{s}_{7,2}, \mathfrak{s}_{7,3}, \mathfrak{s}_{7,4}, \mathfrak{s}_{7,5}, \mathfrak{s}_{7,6}(\epsilon), \mathfrak{s}_{7,7}, \mathfrak{s}_{7,8}(1, a_3), \mathfrak{s}_{7,8}(0, \epsilon), \mathfrak{s}_{7,9}, \mathfrak{s}_8$ where $\epsilon = 1$ over $\mathbb{C}$ and $\epsilon = \pm 1$ over $\mathbb{R}$, whose structure is as described in Theorem 13.8 above and one additional class of algebras, differing from $\mathfrak{s}_{7,9}$ by one additional nonvanishing parameter α

- $\mathfrak{s}_{7,10}(\alpha)$, $\alpha \neq 0$:

$$\begin{array}{lll} [f_1, e_1] = 3e_1, & [f_1, e_2] = 2e_2, & [f_1, e_3] = 2e_3 - e_2, \\ [f_1, e_4] = e_4, & [f_1, e_5] = e_5 + e_4, & [f_1, e_6] = e_6 + e_5 + \alpha e_4, \end{array}$$

$$\mathrm{US} = [0], \quad \mathrm{CS} = [7, 6], \quad \mathrm{DS} = [7, 6, 3, 0].$$

13.4.4. Dimension $n = 5$. When $n = 5$, the investigation must be performed in a different way. The nilradical has the following nonvanishing Lie brackets

$$[x_2, x_5] = x_1, \quad [x_3, x_4] = x_1, \quad [x_4, x_5] = x_2, \tag{13.68}$$

i.e., becomes the algebra $\mathfrak{n}_{5,4}$ presented in the list of 5-dimensional nilpotent algebras in Part 4, Section 18.1. The algebra $\mathfrak{n}_{5,4}$ contains no $\tilde{\mathfrak{n}}_{3,1}$ subalgebra — it has collapsed to the Heisenberg algebra which has different properties. Nevertheless, by a rather straightforward computation resembling those in Sections 13.1 and 13.2

one can construct all its solvable extensions. We only list the results and compare them to their higher-dimensional analogues. In order to make our comparison as simple as possible we work in a basis analogous to (13.59), namely

$$e_1 = x_1, \quad e_2 = x_3, \quad e_3 = x_2, \quad e_4 = x_5, \quad e_5 = x_4. \tag{13.69}$$

(Notice that the basis chosen for presentation in Section 18.1 below corresponds to the basis $(x_1, \ldots x_5)$ here.) The nonvanishing Lie brackets in the new basis $\{e_1, \ldots, e_5\}$ become

$$[e_2, e_5] = e_1, \quad [e_3, e_4] = e_1, \quad [e_4, e_5] = -e_3. \tag{13.70}$$

Although the structure of the nilradical is quite different from the other elements of the series considered in this section, the set of solvable extensions is rather similar. We get analogues of all solvable algebras in Theorem 13.8 with some changes in the structure of $\mathfrak{s}_{n+1,6}$, $\mathfrak{s}_{n+1,8}$, $\mathfrak{s}_{n+1,9}$; in addition, the two algebras $\mathfrak{s}_{n+1,2}$ and $\mathfrak{s}_{n+1,3}$ become identical when $n = 5$. The fact that the algebras $\mathfrak{s}_{n+1,6}$, $\mathfrak{s}_{n+1,8}$, $\mathfrak{s}_{n+1,9}$ must be modified when $n = 5$ can be inferred already from Theorem 13.8 since the Lie brackets as presented there cannot be made sense of if $n = 5$. These structurally different analogues are distinguished by primes below.

Explicitly, assuming the structure of $\mathfrak{n}_{5,4}$ in the form (13.70), we have the following Lie brackets with nonnilpotent element(s) and dimensions of the characteristic series. The 6-dimensional algebras are presented using two notations: the first one corresponds to the notation used in this section in any dimension, the second one refers to the list presented in Section 19.8. We use a hat over $\mathfrak{s}$ to distinguish these two notations.

- $\hat{\mathfrak{s}}_{6,1}(\beta) \simeq \mathfrak{s}_{6,184}$, $\beta \in \mathbb{F} \backslash \{0, -\frac{1}{2}\}$

$$[f_1, e_1] = (1+2\beta)e_1, \quad [f_1, e_2] = 2\beta e_2, \quad [f_1, e_3] = (\beta+1)e_3,$$
$$[f_1, e_4] = \beta e_4, \quad [f_1, e_5] = e_5,$$
$$\text{US} = [0], \quad \text{CS} = [6, 5], \quad \text{DS} = [6, 5, 2, 0].$$

- $\hat{\mathfrak{s}}_{6,2} \simeq \mathfrak{s}_{6,189}$

$$[f_1, e_1] = e_1, \quad [f_1, e_2] = 0, \quad [f_1, e_3] = e_3, \quad [f_1, e_4] = 0, \quad [f_1, e_5] = e_5,$$
$$\text{US} = [0], \quad \text{CS} = [6, 3], \quad \text{DS} = [6, 3, 0].$$

- $\hat{\mathfrak{s}}_{6,4} \simeq \mathfrak{s}_{6,183}$

$$[f_1, e_1] = 0, \quad [f_1, e_2] = -e_2, \quad [f_1, e_3] = \tfrac{1}{2}e_3, \quad [f_1, e_4] = -\tfrac{1}{2}e_4, \quad [f_1, e_5] = e_5,$$
$$\text{US} = [1], \quad \text{CS} = [6, 5], \quad \text{DS} = [6, 5, 2, 0].$$

- $\hat{\mathfrak{s}}_{6,5} \simeq \mathfrak{s}_{6,187,\sigma=0}$

$$[f_1, e_1] = 2e_1, \quad [f_1, e_2] = 2e_2, \quad [f_1, e_3] = e_3, \quad [f_1, e_4] = e_4, \quad [f_1, e_5] = 0,$$
$$\text{US} = [0], \quad \text{CS} = [6, 4], \quad \text{DS} = [6, 4, 1, 0].$$

- $\hat{\mathfrak{s}}'_{6,6} \simeq \mathfrak{s}_{6,186}$

$$[f_1, e_1] = 2e_1, \quad [f_1, e_2] = e_2, \quad [f_1, e_3] = \tfrac{3}{2}e_3, \quad [f_1, e_4] = \tfrac{1}{2}e_4, \quad [f_1, e_5] = e_5 + e_2,$$
$$\text{US} = [0], \quad \text{CS} = [6, 5], \quad \text{DS} = [6, 5, 2, 0].$$

• $\hat{\mathfrak{s}}_{6,7} \simeq \mathfrak{s}_{6,188}$

$$[f_1, e_1] = e_1, \quad [f_1, e_2] = 0, \quad [f_1, e_3] = e_3 - e_1, \quad [f_1, e_4] = e_2, \quad [f_1, e_5] = e_5,$$
$$\mathrm{US} = [0], \quad \mathrm{CS} = [6, 4, 3], \quad \mathrm{DS} = [6, 4, 1, 0].$$

• $\hat{\mathfrak{s}}'_{6,8} \simeq \mathfrak{s}_{6,187,\sigma=1}$

$$[f_1, e_1] = 2e_1, \quad [f_1, e_2] = 2e_2, \quad [f_1, e_3] = e_3, \quad [f_1, e_4] = -e_3 + e_4, \quad [f_1, e_5] = 0,$$
$$\mathrm{US} = [0], \quad \mathrm{CS} = [6, 4], \quad \mathrm{DS} = [6, 4, 1, 0].$$

• $\hat{\mathfrak{s}}'_{6,9} \simeq \mathfrak{s}_{6,185}$

$$[f_1, e_1] = 3e_1, \quad [f_1, e_2] = 2e_2 - e_3, \quad [f_1, e_3] = 2e_3,$$
$$[f_1, e_4] = e_4 + e_5, \quad [f_1, e_5] = e_5,$$
$$\mathrm{US} = [0], \quad \mathrm{CS} = [6, 5], \quad \mathrm{DS} = [6, 5, 2, 0].$$

• $\hat{\mathfrak{s}}_7$

$$[f_1, e_1] = e_1, \quad [f_1, e_2] = 0, \quad [f_1, e_3] = e_3, \quad [f_1, e_4] = 0, \quad [f_1, e_5] = e_5,$$
$$[f_2, e_1] = 2e_1, \quad [f_2, e_2] = 2e_2, \quad [f_2, e_3] = e_3, \quad [f_2, e_4] = e_4, \quad [f_2, e_5] = 0,$$
$$[f_1, f_2] = 0,$$
$$\mathrm{US} = [0], \quad \mathrm{CS} = [7, 5], \quad \mathrm{DS} = [7, 5, 2, 0].$$

We note that in several cases the characteristic series are different from the ones in Theorem 13.8. This difference in behavior is due to the structural difference between $\mathfrak{n}_{n-2,1}$ and the Heisenberg algebra.

13.5. Generalized Casimir invariants of $\mathfrak{n}_{n,3}$ and of its solvable extensions

We proceed to construct generalized Casimir invariants, i.e., invariants of the coadjoint representation, of the nilpotent algebra $\mathfrak{n}_{n,3}$ and its solvable extensions.

We shall reduce the equations (3.2) to the ones encountered and solved in Section 13.1.4 for the subalgebra $\tilde{\mathfrak{n}}_{n-2,1}$ and its solvable extensions.

Considering first the nilpotent algebra $\mathfrak{n}_{n,3}$ we have the operators (3.1) in the form

$$\begin{gathered}
\widehat{E}_1 = 0, \qquad \widehat{E}_2 = e_1 \frac{\partial}{\partial e_n}, \qquad \widehat{E}_3 = e_1 \frac{\partial}{\partial e_{n-1}}, \qquad \widehat{E}_4 = e_2 \frac{\partial}{\partial e_n}, \\
\widehat{E}_k = e_{k-1} \frac{\partial}{\partial e_n}, \quad 5 \leq k \leq n-2, \qquad \widehat{E}_{n-1} = -e_1 \frac{\partial}{\partial e_3} - e_3 \frac{\partial}{\partial e_n}, \\
\widehat{E}_n = -e_1 \frac{\partial}{\partial e_2} - e_2 \frac{\partial}{\partial e_4} - \sum_{k=5}^{n-2} e_{k-1} \frac{\partial}{\partial e_k} + e_3 \frac{\partial}{\partial e_{n-1}}.
\end{gathered} \tag{13.71}$$

It is evident that no solution I of (3.2) can depend on e_3, e_{n-1}, e_n because of the relations $\widehat{E}_{n-1} I = \widehat{E}_3 I = \widehat{E}_2 I = 0$. Consequently, all considered operators $\widehat{E}_j$ can be truncated to act on functions of $\tilde{e}_1 = e_1$, $\tilde{e}_2 = e_2$, $\tilde{e}_3 = e_4, \ldots, \tilde{e}_{n-3} = e_{n-2}$, $\tilde{e}_{n-2} = e_n$ only. Then $\widehat{E}_{3T}, \widehat{E}_{n-1T}$ vanish and the remaining operators are exactly those present in the investigation of invariants of $\mathfrak{n}_{n-2,1}$ in Section 13.1.4. Therefore, the generalized Casimir invariants of $\mathfrak{n}_{n,3}$ are the same as the ones for $\mathfrak{n}_{n-2,1}$ once written in an appropriate basis.

Similarly, when we consider the solvable extensions of $\mathfrak{n}_{n,3}$, the operators $\widehat{E}_j$ in (13.71) get additional $\partial/\partial f_1$ or $\partial/\partial f_1, \partial/\partial f_2$ terms and one $(\widehat{F}_1)$ or two $(\widehat{F}_1, \widehat{F}_2)$ additional operators are present in (3.2).

Let us first consider the case with $\widehat{F}_1$ only. When the derivation D defining f_1 is such that $2c_{n-1} + d_n \neq 0$, we have $\widehat{E}_1 = (2c_{n-1} + d_n)e_1\partial/\partial f_1$ which excludes the dependence of I on f_1. When $2c_{n-1} + d_n = 0$ the situation is only slightly more complicated — the operators $\widehat{E}_2, \widehat{E}_4$ together again exclude the dependence of I on both f_1 and e_n. In both cases, we can restrict all operators (13.71) and $\widehat{F}_1$ to $\mathfrak{n}_{n,3}$ and then to $\tilde{\mathfrak{n}}_{n-2,1}$, reducing the computation to the corresponding solvable extension of $\mathfrak{n}_{n-2,1}$.

In the second case we have two additional operators $\widehat{F}_1, \widehat{F}_2$ and $\partial/\partial f_1, \partial/\partial f_2$ terms in $\widehat{E}_j$. Now the operators $\widehat{E}_1, \widehat{E}_2, \widehat{E}_3, \widehat{E}_4$ are used in the same way to show that an invariant I cannot depend on f_1, f_2.

Altogether, the construction of generalized Casimir invariants was fully reduced to the one for the nilradical $\mathfrak{n}_{n-2,1}$. The results are:

Theorem 13.9. *Let $n \geq 6$. The nilpotent Lie algebra $\mathfrak{n}_{n,3}$ has $n-4$ functionally independent invariants. They can be chosen to be the following polynomials*

$$\xi_0 = e_1,$$

$$\xi_k = \frac{(-1)^k k}{(k+1)!} e_2^{k+1} + \sum_{j=0}^{k-1} (-1)^j \frac{e_2^j\, e_{k+3-j}\, e_1^{k-j}}{j!}, \qquad 1 \leq k \leq n-5. \tag{13.72}$$

The algebras $\mathfrak{s}_{n+1,1}(\beta), \dots, \mathfrak{s}_{n+1,9}$ have $n-5$ invariants each. Their form is

(1) $\mathfrak{s}_{n+1,1}(\beta)$, $\mathfrak{s}_{n+1,2}$, $\mathfrak{s}_{n+1,3}$, $\mathfrak{s}_{n+1,6}$, $\mathfrak{s}_{n+1,7}$ *and* $\mathfrak{s}_{n+1,9}$:

$$\chi_k = \frac{\xi_k^{1+2\beta}}{\xi_0^{2\beta(k+1)}}, \quad 1 \leq k \leq n-5.$$

For $\mathfrak{s}_{n+1,2}$ is $\beta = (n-5)/2$, for $\mathfrak{s}_{n+1,3}$ and $\mathfrak{s}_{n+1,7}$ we have $\beta = 0$, for $\mathfrak{s}_{n+1,6}(\epsilon)$ we have $\beta = (n-4)/2$ and for $\mathfrak{s}_{n+1,9}$ is $\beta = 1$.

(2) $\mathfrak{s}_{n+1,4}$:

$$\chi_1 = \xi_0, \qquad \chi_k = \frac{\xi_k^2}{\xi_1^{k+1}}, \quad 2 \leq k \leq n-5.$$

(3) $\mathfrak{s}_{n+1,5}$:

$$\chi_k = \frac{\xi_k}{\xi_0^{k+1}}, \quad 1 \leq k \leq n-5.$$

(4) $\mathfrak{s}_{n+1,8}(a_2, a_3, \dots, a_{n-3})$:

$$\chi_k = \sum_{q=0}^{[(k+1)/2]} (-1)^q \frac{(\ln \xi_0)^q}{q!} \Bigg(\sum_{i_1+\cdots+i_q=k-2q+1} a_{i_1+3}a_{i_2+3}\cdots a_{i_q+3} + \sum_{j+i_1+\cdots+i_q=k-2q-1} \frac{\xi_{j+1}}{\xi_0^{j+2}}\, a_{i_1+3}a_{i_2+3}\cdots a_{i_q+3} \Bigg),$$
$$1 \leq k \leq n-5.$$

The summation indices take the values $0 \leq j, i_1, \dots, i_q \leq k+1$.

When $n = 6$ the Lie algebra $\mathfrak{s}_{7,10}(\alpha)$ has one invariant which can be chosen in the form $(2e_4e_1 - e_2^2)^3/e_1^4$, i.e., coincides with the one for $\mathfrak{s}_{7,9}$.

The Lie algebra $\mathfrak{s}_{n+2}$ has $n-6$ functionally independent invariants that can be chosen to be

$$\chi_k = \frac{\xi_{k+1}^2}{\xi_1^{k+2}}, \quad 1 \le k \le n-6.$$

When $n = 5$ we have solvable algebras $\hat{\mathfrak{s}}_{6,1}(\beta)$, $\hat{\mathfrak{s}}_{6,2}$, $\hat{\mathfrak{s}}_{6,5}$, $\hat{\mathfrak{s}}'_{6,6}$, $\hat{\mathfrak{s}}_{6,7}$, $\hat{\mathfrak{s}}'_{6,8}$, $\hat{\mathfrak{s}}'_{6,9}$ with no invariants and $\hat{\mathfrak{s}}_{6,4}$ which has two invariants. They can be chosen in the polynomial form

$$e_1, \quad 2e_1^2 f_1 - 2e_1e_2e_5 + e_1e_3e_4 + e_2e_3^2. \tag{13.73}$$

The algebra $\hat{\mathfrak{s}}_7$ has one invariant

$$\frac{(f_2 - 2f_1)e_1^2 + (2e_2e_5 - e_3e_4)e_1 - e_2e_3^2}{e_1^2}. \tag{13.74}$$

We observe that invariants (13.73) and (13.74) of the solvable Lie algebras with the nilradical $\mathfrak{n}_{5,3}$ can depend on elements outside of $\mathfrak{n}_{5,3}$, i.e., f_1 or f_1, f_2. This is related to the fact that there is no $\tilde{\mathfrak{n}}_{3,1}$ subalgebra (it degenerates into the Heisenberg algebra, the properties of which are markedly different).

CHAPTER 14

Levi Decomposable Algebras

Let us now investigate Levi decomposable algebras. In order to construct such algebras one may assume that one already knows all semisimple and solvable Lie algebras of lower dimensions. They will be combined to give rise to Levi decomposable algebras. One then investigates the compatibility of these two algebraic structures.

Our main concern will be the formulation of necessary conditions for the existence of nontrivial Levi factors compatible with a given nilradical. We formulate these conditions in terms of the characteristic ideals of the nilradical, in particular in terms of the lower central series.

Let us consider a nonsolvable Lie algebra $\mathfrak{g}$ with the nonvanishing nilradical $\mathfrak{n}$. Any such algebra $\mathfrak{g}$ can be written in the form (1.2)

$$\mathfrak{g} = \mathfrak{p} \niplus \mathfrak{r}, \qquad [\mathfrak{r}, \mathfrak{g}] \subseteq \mathfrak{n} \subseteq \mathfrak{r}, \quad [\mathfrak{p}, \mathfrak{p}] = \mathfrak{p}$$

where $\mathfrak{r}$ is the *radical*, i.e., the maximal solvable ideal of $\mathfrak{g}$, and $\mathfrak{p}$ is a semisimple Lie algebra, called the *Levi factor*, unique up to automorphisms of $\mathfrak{g}$. We recall that $\mathrm{ad}(\mathfrak{p})|_{\mathfrak{r}}$ provides a representation of $\mathfrak{p}$ on $\mathfrak{r}$ and this fact is the cornerstone in the construction and classification of algebras of this type. We shall consider only the case where $\mathfrak{g}$ is *indecomposable* in the sense that it cannot be decomposed into a direct sum of ideals. This assumption in particular implies that the representation $\mathrm{ad}(\mathfrak{p})|_{\mathfrak{r}}$ of $\mathfrak{p}$ on $\mathfrak{r}$ is faithful, i.e., a monomorphism into $\mathfrak{Der}(\mathfrak{r})$; and that the considered Levi decomposable Lie algebras are *nontrivial* in the sense that $\mathfrak{g} \neq \mathfrak{p} \oplus \mathfrak{r}$.

The classification of algebras of the type (1.2) with $\mathfrak{p} \neq 0, \mathfrak{r} \neq 0$, i.e., of *Levi decomposable* algebras, was considered in some detail by P. Turkowski in [128, 130]. The approach used in [128] was to consider a given semisimple algebra $\mathfrak{p}$ and all its possible representations ρ on a vector space V of chosen dimension. For each ρ, all solvable algebras $\mathfrak{r}$ compatible with the representation ρ were found by an explicit evaluation of the Jacobi identities with unspecified structure constants ${c_{ij}}^k$ of the radical $\mathfrak{r}$. The resulting algebras $\mathfrak{p} \niplus \mathfrak{r}$ were then classified into equivalence classes. Some general properties of Levi decomposable algebras were found in [130] and used to construct all 9-dimensional Levi decomposable algebras. The result was verified in [17] and one missing algebra was found.

The general properties stated by Turkowski are a direct consequence of the complete reducibility of representations of semisimple Lie algebras. Namely, we have

Proposition 14.1. *Let $\mathfrak{g} = \mathfrak{p} \niplus \mathfrak{r}$ be a Levi decomposable Lie algebra with both factors nonvanishing.*

(1) *If the representation $\mathrm{ad}(\mathfrak{p})|_{\mathfrak{r}}$ of $\mathfrak{p}$ on $\mathfrak{r}$ is irreducible then $\mathfrak{r}$ is Abelian.*

(2) *If* $\mathfrak{r}$ *is solvable nonnilpotent, then there exists a complement* $\mathfrak{q}$ *of* $\mathfrak{n}$ *in* $\mathfrak{r}$, *i.e.,*

$$\mathfrak{r} = \mathfrak{n} \dot{+} \mathfrak{q}$$

such that $\mathrm{ad}(p)|_{\mathfrak{q}} = 0$ *for all* $p \in \mathfrak{p}$, *i.e.,* $\mathrm{ad}(\mathfrak{p})|_{\mathfrak{r}}$ *must necessarily contain a copy of the trivial representation.*

(3) *The set of all elements belonging to the trivial representation*

$$\{x \in \mathfrak{n} \mid \mathrm{ad}(p)x = 0, \ \forall p \in \mathfrak{p}\}$$

is a subalgebra of $\mathfrak{n}$.

For a proof of Proposition 14.1 see [130] and also Section 14.1 below.

In view of the second property, it is of interest to study and classify Levi decomposable algebras with nilpotent radicals first.

14.1. Levi decomposable algebras with a nilpotent radical

In this section we provide several more stringent, yet easy to verify, restrictions on the structure of $\mathfrak{n}$ obtained from the compatibility of the nilradical structure with the given representation of $\mathfrak{p}$. These results were originally presented in [113] which we shall closely follow here.

We call any Levi decomposable algebra $\mathfrak{g}$ with the radical $\mathfrak{r}$ a *Levi extension* of $\mathfrak{r}$. In most of this Section we suppose that the nilradical $\mathfrak{n}$ coincides with the radical $\mathfrak{r}$ and that the Levi factor acts faithfully on $\mathfrak{n}$, i.e., $\mathfrak{g}$ is indecomposable.

Let us recall that characteristic ideals are invariant with respect to all derivations. In particular, characteristic ideals of the radical are invariant with respect to $\mathrm{ad}(\mathfrak{p})|_{\mathfrak{r}}$. A simple observation that any subalgebra of the Lie algebra of triangular matrices is necessarily solvable leads to the following proposition.

Proposition 14.2. *If a complete flag*

$$0 \subset V_1 \subset V_2 \subset \cdots \subset V_n = \mathfrak{n}$$

of codimension 1 *subspaces can be built out of characteristic ideals, then no Levi decomposable algebra*

$$\mathfrak{g} = \mathfrak{p} \niplus \mathfrak{n}$$

such that $[\mathfrak{p}, \mathfrak{n}] \neq 0$ *exists.*

When using this criterion we can start by considering ideals in the characteristic series and their centralizers in $\mathfrak{n}$. For example, one can immediately establish without further considerations of the structure of the representations of $\mathfrak{p}$, that out of low-dimensional nilpotent algebras (dimension at most 5), the following can never appear as the nilradical of a Levi decomposable algebra (for notation see Part 4 of this book, Sections 17.1 and 18.1):

- $\dim \mathfrak{n} = 4$

  $\mathfrak{n}_{4,1}$: $[e_2, e_4] = e_1$, $[e_3, e_4] = e_2$; the characteristic flag is

$$0 \subset \mathfrak{n}^3 \subset \mathfrak{n}^2 \subset \mathrm{cent}(\mathfrak{n}^2) \subset \mathfrak{n}.$$

- $\dim \mathfrak{n} = 5$

  $\mathfrak{n}_{5,4}$: $[e_3, e_4] = e_1$, $[e_2, e_5] = e_1$, $[e_4, e_5] = e_2$; the characteristic flag is

$$0 \subset \mathfrak{n}^3 \subset \mathfrak{n}^2 \subset \mathfrak{z}_2 \subset \mathrm{cent}(\mathfrak{n}^2) \subset \mathfrak{n}.$$

$\mathfrak{n}_{5,5}$: $[e_2, e_5] = e_1$, $[e_3, e_5] = e_2$, $[e_4, e_5] = e_3$; the characteristic flag is

$$0 \subset \mathfrak{n}^4 \subset \mathfrak{n}^3 \subset \mathfrak{n}^2 \subset \mathrm{cent}(\mathfrak{n}^3) \subset \mathfrak{n}.$$

$\mathfrak{n}_{5,6}$: $[e_2, e_5] = e_1$, $[e_3, e_4] = e_1$, $[e_3, e_5] = e_2$, $[e_4, e_5] = e_3$; with the same flag as $\mathfrak{n}_{5,5}$.

All other indecomposable nilpotent algebras $\mathfrak{n}_{1,1}, \mathfrak{n}_{3,1}, \mathfrak{n}_{5,1}, \mathfrak{n}_{5,2}, \mathfrak{n}_{5,3}$ of dimension $\leq$ 5 do admit Levi extensions.

We can construct further characteristic ideals as follows. Let $\mathfrak{i}, \mathfrak{j}$ be characteristic ideals in the Lie algebra $\mathfrak{g}$ and let $\mathfrak{k} = \{x \in \mathfrak{g} \mid [x, y] \in \mathfrak{j}, \forall y \in \mathfrak{i}\}$ be a subspace in $\mathfrak{g}$. Then we have

$$[Dx, y] = D[x, y] - [x, Dy] \in \mathfrak{j}$$

by virtue of the definition of $\mathfrak{k}$ and the characteristic property of $\mathfrak{i}, \mathfrak{j}$, i.e., $Dy \in \mathfrak{i}$, $D[x, y] \in \mathfrak{j}$; consequently, $\mathfrak{k}$ is closed under every derivation, i.e., is also a characteristic ideal. Ideals constructed in this manner can be used to refine the sequence of characteristic ideals used in Proposition 14.2.

The complete reducibility of representations of semisimple Lie algebras allows us to deduce the existence of complementary $\mathrm{ad}(\mathfrak{p})$-invariant subspaces $\widetilde{\mathfrak{m}}_j$ of $\mathfrak{n}^{j+1}$ in $\mathfrak{n}^j$

$$\mathfrak{n}^j = \widetilde{\mathfrak{m}}_j \dotplus \mathfrak{n}^{j+1}, \quad \mathrm{ad}(\mathfrak{p})\widetilde{\mathfrak{m}}_j \subseteq \widetilde{\mathfrak{m}}_j, \qquad j = 1, \dots, K. \tag{14.1}$$

Let us now explore what further requirements we can impose on the subspaces $\widetilde{\mathfrak{m}}_j$ without loss of generality. We can take

$$\mathfrak{m}_1 = \widetilde{\mathfrak{m}}_1.$$

Now the commutator of two $\mathrm{ad}(\mathfrak{p})$-invariant subspaces is again an $\mathrm{ad}(\mathfrak{p})$-invariant subspace (see the definition of a derivation, (2.26)). In particular, $[\mathfrak{m}_1, \mathfrak{m}_1]$ is an $\mathrm{ad}(\mathfrak{p})$-invariant subspace of $\mathfrak{n}^2$ and we have

$$\mathfrak{n}^2 = [\mathfrak{m}_1, \mathfrak{m}_1] + \mathfrak{n}^3.$$

Since both $[\mathfrak{m}_1, \mathfrak{m}_1]$ and $\mathfrak{n}^3$ are $\mathrm{ad}(\mathfrak{p})$-invariant, so is their intersection $[\mathfrak{m}_1, \mathfrak{m}_1] \cap \mathfrak{n}^3$. By the complete reducibility of $\mathrm{ad}(\mathfrak{p})$, there is an $\mathrm{ad}(\mathfrak{p})$-invariant complement of $[\mathfrak{m}_1, \mathfrak{m}_1] \cap \mathfrak{n}^3$ in $[\mathfrak{m}_1, \mathfrak{m}_1]$ which we denote $\mathfrak{m}_2$. Altogether, we have

$$\mathfrak{n}^2 = \mathfrak{m}_2 \dotplus \mathfrak{n}^3, \quad \mathfrak{m}_2 \subseteq [\mathfrak{m}_1, \mathfrak{m}_1], \quad \mathrm{ad}(\mathfrak{p})\mathfrak{m}_2 \subseteq \mathfrak{m}_2.$$

Proceeding in the same way, we can construct a sequence of subspaces $\mathfrak{m}_j$ such that

$$\mathfrak{n} = \mathfrak{m}_K \dotplus \mathfrak{m}_{K-1} \dotplus \cdots \dotplus \mathfrak{m}_1 \tag{14.2}$$

where

$$\mathfrak{n}^j = \mathfrak{m}_j \dotplus \mathfrak{n}^{j+1}, \quad \mathfrak{m}_j \subset [\mathfrak{m}_{j-1}, \mathfrak{m}_1], \quad \mathrm{ad}(\mathfrak{p})\mathfrak{m}_j \subset \mathfrak{m}_j. \tag{14.3}$$

We have thus established that in any basis of the nilradical $\mathfrak{n}$ which respects the decomposition (14.2), the matrices of $\mathrm{ad}(\mathfrak{p})|_{\mathfrak{n}}$ have block diagonal form. If any of the blocks is 1-dimensional then it necessarily corresponds to the trivial representation $\rho(p) = 0$, $\forall p \in \mathfrak{p}$. The $\mathfrak{m}_j \times \mathfrak{m}_j$-submatrices of the representation $\mathrm{ad}(\mathfrak{p})|_{\mathfrak{n}}$, $j > 1$, i.e., the matrices of $\mathrm{ad}(\mathfrak{p})|_{\mathfrak{m}_j}$, are fully determined by $\mathrm{ad}(\mathfrak{p})|_{\mathfrak{m}_1}$ through linear relations coming from the definition of a derivation (2.26).

For the same reason, the kernel of $\mathrm{ad}(\mathfrak{p})|_{\mathfrak{m}_1}$ is also the kernel of $\mathrm{ad}(\mathfrak{p})|_{\mathfrak{n}}$. Therefore, the representation $\mathrm{ad}(\mathfrak{p})|_{\mathfrak{n}}$ of the Levi factor $\mathfrak{p}$ is faithful if and only if $\mathrm{ad}(\mathfrak{p})|_{\mathfrak{m}_1}$ is faithful.

In the particular case when $\mathfrak{m}_j$ is one-dimensional we can easily find a simple relation between the representations $\mathrm{ad}(\mathfrak{p})|_{\mathfrak{m}_1}$ and $\mathrm{ad}(\mathfrak{p})|_{\mathfrak{m}_{j+1}}$. Let $\mathfrak{m}_j = \mathrm{span}\{x\}$. We have

$$\mathrm{ad}(p)[x,y] = [x, \mathrm{ad}(p)y], \quad \forall y \in \mathfrak{m}_1,\ \forall p \in \mathfrak{p}. \tag{14.4}$$

Now let us assume that y belongs to a nontrivial irreducible representation ρ contained in $\mathrm{ad}(\mathfrak{p})|_{\mathfrak{m}_1}$ and that $[x,y] \neq 0$. Let V be the space generated by repeated applications of $\mathrm{ad}(p)$, $p \in \mathfrak{p}$ on y, i.e., V is the representation space of the representation ρ. Consider $\mathrm{ad}(x)V \subset [\mathfrak{m}_j, \mathfrak{m}_1]$. Due to (14.4) the subspace $\mathrm{ad}(x)V$ is by construction invariant with respect to $\mathrm{ad}(\mathfrak{p})$. The kernel of $\mathrm{ad}(x)\colon V \to \mathrm{ad}(x)V$ is an invariant subspace of V. By assumption the representation ρ is irreducible and $\mathrm{ad}(x)V \neq 0$. Therefore, $\mathrm{ad}(x)V$ is a vector space isomorphic to V and an irreducible representation ρ' equivalent to ρ is contained in the decomposition of the representation $\mathrm{ad}(\mathfrak{p})|_{[\mathfrak{m}_j,\mathfrak{m}_1]}$ into irreducible representations.

To sum up the case $\dim \mathfrak{m}_j = 1$, let $\rho_1 \oplus \rho_2 \oplus \cdots \oplus \rho_L$ be the decomposition of $\mathrm{ad}(\mathfrak{p})|_{\mathfrak{m}_1}$ into irreducible representations. Then the induced representation on $[\mathfrak{m}_j, \mathfrak{m}_1]$ is a direct sum of irreducible representations equivalent to the ones contained in some subset $\{\rho_a, a \in J \subset \{1, \ldots, L\}\}$. The same necessarily holds also for $\mathfrak{m}_{j+1} \subseteq [\mathfrak{m}_j, \mathfrak{m}_1]$. The reason it may be only a subset is that some of the commutators in (14.4) may vanish.

If $\dim \mathfrak{m}_j > 1$ then the relation between the representations on $\mathfrak{m}_1, \mathfrak{m}_j$ and $\mathfrak{m}_{j+1}$ takes a more complicated form. The action of a block-diagonal derivation D on the commutator of $e_a \in \mathfrak{m}_j$ and $e_b \in \mathfrak{m}_1$ is

$$D[e_a, e_b] = \sum_{c=n+1-\sum_{i=1}^{j} m_i}^{n-\sum_{i=1}^{j-1} m_i} D^c{}_a[e_c, e_b] + \sum_{d=n+1-m_1}^{n} D^d{}_b[e_a, e_d] \tag{14.5}$$

where $D^c{}_a$ are components of the matrix of $D|_{\mathfrak{m}_j} : \mathfrak{m}_j \to \mathfrak{m}_j$ and $D^d{}_b$ are components of $D|_{\mathfrak{m}_1} : \mathfrak{m}_1 \to \mathfrak{m}_1$. Hence the commutator subspace $[\mathfrak{m}_j, \mathfrak{m}_1]$ transforms under a certain subset of irreducible factors in the tensor representation $\mathrm{ad}(\mathfrak{p})|_{\mathfrak{m}_j} \otimes \mathrm{ad}(\mathfrak{p})|_{\mathfrak{m}_1}$ and the same is true also for $\mathfrak{m}_{j+1} \subseteq [\mathfrak{m}_j, \mathfrak{m}_1]$.

We sum up these conclusions in the following theorem

Theorem 14.3. *Let $\mathfrak{g}$ be an indecomposable Lie algebra with a nilpotent radical $\mathfrak{n}$ and the Levi decomposition*

$$\mathfrak{g} = \mathfrak{p} \niplus \mathfrak{n}.$$

There exists a decomposition of $\mathfrak{n}$ into a direct sum of $\mathrm{ad}(\mathfrak{p})$-invariant subspaces

$$\mathfrak{n} = \mathfrak{m}_K \dotplus \mathfrak{m}_{K-1} \dotplus \cdots \dotplus \mathfrak{m}_1$$

where

$$\mathfrak{n}^j = \mathfrak{m}_j \dotplus \mathfrak{n}^{j+1}, \quad \mathfrak{m}_j \subseteq [\mathfrak{m}_{j-1}, \mathfrak{m}_1], \quad \mathrm{ad}(\mathfrak{p})\mathfrak{m}_j \subseteq \mathfrak{m}_j$$

such that $\mathrm{ad}(\mathfrak{p})|_{\mathfrak{m}_1}$ is a faithful representation of $\mathfrak{p}$ on $\mathfrak{m}_1$. For $j = 2, \ldots, K$ the representation $\mathrm{ad}(\mathfrak{p})|_{\mathfrak{m}_j}$ of $\mathfrak{p}$ on the subspace $\mathfrak{m}_j$ can be decomposed into some subset of irreducible components of the tensor representation $\mathrm{ad}(\mathfrak{p})|_{\mathfrak{m}_{j-1}} \otimes \mathrm{ad}(\mathfrak{p})|_{\mathfrak{m}_1}$.

If any of the subspaces $\mathfrak{m}_j$ is one-dimensional then $\mathrm{ad}(\mathfrak{p})|_{\mathfrak{n}}$ must contain a copy of the trivial representation corresponding to the subspace $\mathfrak{m}_j$. If in addition $j < K$ holds, the representation of $\mathfrak{p}$ on $\mathfrak{m}_{j+1}$ can be decomposed into a *sum of irreducible representations, each of which is equivalent to an irreducible representation contained in the decomposition of* $\mathfrak{m}_1$.

In particular, when $\mathrm{ad}(\mathfrak{p})|_{\mathfrak{m}_1}$ *is irreducible and* $\dim \mathfrak{m}_j = 1$ *for some* j *such that* $1 < j < K$ *then the representation* $\mathrm{ad}(\mathfrak{p})|_{\mathfrak{m}_{j+1}}$ *on* $\mathfrak{m}_{j+1}$ *is equivalent to* $\mathrm{ad}(\mathfrak{p})|_{\mathfrak{m}_1}$.

We remark that it was shown in [14] that when the radical $\mathfrak{r}$ of a Levi decomposable algebra $\mathfrak{g}$ has a one-dimensional center then the representation $\mathrm{ad}(\mathfrak{p})|_{\mathfrak{r}}$ contains a copy of the trivial representation. This result is contained in Theorem 14.3 as a particular subcase when $\dim \mathfrak{m}_K = 1$.

Theorem 14.3 gives us a simple dimensional criterion necessary for the existence of Levi extensions of $\mathfrak{n}$.

Corollary 14.4. *A necessary condition for the existence of a nontrivial Levi extension of the nilpotent Lie algebra* $\mathfrak{n}$ *is that a faithful representation of dimension* $m_1 = \dim \mathfrak{n} - \dim \mathfrak{n}^2$ *must exist.*

For example, there cannot be any Levi extension of the Heisenberg algebra $\mathfrak{n}_{3,1}$ ($[e_2, e_3] = e_1$) with a Levi factor other than $\mathfrak{sl}(2, \mathbb{F})$ and similarly for any other nilradical such that $\dim \mathfrak{n} - \dim \mathfrak{n}^2 = 2$.

In particular, if $\mathfrak{n}$ is filiform we have $m_1 = \dim \mathfrak{n} - \dim \mathfrak{n}^2 = 2$ and $m_j = \dim \mathfrak{n}^j - \dim \mathfrak{n}^{j+1} = 1$ for $j = 2, \ldots, n-1$. When $\dim \mathfrak{n} = 3$ we have the Heisenberg algebra which possesses a Levi extension. When $\dim \mathfrak{n} \geq 4$ the existence of a Levi extension would imply that the 1-dimensional subspace $\mathfrak{m}_3$ must carry an equivalent copy of the 2-dimensional irreducible representation of $\mathfrak{sl}(2, \mathbb{F})$ on $\mathfrak{m}_1$, i.e., a clear contradiction. Therefore, we have

Corollary 14.5. *No nontrivial Levi decomposable algebra with a filiform nilradical* $\mathfrak{n}$ *exists when* $\dim \mathfrak{n} \geq 4$.

This conclusion was derived by other means in [53, Lemma 25] and independently in [4, Corollary 1].

14.2. Levi decomposable algebras with nonnilpotent radicals

Let us now turn our attention to nonnilpotent radicals. Let us assume that $\mathfrak{r}$ is a solvable Lie algebra with the nilradical $\mathfrak{n}$. The existence of a Levi extension $\mathfrak{g}$ of the nonnilpotent radical $\mathfrak{r}$ by the Levi factor $\mathfrak{p}$ imposes restrictions that go beyond the ones originating from the existence of Levi extension $\mathfrak{p} \ni\!\!\!\!+ \mathfrak{n} \subset \mathfrak{g}$ of the nilradical $\mathfrak{n}$. (On the other hand, the ideal $\mathfrak{p} \ni\!\!\!\!+ \mathfrak{n}$ of an indecomposable algebra $\mathfrak{g}$ may be decomposable.)

It follows from Proposition 14.1 that we can always put $\mathfrak{r} = \mathfrak{q} \ni\!\!\!\!+ \mathfrak{n}$ where $\mathfrak{q}$ is some vector subspace of $\mathfrak{r}$ and $\mathrm{ad}(\mathfrak{p})|_{\mathfrak{q}} = 0$ holds. Hence we have

$$\mathrm{ad}(p)[x, y] = [\mathrm{ad}(p)x, y] + [x, \mathrm{ad}(p)y] = 0$$

for any $x, y \in \mathfrak{q}$ and $p \in \mathfrak{p}$, i.e., the subspace $[\mathfrak{q}, \mathfrak{q}] \subset \mathfrak{n}$ must be a representation space of the trivial representation (if $[\mathfrak{q}, \mathfrak{q}]$ is nonvanishing). Furthermore, for any $z \in \mathfrak{n}$, $y \in \mathfrak{q}$, $p \in \mathfrak{p}$ we have

$$\mathrm{ad}(p)[y, z] = [y, \mathrm{ad}(p)z]$$

i.e., similarly as in the proof of Theorem 14.3 we see that $\mathrm{ad}(y)$, $y \in \mathfrak{q}$ maps any representation subspace $V \subset \mathfrak{n}$ of an irreducible representation of $\mathfrak{p}$ either to a representation space of an equivalent representation (including V itself) or to zero.

Another restriction comes from the fact that $[\mathfrak{p}, \mathfrak{q}] \subset \mathfrak{n}$ which for the corresponding derivations acting on $\mathfrak{n}$ means that

$$[\mathrm{ad}(p)|_{\mathfrak{n}}, \mathrm{ad}(x)|_{\mathfrak{n}}] \in \mathfrak{Inn}(\mathfrak{n}), \quad \forall p \in \mathfrak{p},\ x \in \mathfrak{q}.$$

This in turn implies that the $\mathfrak{m}_1\mathfrak{m}_1$-blocks of derivations $\mathrm{ad}(p)|_{\mathfrak{n}}$ and $\mathrm{ad}(x)|_{\mathfrak{n}}$ commute.

We collect some of these results into the following theorem

Theorem 14.6. *Let $\mathfrak{g}$ be a Levi decomposable Lie algebra which cannot be decomposed into a direct sum of ideals, $\mathfrak{p}$ its Levi factor, $\mathfrak{r}$ its radical, $\mathfrak{n}$ its nilradical. Let $\mathfrak{n} = \sum_{k=1}^{K} \mathfrak{m}_k$ be the decomposition* (14.2) *of the nilradical $\mathfrak{n}$. Then for any $p \in \mathfrak{p}$ and $x \in \mathfrak{r}$ the submatrices $\big(\mathrm{ad}(p)\big)_{\mathfrak{m}_1\mathfrak{m}_1}$ and $\big(\mathrm{ad}(x)\big)_{\mathfrak{m}_1\mathfrak{m}_1}$ of derivations $\mathrm{ad}(p)|_{\mathfrak{n}}$ and $\mathrm{ad}(x)|_{\mathfrak{n}}$, respectively, commute*

$$\left[\big(\mathrm{ad}(p)\big)_{\mathfrak{m}_1\mathfrak{m}_1}, \big(\mathrm{ad}(x)\big)_{\mathfrak{m}_1\mathfrak{m}_1}\right] = 0.$$

In particular, if the restriction of $\mathrm{ad}(\mathfrak{p})$ to $\mathfrak{m}_1$ is irreducible and $\mathfrak{g}$ is an algebra over $\mathbb{C}$ then $\dim \mathfrak{r} - \dim \mathfrak{n} \leq 1$. When equality holds the $\mathfrak{m}_1\mathfrak{m}_1$-block of the derivation $\mathrm{ad}(f_1)$ is a nonvanishing multiple of the unit operator.

The proof of the statements in the particular case when $\mathrm{ad}(\mathfrak{p})|_{\mathfrak{m}_1}$ is irreducible is a direct consequence of the Schur lemma.

14.3. Levi decomposable algebras of low dimensions

All Levi decomposable Lie algebras up to dimension 9 were classified by P. Turkowski [128, 130], with some small omissions found by R. Campoamor-Stursberg [17]. The generalized Casimir invariants of all these algebras were computed in [18].

As an example of the use of Theorems 14.3 and 14.6 let us find all Levi decomposable algebras up to dimension 7. We proceed differently from [128] but the obtained results are equivalent. We split our computation into three cases

- Levi decomposable algebras with Abelian radical,
- Levi decomposable algebras with non-Abelian radical and Abelian nilradical,
- Levi decomposable algebras with non-Abelian nilradical.

The algebras of dimension at most 6 constructed here are also reproduced in Part 4, Sections 18.7 and 19.15, in relabelled and rescaled bases.

14.3.1. Levi extensions of Abelian radicals. In this case the structure of the algebra $\mathfrak{g}$ is fully determined by the structure of the Levi factor $\mathfrak{p}$ and its representation ρ on the vector space of dimension $\dim \mathfrak{r}$. Any two algebras $\mathfrak{g}$ and $\mathfrak{g}'$ with Abelian radical are isomorphic if and only if the Levi factors $\mathfrak{p}, \mathfrak{p}'$ are isomorphic and the representations ρ, ρ' are equivalent.

We have only two choices for the Levi factor, namely $\mathfrak{sl}(2, \mathbb{R})$ and $\mathfrak{so}(3, \mathbb{R})$ over the field of real numbers, because we are considering Levi decomposable algebras up to dimension 7. Over the field of complex numbers the complexifications of these two algebras coincide, $\mathfrak{sl}(2, \mathbb{C}) \simeq \mathfrak{so}(3, \mathbb{C})$. The 6-dimensional semisimple algebra $\mathfrak{so}(4, \mathbb{C}) = \mathfrak{sl}(3, \mathbb{C}) \oplus \mathfrak{sl}(2, \mathbb{C})$ and its real forms cannot define an indecomposable 7-dimensional Levi decomposable algebra on dimensional grounds. The requirement of indecomposability excludes from the representation ρ any copy of the trivial 1-dimensional representation ρ_1.

Let us start with $\mathfrak{p} = \mathfrak{sl}(2,\mathbb{F})$. The irreducible representations ρ_j of $\mathfrak{sl}(2,\mathbb{F})$ are characterized by their dimension j which we shall use as a distinguishing label. Their structure is as follows:

$$\begin{gathered} \mathfrak{sl}(2,\mathbb{F}) = \mathrm{span}\{h, p_+, p_-\}, \quad [h, p_\pm] = \pm 2p_\pm, \quad [p_+, p_-] = h, \\ V_j = \mathrm{span}\{v_0, \ldots, v_{j-1}\}, \quad \rho_j(h)v_k = (j-1-2k)v_k, \\ \rho_j(p_-)v_k = v_{k+1}, \quad \rho_j(p_+)v_k = k(j-k)v_{k-1}, \qquad k \in \{0, \ldots, j-1\}, \end{gathered} \tag{14.6}$$

where $v_{-1} \equiv v_j \equiv 0$.

We have the following possibilities and algebras

$\dim \mathfrak{g} = 5$: $\rho = \rho_2$: $\mathfrak{sl}(2,\mathbb{F}) \niplus_{\rho_2} 2\mathfrak{a}_1$

	h	p_+	p_-	e_1	e_2
h	0	$2p_+$	$-2p_-$	e_1	$-e_2$
p_+	$-2p_+$	0	h	0	e_1
p_-	$2p_-$	$-h$	0	e_2	0
e_1	$-e_1$	0	$-e_2$	0	0
e_2	e_2	$-e_1$	0	0	0

US $= [0]$
CS $= [5]$
DS $= [5]$,

$\dim \mathfrak{g} = 6$: $\rho = \rho_3$: $\mathfrak{sl}(2,\mathbb{F}) \niplus_{\rho_3} 3\mathfrak{a}_1$

	h	p_+	p_-	e_1	e_2	e_3
h	0	$2p_+$	$-2p_-$	$2e_1$	0	$-2e_3$
p_+	$-2p_+$	0	h	0	$2e_1$	$2e_2$
p_-	$2p_-$	$-h$	0	e_2	e_3	0
e_1	$-2e_1$	0	$-e_2$	0	0	0
e_2	0	$-2e_1$	$-e_3$	0	0	0
e_3	$2e_3$	$-2e_2$	0	0	0	0

US $= [0]$
CS $= [6]$
DS $= [6]$,

$\dim \mathfrak{g} = 7$: There are two possibilities, $\rho = \rho_4$: $\mathfrak{sl}(2,\mathbb{F}) \niplus_{\rho_4} 4\mathfrak{a}_1$

	h	p_+	p_-	e_1	e_2	e_3	e_4
h	0	$2p_+$	$-2p_-$	$3e_1$	e_2	$-e_3$	$-3e_4$
p_+	$-2p_+$	0	h	0	$3e_1$	$4e_2$	$3e_3$
p_-	$2p_-$	$-h$	0	e_2	e_3	e_4	0
e_1	$-3e_1$	0	$-e_2$	0	0	0	0
e_2	$-e_2$	$-3e_1$	$-e_3$	0	0	0	0
e_3	e_3	$-4e_2$	$-e_4$	0	0	0	0
e_4	$3e_4$	$-3e_3$	0	0	0	0	0

US $= [0]$
CS $= [7]$
DS $= [7]$,

and $\rho = \rho_2 \oplus \rho_2$: $\mathfrak{sl}(2,\mathbb{F}) \niplus_{\rho_2 \oplus \rho_2} 4\mathfrak{a}_1$

	h	p_+	p_-	e_1	e_2	e_3	e_4
h	0	$2p_+$	$-2p_-$	e_1	$-e_2$	e_3	$-e_4$
p_+	$-2p_+$	0	h	0	e_1	0	e_3
p_-	$2p_-$	$-h$	0	e_2	0	e_4	0
e_1	$-e_1$	0	$-e_2$	0	0	0	0
e_2	e_2	$-e_1$	0	0	0	0	0
e_3	$-e_3$	0	$-e_4$	0	0	0	0
e_4	e_4	$-e_3$	0	0	0	0	0

US $= [0]$
CS $= [7]$
DS $= [7]$.

It is worth noting that the Levi decomposable algebra $\mathfrak{sl}(2,\mathbb{F}) \niplus_{\rho_2 \oplus \rho_2} 4\mathfrak{a}_1$ is the lowest dimensional one allowing an extension of its Abelian nilradical to a solvable radical by two nonnilpotent elements f_1, f_2. This is possible because two copies of the same irreducible representation ρ_2 of $\mathfrak{p} = \mathfrak{sl}(2, \mathbb{F})$ are present and give rise to two nonisomorphic real 9-dimensional indecomposable Lie algebras with the additional nonvanishing Lie brackets

$$[f_1, e_1] = e_1, \quad [f_1, e_2] = e_2, \quad [f_2, e_3] = e_3, \quad [f_2, e_4] = e_4$$

or

$$[f_1, e_i] = e_i, \quad [f_1, e_{i+2}] = e_{i+2}, \quad [f_2, e_i] = e_{i+2}, \quad [f_2, e_{i+2}] = -e_i, \qquad i = 1, 2.$$

These two 9-dimensional algebras are isomorphic over the field $\mathbb{C}$.

The Levi factor $\mathfrak{p} = \mathfrak{so}(3, \mathbb{F})$ leads to new algebras only over the field $\mathbb{R}$ because $\mathfrak{so}(3, \mathbb{C}) \simeq \mathfrak{sl}(2, \mathbb{C})$. For $\mathfrak{so}(3, \mathbb{R})$ there is no nontrivial 2-dimensional representation but there is a three-dimensional irreducible representation $\tilde{\rho}_3$ in terms of matrices defining $\mathfrak{so}(3, \mathbb{R})$ and a four-dimensional irreducible spinor representation $\tilde{\rho}_4$. We have the following possibilities

dim $\mathfrak{g} = 6$: $\rho = \tilde{\rho}_3$: $\mathfrak{so}(3, \mathbb{R}) \niplus_{\tilde{\rho}_3} 3\mathfrak{a}_1$ (isomorphic to $\mathfrak{sl}(2, \mathbb{C}) \niplus_{\rho_3} 3\mathfrak{a}_1$ over the field $\mathbb{C}$)

	p_1	p_2	p_3	e_1	e_2	e_3
p_1	0	p_3	$-p_2$	0	e_3	$-e_2$
p_2	$-p_3$	0	p_1	$-e_3$	0	e_1
p_3	p_2	$-p_1$	0	e_2	$-e_1$	0
e_1	0	e_3	$-e_2$	0	0	0
e_2	$-e_3$	0	e_1	0	0	0
e_3	e_2	$-e_1$	0	0	0	0

US = [0]
CS = [6]
DS = [6],

dim $\mathfrak{g} = 7$: $\rho = \tilde{\rho}_4$: $\mathfrak{so}(3, \mathbb{R}) \niplus_{\tilde{\rho}_4} 4\mathfrak{a}_1$ (isomorphic to $\mathfrak{sl}(2, \mathbb{C}) \niplus_{\rho_2 \oplus \rho_2} 4\mathfrak{a}_1$ over the field $\mathbb{C}$)

	p_1	p_2	p_3	e_1	e_2	e_3	e_4
p_1	0	p_3	$-p_2$	$\frac{1}{2}e_4$	$\frac{1}{2}e_3$	$-\frac{1}{2}e_2$	$-\frac{1}{2}e_1$
p_2	$-p_3$	0	p_1	$\frac{1}{2}e_2$	$-\frac{1}{2}e_1$	$\frac{1}{2}e_4$	$-\frac{1}{2}e_3$
p_3	p_2	$-p_1$	0	$\frac{1}{2}e_3$	$-\frac{1}{2}e_4$	$-\frac{1}{2}e_1$	$\frac{1}{2}e_2$
e_1	$-\frac{1}{2}e_4$	$-\frac{1}{2}e_2$	$-\frac{1}{2}e_3$	0	0	0	0
e_2	$-\frac{1}{2}e_3$	$\frac{1}{2}e_1$	$\frac{1}{2}e_4$	0	0	0	0
e_3	$\frac{1}{2}e_2$	$-\frac{1}{2}e_4$	$\frac{1}{2}e_1$	0	0	0	0
e_4	$\frac{1}{2}e_1$	$\frac{1}{2}e_3$	$-\frac{1}{2}e_2$	0	0	0	0

US = [0]
CS = [7, 6]
DS = [7, 6].

14.3.2. Levi extensions of non-Abelian radicals with Abelian nilradicals. When we consider Levi extensions of non-Abelian radicals where the nilradical $\mathfrak{n}$ is Abelian, we use Theorem 14.6, i.e., we construct all operators C on the Abelian nilradical $\mathfrak{n}$ which commute with all operators in the representation ρ of the Levi factor $\mathfrak{p}$ on the nilradical $\mathfrak{n}$,

$$[C, \rho(p)] = 0, \quad p \in \mathfrak{p}. \tag{14.7}$$

When the representation ρ is irreducible and we work over the complex numbers such an operator C is unique up to a scalar multiple and consequently defines a unique Levi decomposable algebra with the brackets

$$\mathfrak{q} = \text{span}\{f_1\}, \qquad [\mathfrak{p}, f_1] = 0, \qquad [f_1, z] = z, \quad z \in \mathfrak{n} \tag{14.8}$$

(the scalar multiple was absorbed into redefinition of f_1). When the representation ρ is reducible, several such operators exist.

Another complication arises from the fact that $\mathfrak{p} \niplus \mathfrak{n}$ may now be decomposable — the indecomposability then must arise from the action of $\mathfrak{q}$ on the nilradical. Therefore, representations $\rho = \rho_2 \oplus \rho_1$, etc., must be taken into account.

Up to dimension $\dim \mathfrak{g} = 7$ we have the following possibilities

(1) $\mathfrak{p} = \mathfrak{sl}(2, \mathbb{F})$

- $\dim \mathfrak{g} = 6$, $\dim \mathfrak{n} = 2$, $\rho = \rho_2$, $\mathfrak{q} = \operatorname{span}\{f_1\}$ as in (14.8):

$$\mathfrak{sl}(2, \mathbb{F}) \niplus \mathfrak{s}_{3,1,a=1} \simeq \mathfrak{sl}(2, \mathbb{F}) \niplus_{\rho_1 \oplus \rho_2} (\mathfrak{a}_1 \niplus 2\mathfrak{a}_1)$$

	h	p_+	p_-	e_1	e_2	f_1
h	0	$2p_+$	$-2p_-$	e_1	$-e_2$	0
p_+	$-2p_+$	0	h	0	e_1	0
p_-	$2p_-$	$-h$	0	e_2	0	0
e_1	$-e_1$	0	$-e_2$	0	0	e_1
e_2	e_2	$-e_1$	0	0	0	e_2
f_1	0	0	0	$-e_1$	$-e_2$	0

US = [0]
CS = [6, 5]
DS = [6, 5],

- $\dim \mathfrak{g} = 7$, $\dim \mathfrak{n} = 3$, $\rho = \rho_3$, $\mathfrak{q} = \operatorname{span}\{f_1\}$ as in (14.8):

$$\mathfrak{sl}(2, \mathbb{F}) \niplus \mathfrak{s}_{4,3,a=b=1} \simeq \mathfrak{sl}(2, \mathbb{F}) \niplus_{\rho_1 \oplus \rho_3} (\mathfrak{a}_1 \niplus 3\mathfrak{a}_1)$$

	h	p_+	p_-	e_1	e_2	e_3	f_1
h	0	$2p_+$	$-2p_-$	$2e_1$	0	$-2e_3$	0
p_+	$-2p_+$	0	h	0	$2e_1$	$2e_2$	0
p_-	$2p_-$	$-h$	0	e_2	e_3	0	0
e_1	$-2e_1$	0	$-e_2$	0	0	0	e_1
e_2	0	$-2e_1$	$-e_3$	0	0	0	e_2
e_3	$2e_3$	$-2e_2$	0	0	0	0	e_3
f_1	0	0	0	$-e_1$	$-e_2$	$-e_3$	0

US = [0]
CS = [7, 6]
DS = [7, 6],

- $\dim \mathfrak{g} = 7$, $\dim \mathfrak{n} = 3$, $\rho = \rho_2 \oplus \rho_1$ with the corresponding decomposition of $\mathfrak{n}$ as $\mathfrak{n} = \operatorname{span}\{e_1, e_2\} \oplus \operatorname{span}\{e_3\}$, $\mathfrak{q} = \operatorname{span}\{f_1\}$:

$$\mathfrak{sl}(2, \mathbb{F}) \niplus \mathfrak{s}_{4,3,b=1} \simeq \mathfrak{sl}(2, \mathbb{F}) \niplus_{\rho_1 \oplus \rho_2 \oplus \rho_1} (\mathfrak{a}_1 \niplus 3\mathfrak{a}_1)$$

	h	p_+	p_-	e_1	e_2	e_3	f_1
h	0	$2p_+$	$-2p_-$	e_1	$-e_2$	0	0
p_+	$-2p_+$	0	h	0	e_1	0	0
p_-	$2p_-$	$-h$	0	e_2	0	0	0
e_1	$-e_1$	0	$-e_2$	0	0	0	e_1
e_2	e_2	$-e_1$	0	0	0	0	e_2
e_3	0	0	0	0	0	0	ae_3
f_1	0	0	0	$-e_1$	$-e_2$	$-ae_3$	0

US = [0]
CS = [7, 6]
DS = [7, 6, 5]
$a \neq 0$,

(2) $\mathfrak{p} = \mathfrak{so}(3, \mathbb{R})$

- $\dim \mathfrak{g} = 7$, $\dim \mathfrak{n} = 3$, $\rho = \tilde{\rho}_3$, $\mathfrak{q} = \operatorname{span}\{f_1\}$ as in (14.8):

$$\mathfrak{so}(3, \mathbb{R}) \niplus \mathfrak{s}_{4,3,a=b=1} \simeq \mathfrak{so}(3, \mathbb{R}) \niplus_{\rho_1 \oplus \tilde{\rho}_3} (\mathfrak{a}_1 \niplus 3\mathfrak{a}_1)$$

(isomorphic to $\mathfrak{sl}(2,\mathbb{F}) \niplus_{\rho_1\oplus\rho_3} (\mathfrak{a}_1 \niplus 3\mathfrak{a}_1)$ over the field $\mathbb{C}$)

	p_1	p_2	p_3	e_1	e_2	e_3	f_1
p_1	0	p_3	$-p_2$	0	e_3	$-e_2$	0
p_2	$-p_3$	0	p_1	$-e_3$	0	e_1	0
p_3	p_2	$-p_1$	0	e_2	$-e_1$	0	0
e_1	0	e_3	$-e_2$	0	0	0	e_1
e_2	$-e_3$	0	e_1	0	0	0	e_2
e_3	e_2	$-e_1$	0	0	0	0	e_3
f_1	0	0	0	$-e_1$	$-e_2$	$-e_3$	0

US = [0]
CS = [7, 6]
DS = [7, 6].

14.3.3. Levi decomposable algebras with non-Abelian nilradicals. Let us now classify low dimensional Levi decomposable algebras with non-Abelian nilradicals. When the dimension of the algebra is at most 7 we have three possibilities for the nilradical: the Heisenberg algebra $\mathfrak{h}(1) = \mathfrak{n}_{3,1}$, the model filiform algebra $\mathfrak{n}_{4,1}$ and the decomposable algebra $\mathfrak{h}(1) \oplus \mathfrak{a}_1$. As was already shown, Theorem 14.3 implies that $\mathfrak{n}_{4,1}$ does not have Levi extensions. According to Theorem 14.3, the Levi factor $\mathfrak{p}$ of any Levi extension of the Heisenberg algebra $\mathfrak{h}(1)$ must have reducible action on $\mathfrak{h}(1)$, with invariant subspaces of dimension 2 and 1. This is possible only for the Levi factor $\mathfrak{p} = \mathfrak{sl}(2,\mathbb{F})$. We have without loss of generality $\mathfrak{m}_1 = \mathrm{span}\{e_2, e_3\}$, $e_1 = [e_2, e_3] \in \mathfrak{m}_2 = (\mathfrak{h}(1))^2$ and the resulting Levi decomposable algebra $\mathfrak{sl}(2,\mathbb{F}) \niplus_{\rho_1\oplus\rho_2} \mathfrak{h}(1)$ is unique

	h	p_+	p_-	e_1	e_2	e_3
h	0	$2p_+$	$-2p_-$	0	e_2	$-e_3$
p_+	$-2p_+$	0	h	0	0	e_2
p_-	$2p_-$	$-h$	0	0	e_3	0
e_1	0	0	0	0	0	0
e_2	$-e_2$	0	$-e_3$	0	0	e_1
e_3	e_3	$-e_2$	0	0	$-e_1$	0

US = [1]
CS = [6]
DS = [6].

This Levi extension of the nilpotent radical $\mathfrak{h}(1)$ admits an extension to a solvable radical, as in Theorem 14.6. By irreducibility of the representation of $\mathfrak{sl}(2,\mathbb{F})$ on $\mathfrak{m}_1$ the derivation of $\mathfrak{h}(1)$ corresponding to the nonnilpotent element $f_1 \in \mathfrak{q} \subset \mathfrak{r}$ must have a multiple of the unit matrix in the $\mathfrak{m}_1 \times \mathfrak{m}_1$-block (the overall factor is irrelevant since it can be arbitrarily rescaled by a redefinition of f_1). The $\mathfrak{m}_2 \times \mathfrak{m}_1$-block can be set to zero through a suitable addition of the central element e_1 to e_2, e_3. Hence, there is a unique Levi decomposable algebra with a nonnilpotent radical and the Heisenberg nilradical $\mathfrak{h}(1)$:

$$\mathfrak{sl}(2,\mathbb{F}) \niplus \mathfrak{s}_{4,8,a=1} \simeq \mathfrak{sl}(2,\mathbb{F}) \niplus_{\rho_1\oplus\rho_1\oplus\rho_2} (\mathfrak{a}_1 \niplus \mathfrak{h}(1))$$

	h	p_+	p_-	e_1	e_2	e_3	f_1
h	0	$2p_+$	$-2p_-$	0	e_2	$-e_3$	0
p_+	$-2p_+$	0	h	0	0	e_2	0
p_-	$2p_-$	$-h$	0	0	e_3	0	0
e_1	0	0	0	0	0	0	$2e_1$
e_2	$-e_2$	0	$-e_3$	0	0	e_1	e_2
e_3	e_3	$-e_2$	0	0	$-e_1$	0	e_3
f_1	0	0	0	$-2e_1$	$-e_2$	$-e_3$	0

US = [0]
CS = [7, 6]
DS = [7, 6].

Finally, the structure of the decomposable Lie algebra $\mathfrak{n} = \mathfrak{h}(1) \oplus \mathfrak{a}_1$ implies the filtration by characteristic ideals

$$\mathfrak{n}^2 = \mathrm{span}\{e_1\} \subset C(\mathfrak{n}) = \mathrm{span}\{e_1, e_4\} \subset \mathfrak{n}, \quad \dim \mathfrak{n}^2 = 1, \quad \dim C(\mathfrak{n}) = 2.$$

Consequently, any Levi extension of $\mathfrak{h}(1) \oplus \mathfrak{a}_1$ can be brought to the following form

$$\begin{aligned} &\mathfrak{h}(1) = \mathrm{span}\{e_1, e_2, e_3\}, \quad \mathfrak{a}_1 = \mathrm{span}\{e_4\}, \quad [e_2, e_3] = e_1, \\ &[\mathfrak{p}, e_1] = 0, \quad [\mathfrak{p}, e_4] = 0, \quad [\mathfrak{p}, e_{2,3}] \in \mathrm{span}\{e_2, e_3\} \end{aligned} \tag{14.9}$$

through addition of suitable multiples of e_1 to the original e_4 and e_1, e_4 to the original e_2 and e_3. Consequently, any Levi decomposable algebra with the nilradical $\mathfrak{h}(1) \oplus \mathfrak{a}_1$ is decomposable. In order to obtain an indecomposable algebra we would have to further extend the radical $\mathfrak{h}(1) \oplus \mathfrak{a}_1$ by one nonnilpotent element f_1. This is indeed possible and gives rise to a one-parameter family of nonisomorphic 8-dimensional Levi decomposable algebras of the following form

	h	p_+	p_-	e_1	e_2	e_3	e_4	f_1
h	0	$2p_+$	$-2p_-$	0	e_2	$-e_3$	0	0
p_+	$-2p_+$	0	h	0	0	e_2	0	0
p_-	$2p_-$	$-h$	0	0	e_3	0	0	0
e_1	0	0	0	0	0	0	0	$2e_1$
e_2	$-e_2$	0	$-e_3$	0	0	e_1	0	e_2
e_3	e_3	$-e_2$	0	0	$-e_1$	0	0	e_3
e_4	0	0	0	0	0	0	0	ae_4
f_1	0	0	0	$-2e_1$	$-e_2$	$-e_3$	$-ae_4$	0

US = [0]

CS = [8, 7]

DS = [8, 7, 6]

$a \neq 0$.

Part 4

Low-Dimensional Lie Algebras

CHAPTER 15

Structure of the Lists of Low-Dimensional Lie Algebras

In the following chapters we present lists of all indecomposable Lie algebras up to dimension 6. In addition to the Lie brackets, each algebra has the following properties presented: dimensions of the ideals in the upper central (US), lower central (CS) and derived (DS) series, and functionally independent (generalized) Casimir invariants.

Conventions.

$$a, b, \ldots \in \mathbb{C}, \qquad \alpha, \beta, \ldots \in \mathbb{R}, \qquad \epsilon = \pm 1, \qquad \sigma = 0, 1,$$
$$\arg(\mathbb{C}) = [0, 2\pi), \qquad \arg(0) = 0.$$

Notice that the labeling used in this part for algebras of low dimensions differs from the one used in Chapter 13 due to a different choice of ordering. Namely, the algebras $\mathfrak{n}_{n,1}$; $\mathfrak{n}_{n,2}$ and $\mathfrak{n}_{n,3}$ of Sections 13.1, 13.2 and 13.4 become algebras $\mathfrak{n}_{4,1}, \mathfrak{n}_{5,5}, \mathfrak{n}_{6,7}$; $\mathfrak{n}_{5,6}, \mathfrak{n}_{6,21}$ and $\mathfrak{n}_{5,4}, \mathfrak{n}_{6,12}$, respectively. In both cases the letter $\mathfrak{n}$ indicates that the algebra is nilpotent and the first subscript denotes the dimension of the algebra. The labeling in Chapter 13 is chosen to be convenient in any dimension. The choice in Part 4 gives more information for low dimensions and is explained below in Section 15.1.

15.1. Ordering of the lists

The order of presentation is chosen to make the identification of any given algebra as straightforward as possible, using the structure of the nilradical and dimensions of the characteristic series as the starting point. On the other hand, this choice necessarily hides, to a certain degree, the way these algebras were constructed and classified.

In this aspect the organization of our lists differs substantially from the lists of S. Lie [72], L. Bianchi [6], G. I. Kruchkovich [65], V. V. Morozov [77], J. Patera, R. T. Sharp, P. Winternitz and H. Zassenhaus [87], G. M. Mubarakzyanov [79, 80] and P. Turkowski [128, 129]. In those original articles Lie algebras were ordered using their structural properties, mainly the canonical Jordan form of the action of the nonnilpotent elements on the nilradical or of the elements of a nilpotent algebra on the maximal Abelian ideal, respectively. These properties are, however, difficult to establish unless the algebra is already written in the canonical basis. Our lists are more refined with respect to the classifications mentioned above. Namely, special cases are split off from classes depending on parameters. The reason for this lies in the different structure of the characteristic series of these special cases. Also, typos, omissions and redundancies (present in particular in [80]) were corrected in preparation of our lists. We mention that recently Mubarakzyanov's results [80] were also independently checked and corrected in [111]. The generalized Casimir invariants of these algebras were calculated in [16, 82, 87, 91]. In preparation of our tables we found it more convenient to calculate them ab initio rather than to convert their results to our choice of conventions and our refinement of the classification.

Dimensions $1-3$, 4, 5 and 6 are presented in separate chapters. For any given dimension, individual sections are devoted to nilpotent Lie algebras, solvable Lie algebras ordered according to their nilradicals, and to Levi decomposable and/or simple Lie algebras, respectively. For solvable Lie algebras the structure of the considered nilradical is given at the beginning of the respective section; only the remaining Lie brackets are listed for each solvable algebra with the given nilradical.

Inside each section we order the algebras according to their upper central series, i.e., algebras with larger centers precede algebras with smaller or vanishing centers. When the centers are of the same dimension we similarly order second and higher centers.

We order algebras with the same dimensions of all ideals in the upper central series using the lower central series; the largest dimension of $\mathfrak{g}^2$ is listed first, etc.

Next, we compare the derived series, in the same manner as the lower central series.

Finally, for Lie algebras which share the same dimensions of all three series US, CS and DS, we compare their generalized Casimir invariants[1]. We order the algebras according to the decreasing number of their invariants. When the numbers of invariants coincide, we proceed by their structure. We choose as simple a basis for the invariants as possible. In the dimensions considered such a basis will contain first as many polynomials as possible, then rational functions, ratios of irrational powers of polynomials, exponentials, trigonometric functions and their inverses. However, the reader should be aware that for classes depending on parameters the structure of the invariants may simplify for certain special values of the parameter(s), e.g., a ratio of real powers of polynomials may become a rational function. In order to avoid further splitting into subcases this was not taken into account in our ordering (but proper care was taken to ensure that the presented invariants form a maximal functionally independent set for all values of the parameters).

For nilpotent algebras another distinguishing criterion employed is the dimension of the algebra of derivations. Its determination is rather tedious and it is computed using computer algebra in all but the simplest cases. Consequently, it is difficult to evaluate it properly when parameters are involved and that is why we don't present its values for the solvable algebras.

15.2. Computer-assisted identification of a given Lie algebra

The identification of a given Lie algebra often involves lengthy calculations, e.g., construction of ideals in the characteristic series or evaluation of the Killing form. Thus, the use of computers can significantly reduce the amount of time and effort needed for such a calculation.

The history of formulation and implementation of necessary algorithms is already quite long. One of the first papers on this topic was [5] by R. E. Beck, B. Kolman and I. N. Stewart. Numerous algorithms were described there, without a direct implementation in any programming language. The algorithms described

---

[1] for the sake of brevity denoted simply as Casimir invariants in the lists

in Part 2 of this book were implemented in PASCAL by D. W. Rand in collaboration with P. Winternitz and H. Zassenhaus in [100, 101]. Next, many new and/or alternative algorithms were proposed by the group working at Technische Universiteit Eindhoven, including A. M. Cohen and W. A. de Graaf. These algorithms were published in a series of papers [30–33, 35, 36] and summed up in the book [34]. They were implemented in the computer algebra package GAP as a library of Lie algebra algorithms called ELIAS. Next, they were also ported to the computer algebra system MAGMA.

The most recent, and probably most readily available, implementation of useful computational algorithms for Lie algebras is the LieAlgebras subpackage of the DifferentialGeometry package in Maplesoft's computer algebra system MAPLE. It is available as a standard package of MAPLE from version 11 up, with new functionalities gradually added in subsequent versions. The author of the package is I. M. Anderson at Utah State University.

The LieAlgebras package presently contains a comprehensive collection of nearly 100 commands for creating Lie algebras, Lie algebra cohomology, representations of Lie algebras and for studying their properties. The LieAlgebras package is fully integrated with the rest of the DifferentialGeometry package and supports general calculations on homogeneous spaces and related geometric structures. Further commands for analyzing Lie algebras with parameters, for finding isomorphisms between Lie algebras, for calculating subalgebras of Lie algebras, for decomposing and classifying representations of semisimple Lie algebras are under development.

Its features relevant for Lie algebra identification include:

- Lie algebras can be created within MAPLE in many ways. The command `LieAlgebraData` calculates the structure equations for a matrix algebra or a Lie algebra expressed in terms of Lie brackets of its basis.
- Lie algebras can be also imported from the library containing algebras presented in the papers by M.-P. Gong [34], V. V. Morozov [77], G. M. Mubarakzyanov [79, 80], P. Turkowski [128, 129] and J. Patera, R. T. Sharp, P. Winternitz and H. Zassenhaus [87]. The tables of Lie algebras found in this book will be available in MAPLE 18. The structure equations for the classical and exceptional simple Lie algebras are also available.
- New Lie algebras can be created from previously defined Lie algebras with the commands `DirectSum`, `Deformation`, `Derivations`, `Extension`, `QuotientAlgebra`, `SemiDirectSum`.
- Distinguished subalgebras and ideals of Lie algebras can be constructed with commands `Center`, `Centralizer`, `DerivedAlgebra`, `Nilradical`, `Radical`, `Series`, `SubalgebraNormalizer`.
- Transformations between Lie algebras can be constructed.
- Many properties of Lie algebras and their subalgebras can be tested with the `Query` command.
- The commands `Decompose`, `LeviDecomposition`, `Nilradical` implement the procedures first published in [102] and described in detail in Chapters 5, 6 and 7 of this book.
- There is an extensive collection of commands for the structure theory of both real and complex semisimple Lie algebras which allows one to calculate Cartan subalgebras, root space decompositions, Cartan decompositions, Chevalley basis, and

Dynkin and Satake diagrams. The algorithm for calculating the Cartan subalgebra is that given by W. A. de Graaf in [34].

In order to provide the reader with a glimpse of how the computer algebra system can be efficiently employed in the identification of a given Lie algebra, let us present an example of calculation using the MAPLE's package LieAlgebras.

Example 15.1. We start by loading all the necessary packages

```
> with(DifferentialGeometry): with(LieAlgebras):
    with(LinearAlgebra):
```

(we have suppressed the output by using colon instead of semicolon to end commands).

Next, we build the structure of a real Lie algebra *Alg* by introducing its Lie brackets, basis vectors and the name for the algebra.

```
> ALG := LieAlgebraData([[e1, e2] = 4*e2-2*e5, [e1, e3] =
      -4*e3-2*e6+2*e8, [e1, e5] = 2*e5, [e1, e6] = -e6+e8,
      [e1, e8] = e6-e8, [e2, e3] = e1-4*e4-2*e5-e6+4*e7+e8,
      [e2, e6] = e4+e5-e7, [e2, e8] = -e4-e5+e7, [e3, e5] =
      2*e4+e6-2*e7-e8, [e4, e6] = e7, [e4, e8] = e7, [e5,
      e6] = e4-e7, [e5, e8] = -e4+e7, [e6, e7] = -e7, [e7,
      e8] = e7], [e1, e2, e3, e4, e5, e6, e7, e8], Alg);
```

$ALG := [[e1, e2] = 4\,e2 - 2\,e5, [e1, e3] = -4\,e3 - 2\,e6 + 2\,e8, [e1, e5] = 2\,e5, [e1, e6] = -e6 + e8, [e1, e8] = e6 - e8, [e2, e3] = e1 - 4\,e4 - 2\,e5 - e6 + 4\,e7 + e8, [e2, e6] = e4 + e5 - e7, [e2, e8] = -e4 - e5 + e7, [e3, e5] = 2\,e4 + e6 - 2\,e7 - e8, [e4, e6] = e7, [e4, e8] = e7, [e5, e6] = e4 - e7, [e5, e8] = -e4 + e7, [e6, e7] = -e7, [e7, e8] = e7]$

We declare the algebra to MAPLE as an object to be used by the DifferentialGeometry package and verify that it satisfies Jacobi identities.

```
> DGsetup(ALG, verbose);
```

The following vector fields have been defined and protected:

$[e1, e2, e3, e4, e5, e6, e7, e8]$

The following differential 1-forms have been defined and protected:

$[\theta 1, \theta 2, \theta 3, \theta 4, \theta 5, \theta 6, \theta 7, \theta 8]$

Lie algebra: Alg

```
Alg> Query("Jacobi");
```

true

Next, we check whether the algebra is indecomposable and if not we decompose it into a direct sum.

```
Alg> Query("Indecomposable");
```

false

```
Alg> Decompose(factoralgebras = true);
```

$$\left[\begin{bmatrix} 1 & 0 & 0 & 0 & 0 & 0 & 0 & 0 \\ 0 & 1 & 0 & 0 & 0 & 0 & 0 & 0 \\ 0 & 0 & 1 & 0 & 0 & 0 & 0 & 0 \\ 0 & 0 & 0 & 1 & 0 & 0 & 0 & 0 \\ 0 & 0 & 0 & 0 & 1 & 0 & 0 & 0 \\ 0 & 0 & 0 & 0 & 0 & \frac{1}{2} & 0 & -\frac{1}{2} \\ 0 & 0 & 0 & 0 & 0 & \frac{1}{2} & 0 & \frac{1}{2} \\ 0 & 0 & 0 & 1 & 0 & 0 & 1 & 0 \end{bmatrix}, [e1, e2, e3, e4 - e7, e5, e6 - e8, e6 + e8, e7],\right.$$

$[[[e1, e2] = 4\,e2 - 2\,e5, [e1, e3] = -4\,e3 - 2\,e6, [e1, e5] = 2\,e5, [e1, e6] = -2\,e6, [e2, e3] = e1 - 4\,e4 - 2\,e5 - e6, [e2, e6] = 2\,e4 + 2\,e5, [e3, e5] = 2\,e4 + e6, [e5, e6] = 2\,e4], [[e1, e2] = -2\,e2]]]$

From the last part of the output we find that the algebra *Alg* decomposes into the direct sum of a 6-dimensional ideal spanned by the vectors $e_1, e_2, e_3, e_4 - e_7, e_5, e_6 - e_8$ and a 2-dimensional ideal spanned by $e_6 + e_8$ and e_7. Before proceeding we check that these are indeed ideals.

```
Alg> Query([e1, e2, e3, e4-e7, e5, e6-e8], "Ideal"),
    Query([e6+e8, e7], "Ideal");
```

true, true

We declare these two subalgebras as new objects in MAPLE and investigate their properties, e.g., their radicals and Levi decompositions. We denote the basis vectors in *SubAlg1* by $f_1, \ldots, f_6$ and the basis of *SubAlg2* by h_1, h_2.

```
Alg> DGsetup(LieAlgebraData([e1,e2,e3,e4-e7,e5,e6-e8], SubAlg1),
     [seq(f || k, k = 1..6)], [alpha]);
```

Lie algebra: SubAlg1

```
SubAlg1> ChangeFrame(Alg): DGsetup(LieAlgebraData([e6+e8, e7],
SubAlg2), [h1, h2], [beta]);
```

Lie algebra: SubAlg2

```
SubAlg2> ChangeFrame(SubAlg1); Radical();
```

$$[f6, f5, f4]$$

```
SubAlg1> LeviDecomposition();
```

$$[[f4, f5, f6], [f1 - 4\,f4 - 2\,f5 - f6, f2 + \frac{1}{2}\,f4, f3 - f4]]$$

We have found that the radical of *SubAlg*1 is spanned by f_4, f_5 and f_6 whereas the Levi factor can be chosen as span of $f_1 - 4f_4 - 2f_5 - f_6, f_2 + \frac{1}{2}f_4$ and $f_3 - f_4$. To find to which 3–dimensional real simple algebra is the Levi factor isomorphic, we calculate its Killing form and find its signature.

```
SubAlg1> K := Killing([f1-4*f4-2*f5-f6, f2+(1/2)*f4, f3-f4]);
```

$$K := \begin{bmatrix} 40 & 0 & 0 \\ 0 & 0 & 10 \\ 0 & 10 & 0 \end{bmatrix}$$

```
SubAlg1> convert(map(sign, Eigenvalues(K)), list);
```

$$[-1, 1, 1]$$

The Killing form is indefinite, so the algebra is noncompact. Hence the Levi factor is isomorphic to the algebra $\mathfrak{sl}(2, \mathbb{R})$.

In order to identify the radical of *SubAlg*1 we find the dimensions of its characteristic series.

```
SubAlg1> US = map(nops, Series([f4, f5, f6], "Upper"));
        CS = map(nops, Series([f4, f5, f6], "Lower"));
        DS = map(nops, Series([f4, f5, f6], "Derived"));
```

$$US = [1, 3]$$
$$CS = [3, 1, 0]$$
$$DS = [3, 1, 0]$$

Thus the radical of the 6-dimensional ideal *SubAlg*1 is nilpotent and isomorphic to the Heisenberg algebra $\mathfrak{n}_{3,1}$. Together with the known information about the Levi factor we identify the subalgebra *SubAlg*1 as $\mathfrak{sl}(2) \ni\!\!\!+ \mathfrak{n}_{3,1}$.

The second ideal *SubAlg*2 is solvable and nonnilpotent, as is easily found using

```
SubAlg1> ChangeFrame(SubAlg2); Radical(); Nilradical();
```

$$[h2, h1]$$
$$[h2]$$

i.e., it is isomorphic to $\mathfrak{s}_{2,1}$.

To sum up, we have found that the given Lie algebra *Alg* has the structure

$$(\mathfrak{sl}(2) \ni\!\!\!+ \mathfrak{n}_{3,1}) \oplus \mathfrak{s}_{2,1}. \tag{15.1}$$

An explicit change of basis performing the decomposition (15.1) is obtained via a composition of the changes of bases found above:

```
SubAlg2> newbas := subs({f1 = e1, f2 = e2, f3 = e3, f4 = e4-e7,
        f5 = e5, f6 = e6-e8, h1 = e6+e8, h2 = e7}, [f1-4*f4-2*f5-f6,
        f2+(1/2)*f4, f3-f4, f4, f5, f6, h1, h2]);
```

$$newbas := [e1 - 4\,e4 - 2\,e5 - e6 + 4\,e7 + e8, e2 + \frac{1}{2}\,e4 - \frac{1}{2}\,e7, e3 - e4 + e7, e4 - e7, e5, e6 - e8, e6 + e8, e7]$$

```
SubAlg2> ChangeFrame(Alg); DGsetup(LieAlgebraData(newbas,
        DecomposedAlg), [s1,s2,s3,r1,r2,r3,q1,q2],[omega], verbose);
```

The following vector fields have been defined and protected:

$$[s1, s2, s3, r1, r2, r3, q1, q2]$$

The following differential 1-forms have been defined and protected:

$$[\omega 1, \omega 2, \omega 3, \omega 4, \omega 5, \omega 6, \omega 7, \omega 8]$$

Lie algebra: DecomposedAlg

The structure of the Lie algebra *Alg* in this new basis is given by the following table of Lie brackets:

```
DecomposedAlg> MultiplicationTable("LieTable");
```

	$s1$	$s2$	$s3$	$r1$	$r2$	$r3$	$q1$	$q2$
$s1$	0	$4\,s2$	$-4\,s3$	0	$2\,r1 + 2\,r2$	$-4\,r1 - 2\,r3$	0	0
$s2$	$-4\,s2$	0	$s1$	0	0	$2\,r1 + 2\,r2$	0	0
$s3$	$4\,s3$	$-s1$	0	0	$2r1 + r3$	0	0	0
$r1$	0	0	0	0	0	0	0	0
$r2$	$-2\,r1 - 2\,r2$	0	$-2\,r1 - r3$	0	0	$2\,r1$	0	0
$r3$	$4\,r1 + 2\,r3$	$-2\,r1 - 2\,r2$	0	0	$-2\,r1$	0	0	0
$q1$	0	0	0	0	0	0	0	$-2\,q2$
$q2$	0	0	0	0	0	0	$2\,q2$	0

which makes the decomposition (15.1) into

$$\mathfrak{sl}(2,\mathbb{R}) = \operatorname{span}\{s_1, s_2, s_3\}, \quad \mathfrak{n}_{3,1} = \operatorname{span}\{r_1, r_2, r_3\}, \quad \mathfrak{s}_{2,1} = \operatorname{span}\{q_1, q_2\} \tag{15.2}$$

clearly visible.

For more examples and tutorials the reader may consult the website `http://digitalcommons.usu.edu/dg/`.

CHAPTER 16

Lie Algebras up to Dimension 3

16.1. One-dimensional Lie algebra

- $\mathfrak{n}_{1,1}$

	e_1
e_1	0

US $= [1]$
CS $= [1, 0]$
DS $= [1, 0]$
$\dim \mathfrak{Der} = 1$

Casimir invariant:

$$e_1$$

16.2. Solvable two-dimensional Lie algebra with the nilradical $\mathfrak{n}_{1,1}$

	e_1
e_1	0

- $\mathfrak{s}_{2,1}$

	e_1
e_2	e_1

US $= [0]$
CS $= [2, 1]$
DS $= [2, 1, 0]$

Casimir invariant:

none

16.3. Nilpotent three-dimensional Lie algebra

- $\mathfrak{n}_{3,1}$

	e_1	e_2	e_3
e_1	0	0	0
e_2		0	e_1

US $= [1, 3]$
CS $= [3, 1, 0]$
DS $= [3, 1, 0]$
$\dim \mathfrak{Der} = 6$

Casimir invariant:

$$e_1$$

16.4. Solvable three-dimensional Lie algebras with the nilradical $2\mathfrak{n}_{1,1}$

	e_1	e_2
e_1	0	0

- $\mathfrak{s}_{3,1}$

	e_1	e_2
e_3	e_1	ae_2

US = [0]
CS = [3, 2]
DS = [3, 2, 0]

where the values of the parameter a are
over the field $\mathbb{C}$: $0 < |a| \leq 1$, if $|a| = 1$ then $\arg(a) \leq \pi$;
over the field $\mathbb{R}$: $0 < |a| \leq 1$.
Casimir invariant:

$$\frac{e_1^a}{e_2}$$

- $\mathfrak{s}_{3,2}$

	e_1	e_2
e_3	e_1	$e_1 + e_2$

US = [0]
CS = [3, 2]
DS = [3, 2, 0]

Casimir invariant:

$$e_1 \exp\left(-\frac{e_2}{e_1}\right)$$

- $\mathfrak{s}_{3,3}$ (over the field $\mathbb{C}$: isomorphic to $\mathfrak{s}_{3,1}$)

	e_1	e_2
e_3	$\alpha e_1 - e_2$	$e_1 + \alpha e_2$

US = [0]
CS = [3, 2]
DS = [3, 2, 0]

where $\alpha \geq 0$.
Casimir invariant:

$$(e_1^2 + e_2^2) \exp\left(2\alpha \arctan \frac{e_1}{e_2}\right)$$

16.5. Simple three-dimensional Lie algebras

- $\mathfrak{sl}(2, \mathbb{F})$

	e_1	e_2	e_3
e_1	0	$2e_1$	$-e_2$
e_2		0	$2e_3$

US = [0]
CS = [3]
DS = [3]

Casimir invariant:

$$4e_1e_3 + e_2^2$$

- $\mathfrak{so}(3, \mathbb{R})$ (over the field $\mathbb{C}$: isomorphic to $\mathfrak{sl}(2, \mathbb{C})$)

	e_1	e_2	e_3
e_1	0	e_3	$-e_2$
e_2		0	e_1

US = [0]
CS = [3]
DS = [3]

Casimir invariant:

$$e_1^2 + e_2^2 + e_3^2$$

CHAPTER 17

Four-Dimensional Lie Algebras

17.1. Nilpotent four-dimensional Lie algebra

• $\mathfrak{n}_{4,1}$

	e_1	e_2	e_3	e_4
e_1	0	0	0	0
e_2		0	0	e_1
e_3			0	e_2

US $= [1, 2, 4]$
CS $= [4, 2, 1, 0]$
DS $= [4, 2, 0]$
$\dim \mathfrak{Der} = 7$

Casimir invariants:

$$e_1, \quad 2e_1e_3 - e_2^2$$

17.2. Solvable four-dimensional algebras with the nilradical $3\mathfrak{n}_{1,1}$

	e_1	e_2	e_3
e_1	0	0	0
e_2		0	0

• $\mathfrak{s}_{4,1}$

	e_1	e_2	e_3
e_4	0	e_1	e_3

US $= [1, 2]$
CS $= [4, 2, 1]$
DS $= [4, 2, 0]$

Casimir invariants:

$$e_1, \quad e_3 \exp\left(-\frac{e_2}{e_1}\right)$$

• $\mathfrak{s}_{4,2}$

	e_1	e_2	e_3
e_4	e_1	$e_1 + e_2$	$e_2 + e_3$

US $= [0]$
CS $= [4, 3]$
DS $= [4, 3, 0]$

Casimir invariants:

$$\frac{2e_1e_3 - e_2^2}{e_1^2}, \quad e_1 \exp\left(-\frac{e_2}{e_1}\right)$$

• $\mathfrak{s}_{4,3}$

	e_1	e_2	e_3
e_4	e_1	ae_2	be_3

US $= [0]$
CS $= [4, 3]$
DS $= [4, 3, 0]$

where the values of the parameters a, b are

$$0 < |b| \leq |a| \leq 1, \quad (a, b) \neq (-1, -1)$$

If one or both equalities hold further restrictions on the parameters are necessary in order to avoid redundancies, as discussed in Section 10.4.

Casimir invariants:

$$\frac{e_1^a}{e_2}, \quad \frac{e_1^b}{e_3}$$

- $\mathfrak{s}_{4,4}$

	e_1	e_2	e_3
e_4	e_1	$e_1 + e_2$	ae_3

US = [0]
CS = [4, 3]
DS = [4, 3, 0]

where $a \neq 0$.

Casimir invariants:

$$\frac{e_1^a}{e_3}, \quad e_1 \exp\left(-\frac{e_2}{e_1}\right)$$

- $\mathfrak{s}_{4,5}$ (over the field $\mathbb{C}$: isomorphic to $\mathfrak{s}_{4,3}$)

	e_1	e_2	e_3
e_4	αe_1	$\beta e_2 - e_3$	$e_2 + \beta e_3$

US = [0]
CS = [4, 3]
DS = [4, 3, 0]

where $\alpha > 0$.

Casimir invariants:

$$\frac{(e_2^2 + e_3^2)^\alpha}{e_1^{2\beta}}, \quad e_1 \exp\left(\alpha \arctan \frac{e_2}{e_3}\right)$$

17.3. Solvable four-dimensional Lie algebras with the nilradical $\mathfrak{n}_{3,1}$

	e_1	e_2	e_3
e_1	0	0	0
e_2		0	e_1

- $\mathfrak{s}_{4,6}$

	e_1	e_2	e_3
e_4	0	e_2	$-e_3$

US = [1]
CS = [4, 3]
DS = [4, 3, 1, 0]

Casimir invariants:

$$e_1, \quad e_2 e_3 + e_1 e_4$$

- $\mathfrak{s}_{4,7}$ (over the field $\mathbb{C}$: isomorphic to $\mathfrak{s}_{4,6}$)

	e_1	e_2	e_3
e_4	0	$-e_3$	e_2

US = [1]
CS = [4, 3]
DS = [4, 3, 1, 0]

Casimir invariants:

$$e_1, \quad e_2^2 + e_3^2 - 2e_1 e_4$$

- $\mathfrak{s}_{4,8}$

	e_1	e_2	e_3
e_4	$(1+a)e_1$	e_2	ae_3

US = [0]
CS = [4, 3]
DS = [4, 3, 1, 0]

where the values of the parameter a are
over the field $\mathbb{C}$: $0 < |a| \leq 1$, if $|a| = 1$ then $\arg(a) < \pi$;
over the field $\mathbb{R}$: $-1 < a \leq 1$, $a \neq 0$.
Casimir invariant:

none

- $\mathfrak{s}_{4,9}$ (over the field $\mathbb{C}$: isomorphic to $\mathfrak{s}_{4,8}$)

	e_1	e_2	e_3
e_4	$2\alpha e_1$	$\alpha e_2 - e_3$	$e_2 + \alpha e_3$

US = [0]
CS = [4, 3]
DS = [4, 3, 1, 0]

where $\alpha > 0$.
Casimir invariant:

none

- $\mathfrak{s}_{4,10}$

	e_1	e_2	e_3
e_4	$2e_1$	e_2	$e_2 + e_3$

US = [0]
CS = [4, 3]
DS = [4, 3, 1, 0]

Casimir invariant:

none

- $\mathfrak{s}_{4,11}$

	e_1	e_2	e_3
e_4	e_1	e_2	0

US = [0]
CS = [4, 2]
DS = [4, 2, 0]

Casimir invariant:

none

17.4. Solvable four-dimensional Lie algebras with the nilradical $2\mathfrak{n}_{1,1}$

	e_1	e_2
e_1	0	0

- $\mathfrak{s}_{4,12}$ (over the field $\mathbb{C}$: decomposable)

	e_1	e_2	e_3
e_3	e_1	e_2	0
e_4	$-e_2$	e_1	0

US = [0]
CS = [4, 2]
DS = [4, 2, 0]

Casimir invariant:

none

CHAPTER 18

Five-Dimensional Lie Algebras

18.1. Nilpotent five-dimensional Lie algebras

- $\mathfrak{n}_{5,1}$

	e_1	e_2	e_3	e_4	e_5
e_1	0	0	0	0	0
e_2		0	0	0	0
e_3			0	0	e_1
e_4				0	e_2

US $= [2, 5]$
CS $= [5, 2, 0]$
DS $= [5, 2, 0]$
$\dim \mathfrak{Der} = 13$

Casimir invariants:

$$e_1, \quad e_2, \quad e_2e_3 - e_1e_4$$

- $\mathfrak{n}_{5,2}$

	e_1	e_2	e_3	e_4	e_5
e_1	0	0	0	0	0
e_2		0	0	0	0
e_3			0	e_2	e_1
e_4				0	e_3

US $= [2, 3, 5]$
CS $= [5, 3, 2, 0]$
DS $= [5, 3, 0]$
$\dim \mathfrak{Der} = 10$

Casimir invariants:

$$e_1, \quad e_2, \quad e_3^2 + 2e_2e_5 - 2e_1e_4$$

- $\mathfrak{n}_{5,3}$

	e_1	e_2	e_3	e_4	e_5
e_1	0	0	0	0	0
e_2		0	0	e_1	0
e_3			0	0	e_1
e_4				0	0

US $= [1, 5]$
CS $= [5, 1, 0]$
DS $= [5, 1, 0]$
$\dim \mathfrak{Der} = 15$

Casimir invariants:

$$e_1$$

- $\mathfrak{n}_{5,4}$

	e_1	e_2	e_3	e_4	e_5
e_1	0	0	0	0	0
e_2		0	0	0	e_1
e_3			0	e_1	0
e_4				0	e_2

US $= [1, 3, 5]$
CS $= [5, 2, 1, 0]$
DS $= [5, 2, 0]$
$\dim \mathfrak{Der} = 10$

Casimir invariants:

$$e_1$$

• $\mathfrak{n}_{5,5}$

	e_1	e_2	e_3	e_4	e_5
e_1	0	0	0	0	0
e_2		0	0	0	e_1
e_3			0	0	e_2
e_4				0	e_3

US = [1, 2, 3, 5]
CS = [5, 3, 2, 1, 0]
DS = [5, 3, 0]
$\dim \mathfrak{Der} = 9$

Casimir invariants:

$$e_1, \quad 2e_1e_3 - e_2^2, \quad e_2^3 + 3e_4e_1^2 - 3e_2e_3e_1$$

• $\mathfrak{n}_{5,6}$

	e_1	e_2	e_3	e_4	e_5
e_1	0	0	0	0	0
e_2		0	0	0	e_1
e_3			0	e_1	e_2
e_4				0	e_3

US = [1, 2, 3, 5]
CS = [5, 3, 2, 1, 0]
DS = [5, 3, 0]
$\dim \mathfrak{Der} = 8$

Casimir invariants:

$$e_1$$

18.2. Solvable five-dimensional Lie algebras with the nilradical $4\mathfrak{n}_{1,1}$

	e_1	e_2	e_3	e_4
e_1	0	0	0	0
e_2		0	0	0
e_3			0	0

• $\mathfrak{s}_{5,1}$

	e_1	e_2	e_3	e_4
e_5	0	e_1	e_2	e_4

US = [1, 2, 3]
CS = [5, 3, 2, 1]
DS = [5, 3, 0]

Casimir invariants:

$$e_1, \quad e_2^2 - 2e_1e_3, \quad e_4 \exp\left(-\frac{e_2}{e_1}\right)$$

• $\mathfrak{s}_{5,2}$

	e_1	e_2	e_3	e_4
e_5	0	e_1	e_3	$e_3 + e_4$

US = [1, 2]
CS = [5, 3, 2]
DS = [5, 3, 0]

Casimir invariants:

$$e_1, \quad \frac{e_1e_4 - e_2e_3}{e_3}, \quad e_3 \exp\left(-\frac{e_2}{e_1}\right)$$

• $\mathfrak{s}_{5,3}$

	e_1	e_2	e_3	e_4
e_5	0	e_1	e_3	ae_4

US = [1, 2]
CS = [5, 3, 2]
DS = [5, 3, 0]

where $0 < |a| \leq 1$. If $|a| = 1$ then $\arg(a) \leq \pi$.

Casimir invariants:

$$e_1, \quad \frac{e_3^a}{e_4}, \quad e_3 \exp\left(-\frac{e_2}{e_1}\right)$$

- $\mathfrak{s}_{5,4}$ (over the field $\mathbb{C}$: isomorphic to $\mathfrak{s}_{5,3}$)

	e_1	e_2	e_3	e_4
e_5	0	e_1	$\alpha e_3 - e_4$	$e_3 + \alpha e_4$

US = [1, 2]
CS = [5, 3, 2]
DS = [5, 3, 0]

where $0 \leq \alpha$.

Casimir invariants:

$$e_1, \quad (e_3^2 + e_4^2)\exp\left(-2\alpha\frac{e_2}{e_1}\right), \quad \frac{e_2}{e_1} + \arctan\frac{e_3}{e_4}$$

- $\mathfrak{s}_{5,5}$

	e_1	e_2	e_3	e_4
e_5	e_1	$e_1 + e_2$	$e_2 + e_3$	$e_3 + e_4$

US = [0]
CS = [5, 4]
DS = [5, 4, 0]

Casimir invariants:

$$\frac{2e_1e_3 - e_2^2}{e_1^2}, \quad \frac{3e_1^2e_4 - 3e_1e_2e_3 + e_2^3}{e_1^3}, \quad e_1 \exp\left(-\frac{e_2}{e_1}\right)$$

- $\mathfrak{s}_{5,6}$

	e_1	e_2	e_3	e_4
e_5	e_1	$e_1 + e_2$	ae_3	$e_3 + ae_4$

US = [0]
CS = [5, 4]
DS = [5, 4, 0]

where $0 < |a| \leq 1$. If $|a| = 1$ then $\arg(a) \leq \pi$.

Casimir invariants:

$$\frac{e_1e_4 - e_2e_3}{e_1e_3}, \quad \frac{e_1^a}{e_3}, \quad e_1 \exp\left(-\frac{e_2}{e_1}\right)$$

- $\mathfrak{s}_{5,7}$

	e_1	e_2	e_3	e_4
e_5	e_1	$e_1 + e_2$	$e_2 + e_3$	ae_4

US = [0]
CS = [5, 4]
DS = [5, 4, 0]

where $a \neq 0$.

Casimir invariants:

$$\frac{2e_1e_3 - e_2^2}{e_1^2}, \quad \frac{e_1^a}{e_4}, \quad e_1 \exp\left(-\frac{e_2}{e_1}\right)$$

- $\mathfrak{s}_{5,8}$ (over the field $\mathbb{C}$: isomorphic to $\mathfrak{s}_{5,6}$)

	e_1	e_2	e_3	e_4
e_5	$\alpha e_1 - e_2$	$e_1 + \alpha e_2$	$e_1 + \alpha e_3 - e_4$	$e_2 + e_3 + \alpha e_4$

US = [0]
CS = [5, 4]
DS = [5, 4, 0]

where $0 \leq \alpha$.

Casimir invariants:

$$\frac{e_1e_4 - e_2e_3}{e_1^2 + e_2^2}, \quad \left(e_1^2 + e_2^2\right)\exp\left(-2\alpha\frac{e_1e_3 + e_2e_4}{e_1^2 + e_2^2}\right), \quad \frac{e_1e_3 + e_4e_2}{e_1^2 + e_2^2} + \arctan\frac{e_1}{e_2}$$

- $\mathfrak{s}_{5,9}$

	e_1	e_2	e_3	e_4
e_5	e_1	ae_2	be_3	ce_4

US = [0]
CS = [5, 4]
DS = [5, 4, 0]

where the values of the parameters a, b, c are

$$0 < |c| \leq |b| \leq |a| \leq 1.$$

If one or more equalities hold further restrictions on the parameters are necessary in order to avoid redundancies, as discussed in Section 10.4.

Casimir invariants:

$$\frac{e_1^a}{e_2}, \quad \frac{e_1^b}{e_3}, \quad \frac{e_1^c}{e_4}$$

- $\mathfrak{s}_{5,10}$

	e_1	e_2	e_3	e_4
e_5	e_1	$e_1 + e_2$	ae_3	be_4

US = [0]
CS = [5, 4]
DS = [5, 4, 0]

where $0 < |b| \leq |a|$, if $|b| = |a|$ then $\arg(a) \leq \arg(b)$.

Casimir invariants:

$$\frac{e_1^a}{e_3}, \quad \frac{e_1^b}{e_4}, \quad e_1\exp\left(-\frac{e_2}{e_1}\right)$$

- $\mathfrak{s}_{5,11}$ (over the field $\mathbb{C}$: isomorphic to $\mathfrak{s}_{5,9}$)

	e_1	e_2	e_3	e_4
e_5	αe_1	βe_2	$\gamma e_3 - e_4$	$e_3 + \gamma e_4$

US = [0]
CS = [5, 4]
DS = [5, 4, 0]

where $\alpha > 0$, $\beta \neq 0$, $\alpha \geq \beta \geq -\alpha$.

Casimir invariants:

$$\frac{e_1^\beta}{e_2^\alpha}, \quad \frac{(e_3^2 + e_4^2)^\beta}{e_2^{2\gamma}}, \quad e_1\exp\left(\alpha\arctan\frac{e_3}{e_4}\right)$$

- $\mathfrak{s}_{5,12}$ (over the field $\mathbb{C}$: isomorphic to $\mathfrak{s}_{5,10}$)

	e_1	e_2	e_3	e_4
e_5	e_1	$e_1 + e_2$	$\alpha e_3 - \beta e_4$	$\beta e_3 + \alpha e_4$

US = [0]
CS = [5, 4]
DS = [5, 4, 0]

where $\beta > 0$.

Casimir invariants:

$$\frac{e_3^2 + e_4^2}{e_1^{2\alpha}}, \quad e_1\exp\left(-\frac{e_2}{e_1}\right), \quad \beta\frac{e_2}{e_1} + \arctan\left(\frac{e_3}{e_4}\right)$$

• $\mathfrak{s}_{5,13}$ (over the field $\mathbb{C}$: isomorphic to $\mathfrak{s}_{5,9}$)

	e_1	e_2	e_3	e_4
e_5	$\alpha e_1 - e_2$	$e_1 + \alpha e_2$	$\beta e_3 - \gamma e_4$	$\gamma e_3 + \beta e_4$

US = [0]
CS = [5, 4]
DS = [5, 4, 0]

where $0 < \gamma \leq 1$, $0 \leq \alpha$. If $\alpha = 0$ then $0 \leq \beta$. If $\gamma = 1$ then $|\alpha| \leq |\beta|$.
Casimir invariants:

$$\frac{(e_1^2 + e_2^2)^\beta}{(e_3^2 + e_4^2)^\alpha}, \quad \arctan \frac{e_3}{e_4} - \gamma \arctan \frac{e_1}{e_2}, \quad (e_1^2 + e_2^2) \exp\left(2\alpha \arctan \frac{e_1}{e_2}\right).$$

When $\alpha \to 0$ the first and third invariant become dependent and one of them should be replaced by

$$(e_3^2 + e_4^2) \exp\left(2\beta \arctan \frac{e_1}{e_2}\right).$$

18.3. Solvable five-dimensional Lie algebras with the nilradical $\mathfrak{n}_{3,1} \oplus \mathfrak{n}_{1,1}$

	e_1	e_2	e_3	e_4
e_1	0	0	0	0
e_2		0	e_1	0
e_3			0	0

• $\mathfrak{s}_{5,14}$

	e_1	e_2	e_3	e_4
e_5	0	0	e_2	e_4

US = [1, 2, 3]
CS = [5, 3, 2, 1]
DS = [5, 3, 0]

Casimir invariant:

$$e_1$$

• $\mathfrak{s}_{5,15}$

	e_1	e_2	e_3	e_4
e_5	0	e_2	$-e_3$	e_1

US = [1, 2]
CS = [5, 3]
DS = [5, 3, 1, 0]

Casimir invariant:

$$e_1$$

• $\mathfrak{s}_{5,16}$ (over the field $\mathbb{C}$: isomorphic to $\mathfrak{s}_{5,15}$)

	e_1	e_2	e_3	e_4
e_5	0	$-e_3$	e_2	e_1

US = [1, 2]
CS = [5, 3]
DS = [5, 3, 1, 0]

Casimir invariant:

$$e_1$$

• $\mathfrak{s}_{5,17}$

	e_1	e_2	e_3	e_4
e_5	0	e_2	$-e_3$	ae_4

US = [1]
CS = [5, 4]
DS = [5, 4, 1, 0]

where the values of the parameter a are
over the field $\mathbb{C}$: $0 \leq \mathrm{Re}(a)$, if $\mathrm{Re}(a) = 0$ then $0 < \mathrm{Im}(a)$;
over the field $\mathbb{C}$: $0 < a$.
Casimir invariant:

$$e_1$$

• $\mathfrak{s}_{5,18}$

	e_1	e_2	e_3	e_4
e_5	0	$-e_2$	$e_3 + e_4$	e_4

US = [1]
CS = [5, 4]
DS = [5, 4, 1, 0]

Casimir invariant:

$$e_1$$

• $\mathfrak{s}_{5,19}$ (over the field $\mathbb{C}$: isomorphic to $\mathfrak{s}_{5,17}$)

	e_1	e_2	e_3	e_4
e_5	0	$-e_3$	e_2	αe_4

US = [1]
CS = [5, 4]
DS = [5, 4, 1, 0]

where $\alpha > 0$.
Casimir invariant:

$$e_1$$

• $\mathfrak{s}_{5,20}$

	e_1	e_2	e_3	e_4
e_5	e_1	e_2	e_4	0

US = [1]
CS = [5, 3, 2]
DS = [5, 3, 0]

Casimir invariant:

$$e_4$$

• $\mathfrak{s}_{5,21}$

	e_1	e_2	e_3	e_4
e_5	$2e_1$	$e_2 + e_3$	$e_3 + e_4$	e_4

US = [0]
CS = [5, 4]
DS = [5, 4, 1, 0]

Casimir invariant:

$$\frac{e_4^2}{e_1}$$

• $\mathfrak{s}_{5,22}$

	e_1	e_2	e_3	e_4
e_5	$(a+1)e_1$	e_2	ae_3	be_4

US = [0]
CS = [5, 4]
DS = [5, 4, 1, 0]

where the values of the parameters a, b are
over the field $\mathbb{C}$: $0 < |a| \leq 1$, $b \neq 0$, if $|a| = 1$ then $\arg(a) < \pi$;

over the field $\mathbb{R}$: $-1 < a \leq 1$, $a, b \neq 0$.
Casimir invariant:

$$\frac{e_4^{a+1}}{e_1^b}$$

- $\mathfrak{s}_{5,23}$

	e_1	e_2	e_3	e_4
e_5	$(a+1)e_1$	ae_2	$e_3 + e_4$	e_4

US = [0]
CS = [5, 4]
DS = [5, 4, 1, 0]

where $a \neq 0, -1$.
Casimir invariant:

$$\frac{e_4^{a+1}}{e_1}$$

- $\mathfrak{s}_{5,24}$

	e_1	e_2	e_3	e_4
e_5	$2e_1$	$e_2 + e_3$	e_3	ae_4

US = [0]
CS = [5, 4]
DS = [5, 4, 1, 0]

where $a \neq 0$.
Casimir invariant:

$$\frac{e_1^a}{e_4^2}$$

- $\mathfrak{s}_{5,25}$ (over the field $\mathbb{C}$: isomorphic to $\mathfrak{s}_{5,22}$)

	e_1	e_2	e_3	e_4
e_5	$2\alpha e_1$	$\alpha e_2 - e_3$	$e_2 + \alpha e_3$	βe_4

US = [0]
CS = [5, 4]
DS = [5, 4, 1, 0]

where $\alpha \neq 0$, $\beta > 0$.
Casimir invariant:

$$\frac{e_4^{2\alpha}}{e_1^{\beta}}$$

- $\mathfrak{s}_{5,26}$

	e_1	e_2	e_3	e_4
e_5	$(a+1)e_1$	e_2	ae_3	$e_1 + (a+1)e_4$

US = [0]
CS = [5, 4]
DS = [5, 4, 1, 0]

where the values of the parameter a are
over the field $\mathbb{C}$: $0 < |a| \leq 1$, if $|a| = 1$ then $\arg(a) < \pi$;
over the field $\mathbb{R}$: $-1 < a \leq 1$, $a \neq 0$.
Casimir invariant:

$$e_1 \exp\left(-(a+1)\frac{e_4}{e_1}\right)$$

- $\mathfrak{s}_{5,27}$

	e_1	e_2	e_3	e_4
e_5	$2e_1$	$e_2 + e_3$	e_3	$e_1 + 2e_4$

US = [0]
CS = [5, 4]
DS = [5, 4, 1, 0]

Casimir invariant:

$$e_1 \exp\left(-2\frac{e_4}{e_1}\right)$$

- $\mathfrak{s}_{5,28}$ (over the field $\mathbb{C}$: isomorphic to $\mathfrak{s}_{5,26}$)

	e_1	e_2	e_3	e_4
e_5	$2\alpha e_1$	$\alpha e_2 + e_3$	$-e_2 + \alpha e_3$	$e_1 + 2\alpha e_4$

US = [0]
CS = [5, 4]
DS = [5, 4, 1, 0]

where $\alpha > 0$.

Casimir invariant:

$$e_1 \exp\left(-2\alpha\frac{e_4}{e_1}\right)$$

- $\mathfrak{s}_{5,29}$

	e_1	e_2	e_3	e_4
e_5	e_1	0	$e_3 + e_4$	e_4

US = [0]
CS = [5, 3]
DS = [5, 3, 0]

Casimir invariant:

$$\frac{e_4}{e_1}$$

- $\mathfrak{s}_{5,30}$

	e_1	e_2	e_3	e_4
e_5	e_1	e_2	0	ae_4

US = [0]
CS = [5, 3]
DS = [5, 3, 0]

where $a \neq 0$.

Casimir invariant:

$$\frac{e_1^a}{e_4}$$

- $\mathfrak{s}_{5,31}$

	e_1	e_2	e_3	e_4
e_5	e_1	e_2	0	$e_1 + e_4$

US = [0]
CS = [5, 3]
DS = [5, 3, 0]

Casimir invariant:

$$e_1 \exp\left(-\frac{e_4}{e_1}\right)$$

- $\mathfrak{s}_{5,32}$

	e_1	e_2	e_3	e_4
e_5	e_1	0	$e_3 + e_4$	$e_1 + e_4$

US = [0]
CS = [5, 3]
DS = [5, 3, 0]

Casimir invariant:

$$e_1 \exp\left(-\frac{e_4}{e_1}\right)$$

18.4. Solvable five-dimensional Lie algebras with the nilradical $\mathfrak{n}_{4,1}$

	e_1	e_2	e_3	e_4
e_1	0	0	0	0
e_2		0	0	e_1
e_3			0	e_2

- $\mathfrak{s}_{5,33}$

	e_1	e_2	e_3	e_4
e_5	0	$-e_2$	$-2e_3$	e_4

US = [1]
CS = [5, 4]
DS = [5, 4, 2, 0]

Casimir invariant

$$e_1$$

- $\mathfrak{s}_{5,34}$

	e_1	e_2	e_3	e_4
e_5	$3e_1$	$2e_2$	e_3	$e_3 + e_4$

US = [0]
CS = [5, 4]
DS = [5, 4, 2, 0]

Casimir invariant:

$$\frac{(2e_1e_3 - e_2^2)^3}{e_1^4}$$

- $\mathfrak{s}_{5,35}$

	e_1	e_2	e_3	e_4
e_5	$(a+2)e_1$	$(a+1)e_2$	ae_3	e_4

US = [0]
CS = [5, 4]
DS = [5, 4, 2, 0]

where $a \neq 0, -2$.

Casimir invariant:

$$\frac{(2e_1e_3 - e_2^2)^{2+a}}{e_1^{2(1+a)}}$$

- $\mathfrak{s}_{5,36}$

	e_1	e_2	e_3	e_4
e_5	$2e_1$	e_2	0	e_4

US = [0]
CS = [5, 3]
DS = [5, 3, 1, 0]

Casimir invariant:

$$\frac{2e_1e_3 - e_2^2}{e_1}$$

- $\mathfrak{s}_{5,37}$

	e_1	e_2	e_3	e_4
e_5	e_1	e_2	e_3	0

US = [0]
CS = [5, 3]
DS = [5, 3, 0]

Casimir invariant:

$$\frac{2e_1e_3 - e_2^2}{e_1^2}$$

• $\mathfrak{s}_{5,38}$

	e_1	e_2	e_3	e_4
e_5	e_1	e_2	$\epsilon e_1 + e_3$	0

US = [0]
CS = [5, 3]
DS = [5, 3, 0]

where the value of the parameter ϵ is
over the field $\mathbb{C}$: $\epsilon = 1$;
over the field $\mathbb{R}$: $\epsilon = \pm 1$.
Casimir invariant:

$$e_1^{-2\epsilon} \exp\left(\frac{2e_1e_3 - e_2^2}{e_1^2}\right)$$

18.5. Solvable five dimensional Lie algebras with the nilradical $3\mathfrak{n}_{1,1}$

	e_1	e_2	e_3
e_1	0	0	0
e_2		0	0

• $\mathfrak{s}_{5,39}$

	e_1	e_2	e_3	e_4
e_4	0	e_2	0	0
e_5	0	0	e_3	e_1

US = [1]
CS = [5, 3, 2]
DS = [5, 3, 0]

Casimir invariant:

$$e_1$$

• $\mathfrak{s}_{5,40}$ (over the field $\mathbb{C}$: isomorphic to $\mathfrak{s}_{5,39}$)

	e_1	e_2	e_3	e_4
e_4	0	e_2	e_3	0
e_5	0	$-e_3$	e_2	e_1

US = [1]
CS = [5, 3, 2]
DS = [5, 3, 0]

Casimir invariant:

$$e_1$$

• $\mathfrak{s}_{5,41}$

	e_1	e_2	e_3	e_4
e_4	e_1	0	ae_3	0
e_5	0	e_2	be_3	0

US = [0]
CS = [5, 3]
DS = [5, 3, 0]

where the values of the parameters a, b are

$$0 < |b| \leq |a| \leq 1.$$

If one or both equalities hold further restrictions on the parameters are necessary in order to avoid redundancies, as discussed in Section 10.4.

Casimir invariant:

$$\frac{e_1^a e_2^b}{e_3}$$

• $\mathfrak{s}_{5,42}$

	e_1	e_2	e_3	e_4
e_4	ae_1	e_2	e_3	0
e_5	e_1	0	e_2	0

US = [0]
CS = [5, 3]
DS = [5, 3, 0]

Casimir invariant:

$$\frac{e_2^a}{e_1}\exp\left(\frac{e_3}{e_2}\right)$$

- $\mathfrak{s}_{5,43}$ (over the field $\mathbb{C}$: isomorphic to $\mathfrak{s}_{5,41}$)

	e_1	e_2	e_3	e_4
e_4	αe_1	e_2	e_3	0
e_5	βe_1	$-e_3$	e_2	0

US = [0]
CS = [5, 3]
DS = [5, 3, 0]

where $\alpha^2 + \beta^2 \neq 0$.

Casimir invariant:

$$\frac{e_1^2}{(e_2^2+e_3^2)^\alpha}\exp\left(2\beta\arctan\frac{e_2}{e_3}\right)$$

18.6. Solvable five-dimensional Lie algebras with the nilradical $\mathfrak{n}_{3,1}$

	e_1	e_2	e_3
e_1	0	0	0
e_2		0	e_1

- $\mathfrak{s}_{5,44}$

	e_1	e_2	e_3	e_4
e_4	e_1	e_2	0	0
e_5	0	e_2	$-e_3$	0

US = [0]
CS = [5, 3]
DS = [5, 3, 1, 0]

Casimir invariant:

$$\frac{e_1e_5 + e_2e_3}{e_1}$$

- $\mathfrak{s}_{5,45}$ (over the field $\mathbb{C}$: isomorphic to $\mathfrak{s}_{5,44}$)

	e_1	e_2	e_3	e_4
e_4	$2e_1$	e_2	e_3	0
e_5	0	e_3	$-e_2$	0

US = [0]
CS = [5, 3]
DS = [5, 3, 1, 0]

Casimir invariant:

$$\frac{2e_1e_5 + e_2^2 + e_3^2}{e_1}$$

18.7. Five-dimensional Levi decomposable Lie algebra

- $\mathfrak{sl}(2,\mathbb{F}) \dotplus 2\mathfrak{n}_{1,1}$

	e_1	e_2	e_3	e_4	e_5
e_1	0	$2e_1$	$-e_2$	e_5	0
e_2		0	$2e_3$	e_4	$-e_5$
e_3			0	0	e_4
e_4				0	0

US = [0]
CS = [5]
DS = [5]

Casimir invariant:

$$e_1e_4^2 - e_2e_4e_5 - e_3e_5^2$$

CHAPTER 19

Six-Dimensional Lie Algebras

19.1. Nilpotent six-dimensional Lie algebras

• $\mathfrak{n}_{6,1}$

	e_1	e_2	e_3	e_4	e_5	e_6
e_1	0	0	0	0	0	0
e_2		0	0	0	0	0
e_3			0	0	0	0
e_4				0	e_2	e_3
e_5					0	e_1

US = [3, 6]
CS = [6, 3, 0]
DS = [6, 3, 0]
$\dim \mathfrak{Der} = 18$

Casimir invariants:

$$e_1, \quad e_2, \quad e_3, \quad e_1e_4 + e_2e_6 - e_3e_5$$

• $\mathfrak{n}_{6,2}$

	e_1	e_2	e_3	e_4	e_5	e_6
e_1	0	0	0	0	0	0
e_2		0	0	0	0	0
e_3			0	0	0	e_1
e_4				0	$-e_1$	e_2
e_5					0	0

US = [2, 6]
CS = [6, 2, 0]
DS = [6, 2, 0]
$\dim \mathfrak{Der} = 17$

Casimir invariants:

$$e_1, \quad e_2$$

• $\mathfrak{n}_{6,3}$ (over the field $\mathbb{R}$ only, decomposable over the field $\mathbb{C}$)

	e_1	e_2	e_3	e_4	e_5	e_6
e_1	0	0	0	0	0	0
e_2		0	0	0	0	0
e_3			0	0	e_1	$-e_2$
e_4				0	e_2	e_1
e_5					0	0

US = [2, 6]
CS = [6, 2, 0]
DS = [6, 2, 0]
$\dim \mathfrak{Der} = 16$

Casimir invariants:

$$e_1, \quad e_2$$

- $\mathfrak{n}_{6,4}$

	e_1	e_2	e_3	e_4	e_5	e_6
e_1	0	0	0	0	0	0
e_2		0	0	0	0	0
e_3			0	0	e_1	e_2
e_4				0	0	e_1
e_5					0	e_3

US $= [2, 4, 6]$
CS $= [6, 3, 2, 0]$
DS $= [6, 3, 0]$
$\dim \mathfrak{Der} = 13$

Casimir invariants:

$$e_1, \quad e_2$$

- $\mathfrak{n}_{6,5}$

	e_1	e_2	e_3	e_4	e_5	e_6
e_1	0	0	0	0	0	0
e_2		0	0	0	0	0
e_3			0	0	e_2	e_1
e_4				0	e_2	0
e_5					0	e_3

US $= [2, 4, 6]$
CS $= [6, 3, 2, 0]$
DS $= [6, 3, 0]$
$\dim \mathfrak{Der} = 12$

Casimir invariants:

$$e_1, \quad e_2$$

- $\mathfrak{n}_{6,6}$ (over the field $\mathbb{C}$: isomorphic to $\mathfrak{n}_{6,5}$)

	e_1	e_2	e_3	e_4	e_5	e_6
e_1	0	0	0	0	0	0
e_2		0	0	0	0	0
e_3			0	0	e_1	$-e_2$
e_4				0	e_2	e_1
e_5					0	e_3

US $= [2, 4, 6]$
CS $= [6, 3, 2, 0]$
DS $= [6, 3, 0]$
$\dim \mathfrak{Der} = 12$

Casimir invariants:

$$e_1, \quad e_2$$

- $\mathfrak{n}_{6,7}$

	e_1	e_2	e_3	e_4	e_5	e_6
e_1	0	0	0	0	0	0
e_2		0	0	0	0	0
e_3			0	0	0	e_1
e_4				0	0	e_2
e_5					0	e_3

US $= [2, 4, 6]$
CS $= [6, 3, 1, 0]$
DS $= [6, 3, 0]$
$\dim \mathfrak{Der} = 15$

Casimir invariants:

$$e_1, \quad e_2, \quad e_1e_4 - e_2e_3, \quad 2e_1e_5 - e_3^2$$

- $\mathfrak{n}_{6,8}$

	e_1	e_2	e_3	e_4	e_5	e_6
e_1	0	0	0	0	0	0
e_2		0	0	0	0	0
e_3			0	0	0	e_1
e_4				0	$-e_1$	e_2
e_5					0	e_3

US $= [2, 4, 6]$
CS $= [6, 3, 1, 0]$
DS $= [6, 3, 0]$
$\dim \mathfrak{Der} = 14$

Casimir invariants:

$$e_1, \quad e_2$$

- $\mathfrak{n}_{6,9}$

	e_1	e_2	e_3	e_4	e_5	e_6
e_1	0	0	0	0	0	0
e_2		0	0	0	0	0
e_3			0	0	0	e_1
e_4				0	e_2	0
e_5					0	e_3

US = [2, 4, 6]
CS = [6, 3, 1, 0]
DS = [6, 3, 0]
$\dim \mathfrak{Der} = 13$

Casimir invariants:

$$e_1, \quad e_2$$

- $\mathfrak{n}_{6,10}$

	e_1	e_2	e_3	e_4	e_5	e_6
e_1	0	0	0	0	0	0
e_2		0	0	0	0	0
e_3			0	0	0	e_1
e_4				0	e_2	e_3
e_5					0	e_4

US = [2, 3, 4, 6]
CS = [6, 4, 3, 1, 0]
DS = [6, 4, 0]
$\dim \mathfrak{Der} = 11$

Casimir invariants:

$$e_1, \quad e_2$$

- $\mathfrak{n}_{6,11}$

	e_1	e_2	e_3	e_4	e_5	e_6
e_1	0	0	0	0	0	0
e_2		0	0	0	0	e_1
e_3			0	0	e_1	0
e_4				0	e_3	0
e_5					0	0

US = [1, 4, 6]
CS = [6, 2, 1, 0]
DS = [6, 2, 0]
$\dim \mathfrak{Der} = 14$

Casimir invariants:

$$e_1, \quad 2e_1e_4 - e_3^2$$

- $\mathfrak{n}_{6,12}$

	e_1	e_2	e_3	e_4	e_5	e_6
e_1	0	0	0	0	0	0
e_2		0	0	0	0	e_1
e_3			0	0	e_1	0
e_4				0	e_3	0
e_5					0	e_2

US = [1, 3, 6]
CS = [6, 3, 1, 0]
DS = [6, 3, 0]
$\dim \mathfrak{Der} = 12$

Casimir invariants:

$$e1, \quad 2e_1e_4 - e_3^2$$

- $\mathfrak{n}_{6,13}$

	e_1	e_2	e_3	e_4	e_5	e_6
e_1	0	0	0	0	0	0
e_2		0	0	0	0	ϵe_1
e_3			0	0	e_1	0
e_4				0	e_3	e_2
e_5					0	0

US $= [1, 3, 6]$
CS $= [6, 3, 1, 0]$
DS $= [6, 3, 0]$
$\dim \mathfrak{Der} = 11$

where the values of the parameter ϵ are
over the field $\mathbb{C}$: $\epsilon = 1$;
over the field $\mathbb{R}$: $\epsilon = \pm 1$.
Casimir invariants:

$$e_1, \quad 2e_1e_4 - \epsilon e_2^2 - e_3^2$$

- $\mathfrak{n}_{6,14}$

	e_1	e_2	e_3	e_4	e_5	e_6
e_1	0	0	0	0	0	0
e_2		0	0	0	0	ϵe_1
e_3			0	0	e_1	0
e_4				0	e_3	e_2
e_5					0	e_4

US $= [1, 3, 4, 6]$
CS $= [6, 4, 3, 1, 0]$
DS $= [6, 4, 0]$
$\dim \mathfrak{Der} = 10$

where the values of the parameter ϵ are
over the field $\mathbb{C}$: $\epsilon = 1$;
over the field $\mathbb{R}$: $\epsilon = \pm 1$.
Casimir invariants:

$$e_1, \quad 2e_1e_4 - \epsilon e_2^2 - e_3^2$$

- $\mathfrak{n}_{6,15}$

	e_1	e_2	e_3	e_4	e_5	e_6
e_1	0	0	0	0	0	0
e_2		0	0	0	0	e_1
e_3			0	0	e_1	0
e_4				0	e_3	e_1
e_5					0	e_4

US $= [1, 3, 4, 6]$
CS $= [6, 3, 2, 1, 0]$
DS $= [6, 3, 0]$
$\dim \mathfrak{Der} = 11$

Casimir invariants:

$$e_1, \quad 2e_1(e_2 - e_4) + e_3^2$$

- $\mathfrak{n}_{6,16}$

	e_1	e_2	e_3	e_4	e_5	e_6
e_1	0	0	0	0	0	0
e_2		0	0	0	0	e_1
e_3			0	0	e_1	0
e_4				0	e_3	0
e_5					0	e_4

US $= [1, 3, 4, 6]$
CS $= [6, 3, 2, 1, 0]$
DS $= [6, 3, 0]$
$\dim \mathfrak{Der} = 12$

Casimir invariants:

$$e_1, \quad 2e_1e_4 - e_3^2$$

- $\mathfrak{n}_{6,17}$

	e_1	e_2	e_3	e_4	e_5	e_6
e_1	0	0	0	0	0	0
e_2		0	0	0	0	e_1
e_3			0	0	e_1	e_2
e_4				0	e_2	e_3
e_5					0	0

$\mathrm{US} = [1, 2, 4, 6]$
$\mathrm{CS} = [6, 3, 2, 1, 0]$
$\mathrm{DS} = [6, 3, 0]$
$\dim \mathfrak{Der} = 10$

Casimir invariants:

$$e_1, \quad 3e_1^2e_4 - 3e_1e_2e_3 + e_2^3$$

- $\mathfrak{n}_{6,18}$

	e_1	e_2	e_3	e_4	e_5	e_6
e_1	0	0	0	0	0	0
e_2		0	0	0	0	e_1
e_3			0	e_1	e_2	0
e_4				0	e_3	0
e_5					0	e_4

$\mathrm{US} = [1, 2, 3, 4, 6]$
$\mathrm{CS} = [6, 4, 3, 2, 1, 0]$
$\mathrm{DS} = [6, 4, 1, 0]$
$\dim \mathfrak{Der} = 9$

Casimir invariants:

$$e_1, \quad 2e_1e_5 - 2e_2e_4 + e_3^2$$

- $\mathfrak{n}_{6,19}$

	e_1	e_2	e_3	e_4	e_5	e_6
e_1	0	0	0	0	0	0
e_2		0	0	0	0	e_1
e_3			0	e_1	e_2	0
e_4				0	e_3	e_2
e_5					0	e_4

$\mathrm{US} = [1, 2, 3, 4, 6]$
$\mathrm{CS} = [6, 4, 3, 2, 1, 0]$
$\mathrm{DS} = [6, 4, 1, 0]$
$\dim \mathfrak{Der} = 8$

Casimir invariants:

$$e_1, \quad 6e_1^2e_5 - 6e_1e_2e_4 + 3e_1e_3^2 + 2e_2^3$$

- $\mathfrak{n}_{6,20}$

	e_1	e_2	e_3	e_4	e_5	e_6
e_1	0	0	0	0	0	0
e_2		0	0	0	0	e_1
e_3			0	0	0	e_2
e_4				0	0	e_3
e_5					0	e_4

$\mathrm{US} = [1, 2, 3, 4, 6]$
$\mathrm{CS} = [6, 4, 3, 2, 1, 0]$
$\mathrm{DS} = [6, 4, 0]$
$\dim \mathfrak{Der} = 11$

Casimir invariants:

$$e_1, \quad 2e_1e_3 - e_2^2, \quad 3e_1^2e_4 - 3e_1e_2e_3 + e_2^3, \quad 8e_1^3e_5 + 4e_1e_2^2e_3 - 8e_1^2e_2e_4 - e_2^4$$

• $\mathfrak{n}_{6,21}$

	e_1	e_2	e_3	e_4	e_5	e_6
e_1	0	0	0	0	0	0
e_2		0	0	0	0	e_1
e_3			0	0	0	e_2
e_4				0	e_1	e_3
e_5					0	e_4

US $= [1, 2, 3, 4, 6]$
CS $= [6, 4, 3, 2, 1, 0]$
DS $= [6, 4, 0]$
$\dim \mathfrak{Der} = 10$

Casimir invariants:

$$e_1, \quad 2e_1e_3 - e_2^2$$

• $\mathfrak{n}_{6,22}$

	e_1	e_2	e_3	e_4	e_5	e_6
e_1	0	0	0	0	0	0
e_2		0	0	0	0	e_1
e_3			0	0	e_1	e_2
e_4				0	e_2	e_3
e_5					0	e_4

US $= [1, 2, 3, 4, 6]$
CS $= [6, 4, 3, 2, 1, 0]$
DS $= [6, 4, 0]$
$\dim \mathfrak{Der} = 9$

Casimir invariants:

$$e_1, \quad 3e_4e_1^2 - 3e_1e_3e_2 + e_2^3$$

19.2. Solvable six-dimensional Lie algebras with the nilradical $5\mathfrak{n}_{1,1}$

	e_1	e_2	e_3	e_4	e_5
e_1	0	0	0	0	0
e_2		0	0	0	0
e_3			0	0	0
e_4				0	0

• $\mathfrak{s}_{6,1}$

	e_1	e_2	e_3	e_4	e_5
e_6	0	0	e_1	e_2	e_5

US $= [2, 4]$
CS $= [6, 3, 1]$
DS $= [6, 3, 0]$

Casimir invariants:

$$e_1, \quad e_2, \quad e_1e_4 - e_2e_3, \quad e_5 \exp\left(-\frac{e_3}{e_1}\right)$$

• $\mathfrak{s}_{6,2}$

	e_1	e_2	e_3	e_4	e_5
e_6	0	e_1	e_2	e_3	e_5

US $= [1, 2, 3, 4]$
CS $= [6, 4, 3, 2, 1]$
DS $= [6, 4, 0]$

Casimir invariants:

$$e_1, \quad 2e_1e_3 - e_2^2, \quad 3e_1^2e_4 - 3e_1e_2e_3 + e_2^3, \quad e_5 \exp\left(-\frac{e_2}{e_1}\right)$$

- $\mathfrak{s}_{6,3}$

	e_1	e_2	e_3	e_4	e_5
e_6	0	e_1	e_2	e_4	$e_4 + e_5$

US $= [1, 2, 3]$
CS $= [6, 4, 3, 2]$
DS $= [6, 4, 0]$

Casimir invariants:

$$e_1, \quad 2e_1e_3 - e_2^2, \quad \frac{e_1e_5 - e_2e_4}{e_4}, \quad e_4 \exp\left(-\frac{e_2}{e_1}\right)$$

- $\mathfrak{s}_{6,4}$

	e_1	e_2	e_3	e_4	e_5
e_6	0	e_1	e_2	e_4	ae_5

US $= [1, 2, 3]$
CS $= [6, 4, 3, 2]$
DS $= [6, 4, 0]$

where $0 < |a| \leq 1$. If $|a| = 1$ then $\arg(a) \leq \pi$.
Casimir invariants:

$$e_1, \quad 2e_1e_3 - e_2^2, \quad \frac{e_4^a}{e_5}, \quad e_4 \exp\left(-\frac{e_2}{e_1}\right)$$

- $\mathfrak{s}_{6,5}$ (over the field $\mathbb{C}$: isomorphic to $\mathfrak{s}_{6,4}$)

	e_1	e_2	e_3	e_4	e_5
e_6	0	e_1	e_2	$\alpha e_4 - e_5$	$e_4 + \alpha e_5$

US $= [1, 2, 3]$
CS $= [6, 4, 3, 2]$
DS $= [6, 4, 0]$

where $0 \leq \alpha$.
Casimir invariants:

$$e_1, \quad 2e_1e_3 - e_2^2, \quad (e_4^2 + e_5^2)\exp\left(-2\alpha\frac{e_2}{e_1}\right), \quad \frac{e_2}{e_1} + \arctan\frac{e_4}{e_5}$$

- $\mathfrak{s}_{6,6}$

	e_1	e_2	e_3	e_4	e_5
e_6	e_1	$e_1 + e_2$	$e_2 + e_3$	0	e_4

US $= [1, 2]$
CS $= [6, 4, 3]$
DS $= [6, 4, 0]$

Casimir invariants:

$$e_4, \quad \frac{e_1e_5 - e_2e_4}{e_1}, \quad \frac{2e_1e_3 - e_2^2}{e_1^2}, \quad e_1 \exp\left(-\frac{e_2}{e_1}\right)$$

- $\mathfrak{s}_{6,7}$

	e_1	e_2	e_3	e_4	e_5
e_6	ae_1	$e_1 + ae_2$	0	e_3	e_5

US $= [1, 2]$
CS $= [6, 4, 3]$
DS $= [6, 4, 0]$

where $a \neq 0$.
Casimir invariants:

$$e_3, \quad \frac{e_1e_4 - e_2e_3}{e_1}, \quad \frac{e_5^a}{e_1}, \quad e_5 \exp\left(-\frac{e_2}{e_1}\right)$$

- $\mathfrak{s}_{6,8}$

	e_1	e_2	e_3	e_4	e_5
e_6	0	e_1	e_3	ae_4	be_5

US = [1, 2]
CS = [6, 4, 3]
DS = [6, 4, 0]

where the values of the parameters a, b are $0 < |b| \leq |a| \leq 1$. If one or both equalities hold further restrictions on the parameters are necessary in order to avoid redundancies, as discussed in Section 10.4.

Casimir invariants:

$$e_1, \quad \frac{e_3^a}{e_4}, \quad \frac{e_3^b}{e_5}, \quad e_3 \exp\left(-\frac{e_2}{e_1}\right)$$

- $\mathfrak{s}_{6,9}$ (over the field $\mathbb{C}$: isomorphic to $\mathfrak{s}_{6,8}$)

	e_1	e_2	e_3	e_4	e_5
e_6	0	e_1	αe_3	$\beta e_4 - e_5$	$e_4 + \beta e_5$

US = [1, 2]
CS = [6, 4, 3]
DS = [6, 4, 0]

where $0 < \alpha$.

Casimir invariants:

$$e_1, \quad \frac{(e_4^2 + e_5^2)^\alpha}{e_3^{2\beta}}, \quad e_3 \exp\left(-\alpha\frac{e_2}{e_1}\right), \quad e_3^2 \exp\left(2\alpha \arctan\frac{e_4}{e_5}\right)$$

- $\mathfrak{s}_{6,10}$

	e_1	e_2	e_3	e_4	e_5
e_6	e_1	$e_1 + e_2$	$e_2 + e_3$	$e_3 + e_4$	$e_4 + e_5$

US = [0]
CS = [6, 5]
DS = [6, 5, 0]

Casimir invariants:

$$\frac{e_2 - e_5}{e_1}, \quad \frac{2e_1e_3 - e_2^2}{e_1^2}, \quad \frac{3e_1^2e_4 - 3e_1e_2e_3 + e_2^3}{e_1^3}, \quad e_1 \exp\left(-\frac{e_2}{e_1}\right)$$

- $\mathfrak{s}_{6,11}$

	e_1	e_2	e_3	e_4	e_5
e_6	e_1	$e_1 + e_2$	$e_2 + e_3$	ae_4	$e_4 + ae_5$

US = [0]
CS = [6, 5]
DS = [6, 5, 0]

where $a \neq 0$.

Casimir invariants:

$$\frac{2e_1e_3 - e_2^2}{e_1^2}, \quad \frac{e_1e_5 - e_2e_4}{e_1e_4}, \quad \frac{e_1^a}{e_4}, \quad e_1 \exp\left(-\frac{e_2}{e_1}\right)$$

- $\mathfrak{s}_{6,12}$

	e_1	e_2	e_3	e_4	e_5
e_6	ae_1	$e_1 + ae_2$	$e_2 + ae_3$	$e_3 + ae_4$	e_5

US = [0]
CS = [6, 5]
DS = [6, 5, 0]

where $a \neq 0$.

Casimir invariants:

$$\frac{2e_1e_3 - e_2^2}{e_1^2}, \quad \frac{3e_1^2e_4 - 3e_1e_2e_3 + e_2^3}{e_1^3}, \quad \frac{e_5^a}{e_1}, \quad e_1 \exp\left(-a\frac{e_2}{e_1}\right).$$

- $\mathfrak{s}_{6,13}$

	e_1	e_2	e_3	e_4	e_5
e_6	ae_1	$e_1 + ae_2$	$e_2 + ae_3$	e_4	be_5

US = [0]
CS = [6, 5]
DS = [6, 5, 0]

where $a \neq 0$, $0 < |b| \leq 1$. If $|b| = 1$ then $\arg(b) \leq \pi$.

Casimir invariants:

$$\frac{2e_1e_3 - e_2^2}{e_1^2}, \quad \frac{e_4^a}{e_1}, \quad \frac{e_5^a}{e_1^b}, \quad e_1 \exp\left(-a\frac{e_2}{e_1}\right)$$

- $\mathfrak{s}_{6,14}$

	e_1	e_2	e_3	e_4	e_5
e_6	ae_1	$e_1 + ae_2$	be_3	$e_3 + be_4$	e_5

US = [0]
CS = [6, 5]
DS = [6, 5, 0]

where the values of the parameters a, b are $0 < |b| \leq |a|$. If $|a| = |b|$ then $\arg(a) \leq \arg(b)$.

Casimir invariants:

$$\frac{e_1e_4 - e_2e_3}{e_1e_3}, \quad \frac{e_5^a}{e_1}, \quad \frac{e_5^b}{e_3}, \quad e_5 \exp\left(-\frac{e_2}{e_1}\right)$$

- $\mathfrak{s}_{6,15}$ (over the field $\mathbb{C}$: isomorphic to $\mathfrak{s}_{6,13}$)

	e_1	e_2	e_3	e_4	e_5
e_6	αe_1	$e_1 + \alpha e_2$	$e_2 + \alpha e_3$	$\beta e_4 - e_5$	$e_4 + \beta e_5$

US = [0]
CS = [6, 5]
DS = [6, 5, 0]

where $0 < \alpha$.

Casimir invariants:

$$\frac{2e_1e_3 - e_2^2}{e_1^2}, \quad \frac{(e_4^2 + e_5^2)^\alpha}{e_1^{2\beta}}, \quad e_1 \exp\left(-\alpha\frac{e_2}{e_1}\right), \quad \frac{e_2}{e_1} + \arctan\frac{e_4}{e_5}$$

- $\mathfrak{s}_{6,16}$ (over the field $\mathbb{C}$: isomorphic to $\mathfrak{s}_{6,14}$)

	e_1	e_2	e_3	e_4	e_5
e_6	$\alpha e_1 - e_2$	$e_1 + \alpha e_2$	$e_1 + \alpha e_3 - e_4$	$e_2 + e_3 + \alpha e_4$	βe_5

US = [0]
CS = [6, 5]
DS = [6, 5, 0]

where $0 < \beta$.

Casimir invariants:

$$\frac{e_1e_4 - e_2e_3}{e_1^2 + e_2^2}, \quad \frac{(e_1^2 + e_2^2)^\beta}{e_5^{2\alpha}}, \quad e_5 \exp\left(-\beta\frac{e_1e_3 + e_2e_4}{e_1^2 + e_2^2}\right), \quad \frac{e_1e_3 + e_2e_4}{e_1^2 + e_2^2} + \arctan\frac{e_1}{e_2}$$

- $\mathfrak{s}_{6,17}$

	e_1	e_2	e_3	e_4	e_5
e_6	e_1	ae_2	be_3	ce_4	de_5

US = [0]
CS = [6, 5]
DS = [6, 5, 0]

where the values of the parameters a, b, c, d are $0 < |d| \leq |c| \leq |b| \leq |a| \leq 1$. If one or more equalities hold further restrictions on the parameters are necessary in order to avoid redundancies, as discussed in Section 10.4.

Casimir invariants:

$$\frac{e_1^a}{e_2}, \quad \frac{e_1^b}{e_3}, \quad \frac{e_1^c}{e_4}, \quad \frac{e_1^d}{e_5}$$

- $\mathfrak{s}_{6,18}$

	e_1	e_2	e_3	e_4	e_5
e_6	ae_1	$e_1 + ae_2$	e_3	be_4	ce_5

US = [0]
CS = [6, 5]
DS = [6, 5, 0]

where the values of the parameters a, b, c are $0 < |c| \leq |b| \leq 1$, $a \neq 0$. If one or both equalities hold further restrictions on the parameters are necessary in order to avoid redundancies, as discussed in Section 10.4.

Casimir invariants:

$$\frac{e_3^a}{e_1}, \quad \frac{e_4^a}{e_1^b}, \quad \frac{e_5^a}{e_1^c}, \quad e_1 \exp\left(-a\frac{e_2}{e_1}\right)$$

- $\mathfrak{s}_{6,19}$ (over the field $\mathbb{C}$: isomorphic to $\mathfrak{s}_{6,17}$)

	e_1	e_2	e_3	e_4	e_5
e_6	αe_1	βe_2	γe_3	$\delta e_4 - e_5$	$e_4 + \delta e_5$

US = [0]
CS = [6, 5]
DS = [6, 5, 0]

where $0 < |\gamma| \leq |\beta| \leq \alpha$.

Casimir invariants:

$$\frac{e_2^\alpha}{e_1^\beta}, \quad \frac{e_3^\alpha}{e_1^\gamma}, \quad \frac{(e_4^2 + e_5^2)^\alpha}{e_1^{2\delta}}, \quad e_1 \exp\left(\alpha \arctan\frac{e_4}{e_5}\right)$$

- $\mathfrak{s}_{6,20}$ (over the field $\mathbb{C}$: isomorphic to $\mathfrak{s}_{6,18}$)

	e_1	e_2	e_3	e_4	e_5
e_6	αe_1	$e_1 + \alpha e_2$	βe_3	$\gamma e_4 - e_5$	$e_4 + \gamma e_5$

US = [0]
CS = [6, 5]
DS = [6, 5, 0]

where $0 < \alpha$, $\beta \neq 0$.

Casimir invariants:

$$\frac{e_3^\alpha}{e_1^\beta}, \quad \frac{(e_4^2 + e_5^2)^\alpha}{e_1^{2\gamma}}, \quad e_1 \exp\left(-\alpha\frac{e_2}{e_1}\right), \quad \frac{e_2}{e_1} + \arctan\frac{e_4}{e_5}$$

- $\mathfrak{s}_{6,21}$ (over the field $\mathbb{C}$: isomorphic to $\mathfrak{s}_{6,17}$)

	e_1	e_2	e_3	e_4	e_5
e_6	$\alpha e_1 - e_2$	$e_1 + \alpha e_2$	$\beta e_3 - \gamma e_4$	$\gamma e_3 + \beta e_4$	δe_5

US = [0]
CS = [6, 5]
DS = [6, 5, 0]

where $0 < \delta$, $0 < \gamma \leq 1$. If $\gamma = 1$ then $\alpha \leq \beta$.

Casimir invariants:

$$\frac{(e_1^2 + e_2^2)^\delta}{e_5^{2\alpha}}, \quad \frac{(e_3^2 + e_4^2)^\delta}{e_5^{2\beta}}, \quad \gamma \arctan\frac{e_1}{e_2} - \arctan\frac{e_3}{e_4}, \quad e_5 \exp\left(\delta \arctan\frac{e_1}{e_2}\right)$$

19.3. Solvable six-dimensional Lie algebras with the nilradical $\mathfrak{n}_{3,1} \oplus 2\mathfrak{n}_{1,1}$

	e_1	e_2	e_3	e_4	e_5
e_1	0	0	0	0	0
e_2		0	e_1	0	0
e_3			0	0	0
e_4				0	0

- $\mathfrak{s}_{6,22}$

	e_1	e_2	e_3	e_4	e_5
e_6	0	0	e_4	0	e_5

US $= [2, 4]$
CS $= [6, 3, 1]$
DS $= [6, 3, 0]$

Casimir invariants:

$$e_1, \quad e_4$$

- $\mathfrak{s}_{6,23}$

	e_1	e_2	e_3	e_4	e_5
e_6	0	e_3	e_4	0	e_5

US $= [2, 3, 4]$
CS $= [6, 4, 3, 1]$
DS $= [6, 4, 0]$

Casimir invariants:

$$e_1, \quad e_4$$

- $\mathfrak{s}_{6,24}$

	e_1	e_2	e_3	e_4	e_5
e_6	0	e_2	$-e_3$	0	e_4

US $= [2, 3]$
CS $= [6, 4, 3]$
DS $= [6, 4, 1, 0]$

Casimir invariants:

$$e_1, \quad e_4$$

- $\mathfrak{s}_{6,25}$ (over the field $\mathbb{C}$: isomorphic to $\mathfrak{s}_{6,24}$)

	e_1	e_2	e_3	e_4	e_5
e_6	0	$-e_3$	e_2	e_5	0

US $= [2, 3]$
CS $= [6, 4, 3]$
DS $= [6, 4, 1, 0]$

Casimir invariants:

$$e_1, \quad e_5$$

- $\mathfrak{s}_{6,26}$

	e_1	e_2	e_3	e_4	e_5
e_6	0	0	0	e_4	e_1

US $= [1, 4]$
CS $= [6, 2, 1]$
DS $= [6, 2, 0]$

Casimir invariants:

$$e_1, \quad e_4 \exp\left(-\frac{e_5}{e_1}\right)$$

- $\mathfrak{s}_{6,27}$

	e_1	e_2	e_3	e_4	e_5
e_6	0	0	e_4	e_1	e_5

US $= [1,3,4]$
CS $= [6,3,2,1]$
DS $= [6,3,0]$

Casimir invariants:

$$e_1, \quad e_5 \exp\left(-\frac{e_4}{e_1}\right)$$

- $\mathfrak{s}_{6,28}$

	e_1	e_2	e_3	e_4	e_5
e_6	0	0	e_2	e_1	e_5

US $= [1,3,4]$
CS $= [6,3,2,1]$
DS $= [6,3,0]$

Casimir invariants:

$$e_1, \quad e_5 \exp\left(-\frac{e_4}{e_1}\right)$$

- $\mathfrak{s}_{6,29}$

	e_1	e_2	e_3	e_4	e_5
e_6	0	e_3	e_4	e_1	e_5

US $= [1,2,3,4]$
CS $= [6,4,3,2,1]$
DS $= [6,4,0]$

Casimir invariants:

$$e_1, \quad e_5 \exp\left(-\frac{e_4}{e_1}\right)$$

- $\mathfrak{s}_{6,30}$

	e_1	e_2	e_3	e_4	e_5
e_6	0	e_2	$-e_3$	e_5	e_1

US $= [1,2,3]$
CS $= [6,4,3]$
DS $= [6,4,1,0]$

Casimir invariants:

$$e_1, \quad 2e_1e_4 - e_5^2$$

- $\mathfrak{s}_{6,31}$ (over the field $\mathbb{C}$: isomorphic to $\mathfrak{s}_{6,30}$)

	e_1	e_2	e_3	e_4	e_5
e_6	0	$-e_3$	e_2	e_5	e_1

US $= [1,2,3]$
CS $= [6,4,3]$
DS $= [6,4,1,0]$

Casimir invariants:

$$e_1, \quad 2e_1e_4 - e_5^2$$

- $\mathfrak{s}_{6,32}$

	e_1	e_2	e_3	e_4	e_5
e_6	0	0	e_2	e_4	ae_5

US $= [1,2,3]$
CS $= [6,4,3,2]$
DS $= [6,4,0]$

where the value of the parameter a is $0 < |a| \le 1$. If $|a| = 1$ then $\arg(a) \le \pi$.

Casimir invariants:

$$e_1, \quad \frac{e_4^a}{e_5}$$

- $\mathfrak{s}_{6,33}$

	e_1	e_2	e_3	e_4	e_5
e_6	0	0	e_2	e_4	$e_4 + e_5$

US = [1, 2, 3]
CS = [6, 4, 3, 2]
DS = [6, 4, 0]

Casimir invariants:

$$e_1, \quad e_4 \exp\left(-\frac{e_5}{e_4}\right)$$

- $\mathfrak{s}_{6,34}$ (over the field $\mathbb{C}$: isomorphic to $\mathfrak{s}_{6,32}$)

	e_1	e_2	e_3	e_4	e_5
e_6	0	0	e_2	$\alpha e_4 - e_5$	$e_4 + \alpha e_5$

US = [1, 2, 3]
CS = [6, 4, 3, 2]
DS = [6, 4, 0]

where $\alpha \geq 0$.

Casimir invariants:

$$e_1, \quad (e_4^2 + e_5^2) \exp\left(2\alpha \arctan \frac{e_4}{e_5}\right)$$

- $\mathfrak{s}_{6,35}$

	e_1	e_2	e_3	e_4	e_5
e_6	0	ae_2	$-ae_3$	e_4	e_1

US = [1, 2]
CS = [6, 4]
DS = [6, 4, 1, 0]

where $\arg(a) < \pi$, $a \neq 0$.

Casimir invariants:

$$e_1, \quad e_4 \exp\left(-\frac{e_5}{e_1}\right)$$

- $\mathfrak{s}_{6,36}$

	e_1	e_2	e_3	e_4	e_5
e_6	0	$-e_2$	$e_3 + e_4$	e_4	e_1

US = [1, 2]
CS = [6, 4]
DS = [6, 4, 1, 0]

Casimir invariants:

$$e_1, \quad e_4 \exp\left(-\frac{e_5}{e_1}\right)$$

- $\mathfrak{s}_{6,37}$ (over the field $\mathbb{C}$: isomorphic to $\mathfrak{s}_{6,35}$)

	e_1	e_2	e_3	e_4	e_5
e_6	0	$-e_3$	e_2	αe_4	e_1

US = [1, 2]
CS = [6, 4]
DS = [6, 4, 1, 0]

where $\alpha \neq 0$.

Casimir invariants:

$$e_1, \quad e_4 \exp\left(-\alpha \frac{e_5}{e_1}\right)$$

- $\mathfrak{s}_{6,38}$

	e_1	e_2	e_3	e_4	e_5
e_6	$(a+1)e_1$	e_2	ae_3	0	e_4

US = [1, 2]
CS = [6, 4, 3]
DS = [6, 4, 1, 0]

where $0 < |a| \leq 1$. If $|a| = 1$ then $\arg(a) < \pi$.

Casimir invariants:

$$e_4, \quad e_1 \exp\left(-(a+1)\frac{e_5}{e_4}\right)$$

- $\mathfrak{s}_{6,39}$

	e_1	e_2	e_3	e_4	e_5
e_6	$2e_1$	e_2	$e_2 + e_3$	0	e_4

US = [1, 2]
CS = [6, 4, 3]
DS = [6, 4, 1, 0]

Casimir invariants:

$$e_4, \quad e_1 \exp\left(-2\frac{e_5}{e_4}\right)$$

- $\mathfrak{s}_{6,40}$ (over the field $\mathbb{C}$: isomorphic to $\mathfrak{s}_{6,38}$)

	e_1	e_2	e_3	e_4	e_5
e_6	$2\alpha e_1$	$\alpha e_2 - e_3$	$e_2 + \alpha e_3$	0	e_4

US = [1, 2]
CS = [6, 4, 3]
DS = [6, 4, 1, 0]

where $\alpha > 0$.

Casimir invariants:

$$e_4, \quad e_1 \exp\left(-2\alpha\frac{e_5}{e_4}\right)$$

- $\mathfrak{s}_{6,41}$

	e_1	e_2	e_3	e_4	e_5
e_6	e_1	e_2	e_4	e_5	0

US = [1, 2]
CS = [6, 4, 3, 2]
DS = [6, 4, 0]

Casimir invariants:

$$e_5, \quad e_1 \exp\left(-\frac{e_4}{e_5}\right)$$

- $\mathfrak{s}_{6,42}$

	e_1	e_2	e_3	e_4	e_5
e_6	e_1	e_2	0	0	e_4

US = [1, 2]
CS = [6, 3, 2]
DS = [6, 3, 0]

Casimir invariants:

$$e_4, \quad e_1 \exp\left(-\frac{e_5}{e_4}\right)$$

- $\mathfrak{s}_{6,43}$

	e_1	e_2	e_3	e_4	e_5
e_6	0	$e_2 + e_4$	$-e_3 + e_5$	e_4	$-e_5$

US = [1]
CS = [6, 5]
DS = [6, 5, 1, 0]

Casimir invariants:

$$e_1, \quad e_4e_5$$

- $\mathfrak{s}_{6,44}$ (over the field $\mathbb{C}$: isomorphic to $\mathfrak{s}_{6,43}$)

	e_1	e_2	e_3	e_4	e_5
e_6	0	$-e_3+e_4$	e_2+e_5	$-e_5$	e_4

US = [1]
CS = [6, 5]
DS = [6, 5, 1, 0]

Casimir invariants:

$$e_1, \quad e_4^2+e_5^2$$

- $\mathfrak{s}_{6,45}$

	e_1	e_2	e_3	e_4	e_5
e_6	0	ae_2	$-ae_3$	e_4	be_5

US = [1]
CS = [6, 5]
DS = [6, 5, 1, 0]

where $a \neq 0$, $\arg(a) < \pi$, $0 < |b| \leq 1$. If $|b| = 1$ then $\arg(b) \leq \pi$.

Casimir invariants:

$$e_1, \quad \frac{e_4^b}{e_5}$$

- $\mathfrak{s}_{6,46}$ (over the field $\mathbb{C}$: isomorphic to $\mathfrak{s}_{6,45}$)

	e_1	e_2	e_3	e_4	e_5
e_6	0	$-e_3$	e_2	αe_4	βe_5

US = [1]
CS = [6, 5]
DS = [6, 5, 1, 0]

where the values of the parameters α, β are $0 < |\beta| \leq |\alpha|$. If $|\alpha| = |\beta|$ then $\beta \leq \alpha$.

Casimir invariants:

$$e_1, \quad \frac{e_4^\beta}{e_5^\alpha}$$

- $\mathfrak{s}_{6,47}$

	e_1	e_2	e_3	e_4	e_5
e_6	0	$-e_2$	e_3+e_4	e_4	ae_5

US = [1]
CS = [6, 5]
DS = [6, 5, 1, 0]

where $a \neq 0$.

Casimir invariants:

$$e_1, \quad \frac{e_4^a}{e_5}$$

- $\mathfrak{s}_{6,48}$

	e_1	e_2	e_3	e_4	e_5
e_6	0	e_2	$-e_3$	ae_4	e_4+ae_5

US = [1]
CS = [6, 5]
DS = [6, 5, 1, 0]

where $a \neq 0$.

Casimir invariants:

$$e_1, \quad e_4 \exp\left(-a\frac{e_5}{e_4}\right)$$

- $\mathfrak{s}_{6,49}$ (over the field $\mathbb{C}$: isomorphic to $\mathfrak{s}_{6,48}$)

	e_1	e_2	e_3	e_4	e_5
e_6	0	$-e_3$	e_2	$\alpha e_4 + e_5$	αe_5

US = [1]
CS = [6, 5]
DS = [6, 5, 1, 0]

where $\alpha \neq 0$.

Casimir invariants:

$$e_1, \quad e_5 \exp\left(-\alpha \frac{e_4}{e_5}\right)$$

- $\mathfrak{s}_{6,50}$

	e_1	e_2	e_3	e_4	e_5
e_6	0	$-e_2$	$e_3 + e_4$	$e_4 + e_5$	e_5

US = [1]
CS = [6, 5]
DS = [6, 5, 1, 0]

Casimir invariants:

$$e_1, \quad e_5 \exp\left(-\frac{e_4}{e_5}\right)$$

- $\mathfrak{s}_{6,51}$ (over the field $\mathbb{C}$: isomorphic to $\mathfrak{s}_{6,45}$)

	e_1	e_2	e_3	e_4	e_5
e_6	0	αe_2	$-\alpha e_3$	$\beta e_4 - e_5$	$e_4 + \beta e_5$

US = [1]
CS = [6, 5]
DS = [6, 5, 1, 0]

where $\alpha > 0$, $\beta \geq 0$.

Casimir invariants:

$$e_1, \quad (e_4^2 + e_5^2) \exp\left(2\beta \arctan \frac{e_4}{e_5}\right)$$

- $\mathfrak{s}_{6,52}$ (over the field $\mathbb{C}$: isomorphic to $\mathfrak{s}_{6,45}$)

	e_1	e_2	e_3	e_4	e_5
e_6	0	$-e_3$	e_2	$\alpha e_4 - \beta e_5$	$\beta e_4 + \alpha e_5$

US = [1]
CS = [6, 5]
DS = [6, 5, 1, 0]

where $\alpha \geq 0$, $\beta > 0$.

Casimir invariants:

$$e_1, \quad (e_4^2 + e_5^2)^{\beta} \exp\left(2\alpha \arctan \frac{e_4}{e_5}\right)$$

- $\mathfrak{s}_{6,53}$

	e_1	e_2	e_3	e_4	e_5
e_6	e_1	$e_2 + e_4$	e_5	e_4	0

US = [1]
CS = [6, 4, 3]
DS = [6, 4, 0]

Casimir invariants:

$$e_5, \quad \frac{e_4}{e_1}$$

- $\mathfrak{s}_{6,54}$

	e_1	e_2	e_3	e_4	e_5
e_6	ae_1	ae_2	e_5	e_4	0

US = [1]
CS = [6, 4, 3]
DS = [6, 4, 0]

where $a \neq 0$.

Casimir invariants:

$$e_5, \quad \frac{e_4^a}{e_1}$$

- $\mathfrak{s}_{6,55}$

	e_1	e_2	e_3	e_4	e_5
e_6	e_1	$e_2 + e_4$	e_5	$e_1 + e_4$	0

US = [1]
CS = [6, 4, 3]
DS = [6, 4, 0]

Casimir invariants:

$$e_5, \quad e_1 \exp\left(-\frac{e_4}{e_1}\right)$$

- $\mathfrak{s}_{6,56}$

	e_1	e_2	e_3	e_4	e_5
e_6	e_1	e_2	e_5	$e_1 + e_4$	0

US = [1]
CS = [6, 4, 3]
DS = [6, 4, 0]

Casimir invariants:

$$e_5, \quad e_1 \exp\left(-\frac{e_4}{e_1}\right)$$

- $\mathfrak{s}_{6,57}$

	e_1	e_2	e_3	e_4	e_5
e_6	$2ae_1$	$ae_2 + e_3$	$ae_3 + e_4$	ae_4	e_5

US = [0]
CS = [6, 5]
DS = [6, 5, 1, 0]

where $a \neq 0$.

Casimir invariants:

$$\frac{e_4^2}{e_1}, \quad \frac{e_5^{2a}}{e_1}$$

- $\mathfrak{s}_{6,58}$

	e_1	e_2	e_3	e_4	e_5
e_6	$(a+1)e_1$	$e_2 + e_4$	$ae_3 + e_5$	e_4	ae_5

US = [0]
CS = [6, 5]
DS = [6, 5, 1, 0]

where $0 < |a| \leq 1$. If $|a| = 1$ then $\arg(a) < \pi$.

Casimir invariants:

$$\frac{e_4 e_5}{e_1}, \quad \frac{e_4^{a+1}}{e_1}$$

- $\mathfrak{s}_{6,59}$

	e_1	e_2	e_3	e_4	e_5
e_6	$2e_1$	$e_2 + e_3$	$e_3 + e_4$	$e_4 + e_5$	e_5

US = [0]
CS = [6, 5]
DS = [6, 5, 1, 0]

Casimir invariants:

$$\frac{e_5^2}{e_1}, \quad e_1 \exp\left(-2\frac{e_4}{e_5}\right)$$

- $\mathfrak{s}_{6,60}$

	e_1	e_2	e_3	e_4	e_5
e_6	$2e_1$	$e_2 + e_3$	$e_3 + e_4$	e_4	$e_1 + 2e_5$

US = [0]
CS = [6, 5]
DS = [6, 5, 1, 0]

Casimir invariants:

$$\frac{e_4^2}{e_1}, \quad e_1 \exp\left(-2\frac{e_5}{e_1}\right)$$

- $\mathfrak{s}_{6,61}$

	e_1	e_2	e_3	e_4	e_5
e_6	$2e_1$	$e_2 + e_3 + e_4$	$e_3 + e_5$	$e_4 + e_5$	e_5

US = [0]
CS = [6, 5]
DS = [6, 5, 1, 0]

Casimir invariants:

$$\frac{e_5^2}{e_1}, \quad e_1 \exp\left(-2\frac{e_4}{e_5}\right)$$

- $\mathfrak{s}_{6,62}$

	e_1	e_2	e_3	e_4	e_5
e_6	$(a+1)e_1$	e_2	ae_3	$(1+a)e_4 + e_5$	$e_1 + (1+a)e_5$

US = [0]
CS = [6, 5]
DS = [6, 5, 1, 0]

where $0 < |a| \leq 1$. If $|a| = 1$ then $\arg(a) < \pi$.

Casimir invariants:

$$\frac{2e_1e_4 - e_5^2}{e_1^2}, \quad e_1 \exp\left(-(a+1)\frac{e_5}{e_1}\right)$$

- $\mathfrak{s}_{6,63}$

	e_1	e_2	e_3	e_4	e_5
e_6	$2e_1$	e_2	$e_2 + e_3$	$2e_4 + e_5$	$e_1 + 2e_5$

US = [0]
CS = [6, 5]
DS = [6, 5, 1, 0]

Casimir invariants:

$$\frac{2e_1e_4 - e_5^2}{e_1^2}, \quad e_1 \exp\left(-2\frac{e_5}{e_1}\right)$$

- $\mathfrak{s}_{6,64}$ (over the field $\mathbb{C}$: isomorphic to $\mathfrak{s}_{6,62}$)

	e_1	e_2	e_3	e_4	e_5
e_6	$2\alpha e_1$	$\alpha e_2 - e_3$	$e_2 + \alpha e_3$	$2\alpha e_4 + e_5$	$e_1 + 2\alpha e_5$

US = [0]
CS = [6, 5]
DS = [6, 5, 1, 0]

where $\alpha > 0$.

Casimir invariants:

$$\frac{2e_1e_4 - e_5^2}{e_1^2}, \quad e_1 \exp\left(-2\alpha\frac{e_5}{e_1}\right)$$

- $\mathfrak{s}_{6,65}$ (over the field $\mathbb{C}$: isomorphic to $\mathfrak{s}_{6,58}$)

	e_1	e_2	e_3	e_4	e_5
e_6	$2\alpha e_1$	$\alpha e_2 - e_3 + e_4$	$e_2 + \alpha e_3 + e_5$	$\alpha e_4 - e_5$	$e_4 + \alpha e_5$

US = [0]
CS = [6, 5]
DS = [6, 5, 1, 0]

where $\alpha > 0$.

Casimir invariants:

$$\frac{e_4^2 + e_5^2}{e_1}, \quad e_1 \exp\left(2\alpha \arctan \frac{e_4}{e_5}\right)$$

- $\mathfrak{s}_{6,66}$

	e_1	e_2	e_3	e_4	e_5
e_6	$(a+b)e_1$	ae_2	be_3	e_4	ce_5

US = [0]
CS = [6, 5]
DS = [6, 5, 1, 0]

where the values of the parameters a, b, c are $0 < |b| \leq |a|$, $0 < |c| \leq 1$. If $|a| = |b|$ then $\arg(a) \leq \arg(b)$, $a \neq -b$. If $|c| = |1|$ then $\arg(c) \leq \pi$.

Casimir invariants:

$$\frac{e_4^{a+b}}{e_1}, \quad \frac{e_5^{a+b}}{e_1^c}$$

- $\mathfrak{s}_{6,67}$

	e_1	e_2	e_3	e_4	e_5
e_6	$(a+1)e_1$	ae_2	$e_3 + e_4$	e_4	be_5

US = [0]
CS = [6, 5]
DS = [6, 5, 1, 0]

where $a \neq -1, 0$, $b \neq 0$.

Casimir invariants:

$$\frac{e_4^{a+1}}{e_1}, \quad \frac{e_5^{a+1}}{e_1^b}$$

- $\mathfrak{s}_{6,68}$

	e_1	e_2	e_3	e_4	e_5
e_6	$2ae_1$	ae_2	$e_2 + ae_3$	e_4	be_5

US = [0]
CS = [6, 5]
DS = [6, 5, 1, 0]

where the values of the parameters a, b are $a \neq 0$, $0 < |b| \leq 1$. If $|b| = 1$ then $\arg(b) \leq \pi$.

Casimir invariants:

$$\frac{e_4^{2a}}{e_1}, \quad \frac{e_5^{2a}}{e_1^b}$$

- $\mathfrak{s}_{6,69}$ (over the field $\mathbb{C}$: isomorphic to $\mathfrak{s}_{6,66}$)

	e_1	e_2	e_3	e_4	e_5
e_6	$2\alpha e_1$	$\alpha e_2 - e_3$	$e_2 + \alpha e_3$	βe_4	γe_5

US = [0]
CS = [6, 5]
DS = [6, 5, 1, 0]

where the values of the parameters α, β, γ are $\alpha \neq 0$, $0 < |\gamma| \leq |\beta|$. If $|\gamma| = |\beta|$ then $\gamma \leq \beta$.

Casimir invariants:

$$\frac{e_4^{2\alpha}}{e_1^{\beta}}, \quad \frac{e_5^{2\alpha}}{e_1^{\gamma}}$$

- $\mathfrak{s}_{6,70}$

	e_1	e_2	e_3	e_4	e_5
e_6	$2e_1$	e_2	e_2+e_3	ae_4	e_4+ae_5

US = [0]
CS = [6, 5]
DS = [6, 5, 1, 0]

where $a \neq 0$.

Casimir invariants:

$$\frac{e_4^2}{e_1^a}, \quad e_1 \exp\left(-2\frac{e_5}{e_4}\right)$$

- $\mathfrak{s}_{6,71}$

	e_1	e_2	e_3	e_4	e_5
e_6	$(a+b)e_1$	ae_2	be_3	e_4	$e_1+(a+b)e_5$

US = [0]
CS = [6, 5]
DS = [6, 5, 1, 0]

where $0 < |b| \leq |a|$, $a \neq -b$. If $|a| = |b|$ then $\arg(a) \leq \arg(b)$.

Casimir invariants:

$$\frac{e_4^{a+b}}{e_1}, \quad e_1 \exp\left(-(a+b)\frac{e_5}{e_1}\right)$$

- $\mathfrak{s}_{6,72}$

	e_1	e_2	e_3	e_4	e_5
e_6	$(a+1)e_1$	ae_2	e_3+e_4	e_4	$e_1+(a+1)e_5$

US = [0]
CS = [6, 5]
DS = [6, 5, 1, 0]

where $a \neq -1, 0$.

Casimir invariants:

$$\frac{e_4^{a+1}}{e_1}, \quad e_1 \exp\left(-(a+1)\frac{e_5}{e_1}\right)$$

- $\mathfrak{s}_{6,73}$

	e_1	e_2	e_3	e_4	e_5
e_6	$2ae_1$	ae_2	e_2+ae_3	e_1+2ae_4	e_5

US = [0]
CS = [6, 5]
DS = [6, 5, 1, 0]

where $a \neq 0$.

Casimir invariants:

$$\frac{e_5^{2a}}{e_1}, \quad e_1 \exp\left(-2a\frac{e_4}{e_1}\right)$$

- $\mathfrak{s}_{6,74}$

	e_1	e_2	e_3	e_4	e_5
e_6	$(a+1)e_1$	e_2	ae_3	be_4	e_4+be_5

US = [0]
CS = [6, 5]
DS = [6, 5, 1, 0]

where $0 < |a| \leq 1$, $b \neq 0$. If $|a| = 1$ then $\arg(a) < \pi$.

Casimir invariants:

$$\frac{e_4^{a+1}}{e_1^{b}}, \quad e_1 \exp\left(-(a+1)\frac{e_5}{e_4}\right)$$

- $\mathfrak{s}_{6,75}$

	e_1	e_2	e_3	e_4	e_5
e_6	$(a+1)e_1$	ae_2	e_3+e_4	e_4+e_5	e_5

US = [0]
CS = [6, 5]
DS = [6, 5, 1, 0]

where $a \neq 0, -1$.

Casimir invariants:

$$\frac{e_5^{a+1}}{e_1}, \quad e_1 \exp\left(-(a+1)\frac{e_4}{e_5}\right)$$

- $\mathfrak{s}_{6,76}$ (over the field $\mathbb{C}$: isomorphic to $\mathfrak{s}_{6,71}$)

	e_1	e_2	e_3	e_4	e_5
e_6	$2\alpha e_1$	$\alpha e_2 - e_3$	$e_2 + \alpha e_3$	βe_4	$e_1 + 2\alpha e_5$

US = [0]
CS = [6, 5]
DS = [6, 5, 1, 0]

where $\alpha > 0$, $\beta \neq 0$.

Casimir invariants:

$$\frac{e_4^{2\alpha}}{e_1^{\beta}}, \quad e_4 \exp\left(-\beta\frac{e_5}{e_1}\right)$$

- $\mathfrak{s}_{6,77}$ (over the field $\mathbb{C}$: isomorphic to $\mathfrak{s}_{6,74}$)

	e_1	e_2	e_3	e_4	e_5
e_6	$2\alpha e_1$	$\alpha e_2 - e_3$	$e_2 + \alpha e_3$	$\beta e_4 + e_5$	βe_5

US = [0]
CS = [6, 5]
DS = [6, 5, 1, 0]

where $\alpha > 0$, $\beta \neq 0$.

Casimir invariants:

$$\frac{e_5^{2\alpha}}{e_1^{\beta}}, \quad e_5 \exp\left(-\beta\frac{e_4}{e_5}\right)$$

- $\mathfrak{s}_{6,78}$ (over the field $\mathbb{C}$: isomorphic to $\mathfrak{s}_{6,66}$)

	e_1	e_2	e_3	e_4	e_5
e_6	$(\alpha+\beta)e_1$	αe_2	βe_3	$\gamma e_4 - e_5$	$e_4 + \gamma e_5$

US = [0]
CS = [6, 5]
DS = [6, 5, 1, 0]

where $0 < |\beta| \leq |\alpha|$, $\beta \neq -\alpha$, $\gamma > 0$.

Casimir invariants:

$$\frac{(e_4^2+e_5^2)^{\alpha+\beta}}{e_1^{2\gamma}}, \quad e_1 \exp\left((\alpha+\beta)\arctan\frac{e_4}{e_5}\right)$$

- $\mathfrak{s}_{6,79}$ (over the field $\mathbb{C}$: isomorphic to $\mathfrak{s}_{6,68}$)

	e_1	e_2	e_3	e_4	e_5
e_6	$2\alpha e_1$	αe_2	$e_2 + \alpha e_3$	$\beta e_4 - e_5$	$e_4 + \beta e_5$

US = [0]
CS = [6, 5]
DS = [6, 5, 1, 0]

where $\alpha \neq 0$, $\beta \geq 0$.

Casimir invariants:

$$\frac{(e_4^2+e_5^2)^\alpha}{e_1^\beta}, \quad e_1 \exp\left(2\alpha \arctan \frac{e_4}{e_5}\right)$$

- $\mathfrak{s}_{6,80}$ (over the field $\mathbb{C}$: isomorphic to $\mathfrak{s}_{6,66}$)

	e_1	e_2	e_3	e_4	e_5
e_6	$2\alpha e_1$	$\alpha e_2 - e_3$	$e_2 + \alpha e_3$	$\beta e_4 - \gamma e_5$	$\gamma e_4 + \beta e_5$

US = [0]
CS = [6, 5]
DS = [6, 5, 1, 0]

where $\alpha, \gamma > 0$.

Casimir invariants:

$$\frac{(e_4^2+e_5^2)^\alpha}{e_1^\beta}, \quad e_1^\gamma \exp\left(2\alpha \arctan \frac{e_4}{e_5}\right)$$

- $\mathfrak{s}_{6,81}$

	e_1	e_2	e_3	e_4	e_5
e_6	e_1	0	$e_3 + e_4$	e_4	ae_5

US = [0]
CS = [6, 4]
DS = [6, 4, 0]

where $a \neq 0$.

Casimir invariants:

$$\frac{e_4}{e_1}, \quad \frac{e_1^a}{e_5}$$

- $\mathfrak{s}_{6,82}$

	e_1	e_2	e_3	e_4	e_5
e_6	e_1	0	$e_3 + e_4$	e_4	$e_1 + e_5$

US = [0]
CS = [6, 4]
DS = [6, 4, 0]

Casimir invariants:

$$\frac{e_4}{e_1}, \quad e_1 \exp\left(-\frac{e_5}{e_1}\right)$$

- $\mathfrak{s}_{6,83}$

	e_1	e_2	e_3	e_4	e_5
e_6	e_1	0	$e_3 + e_4$	$e_4 + e_5$	e_5

US = [0]
CS = [6, 4]
DS = [6, 4, 0]

Casimir invariants:

$$\frac{e_5}{e_1}, \quad e_1 \exp\left(-\frac{e_4}{e_5}\right)$$

- $\mathfrak{s}_{6,84}$

	e_1	e_2	e_3	e_4	e_5
e_6	e_1	e_2	0	$e_4 + e_5$	$e_1 + e_5$

US = [0]
CS = [6, 4]
DS = [6, 4, 0]

Casimir invariants:

$$\frac{2e_1e_4 - e_5^2}{e_1^2}, \quad e_1 \exp\left(-\frac{e_5}{e_1}\right)$$

- $\mathfrak{s}_{6,85}$

	e_1	e_2	e_3	e_4	e_5
e_6	e_1	0	$e_3 + e_4$	$e_4 + e_5$	$e_1 + e_5$

US = [0]
CS = [6, 4]
DS = [6, 4, 0]

Casimir invariants:

$$\frac{2e_1e_4 - e_5^2}{e_1^2}, \quad e_1 \exp\left(-\frac{e_5}{e_1}\right)$$

- $\mathfrak{s}_{6,86}$

	e_1	e_2	e_3	e_4	e_5
e_6	ae_1	ae_2	0	e_4	be_5

US = [0]
CS = [6, 4]
DS = [6, 4, 0]

where $a \neq 0$, $0 < |b| \leq 1$. If $|b| = 1$ then $\arg(b) \leq \pi$.

Casimir invariants:

$$\frac{e_4^a}{e_1}, \quad \frac{e_5^a}{e_1^b}$$

- $\mathfrak{s}_{6,87}$

	e_1	e_2	e_3	e_4	e_5
e_6	ae_1	ae_2	0	e_4	$e_1 + ae_5$

US = [0]
CS = [6, 4]
DS = [6, 4, 0]

where $a \neq 0$.

Casimir invariants:

$$\frac{e_4^a}{e_1}, \quad e_1 \exp\left(-a\frac{e_5}{e_1}\right)$$

- $\mathfrak{s}_{6,88}$

	e_1	e_2	e_3	e_4	e_5
e_6	e_1	0	$e_3 + e_4$	$e_1 + e_4$	ae_5

US = [0]
CS = [6, 4]
DS = [6, 4, 0]

where $a \neq 0$.

Casimir invariants:

$$\frac{e_1^a}{e_5}, \quad e_1 \exp\left(-\frac{e_4}{e_1}\right)$$

- $\mathfrak{s}_{6,89}$

	e_1	e_2	e_3	e_4	e_5
e_6	e_1	e_2	0	ae_4	$e_4 + ae_5$

US = [0]
CS = [6, 4]
DS = [6, 4, 0]

where $a \neq 0$.

Casimir invariants:

$$\frac{e_1^a}{e_4}, \quad e_1 \exp\left(-\frac{e_5}{e_4}\right)$$

• $\mathfrak{s}_{6,90}$ (over the field $\mathbb{C}$: isomorphic to $\mathfrak{s}_{6,86}$)

	e_1	e_2	e_3	e_4	e_5
e_6	αe_1	αe_2	0	$\beta e_4 - e_5$	$e_4 + \beta e_5$

US = [0]
CS = [6, 4]
DS = [6, 4, 0]

where $\alpha \neq 0$, $\beta \geq 0$.

Casimir invariants:

$$\frac{(e_4^2 + e_5^2)^\alpha}{e_1^{2\beta}}, \quad e_1 \exp\left(\alpha \arctan \frac{e_4}{e_5}\right)$$

19.4. Solvable six-dimensional Lie algebras with the nilradical $\mathfrak{n}_{4,1} \oplus \mathfrak{n}_{1,1}$

	e_1	e_2	e_3	e_4	e_5
e_1	0	0	0	0	0
e_2		0	0	e_1	0
e_3			0	e_2	0
e_4				0	0

• $\mathfrak{s}_{6,91}$

	e_1	e_2	e_3	e_4	e_5
e_6	0	0	ϵe_1	0	e_5

US = [1, 2, 4]
CS = [6, 3, 2, 1]
DS = [6, 3, 0]

Casimir invariants:

$$e_1, \quad e_5^2 \exp\left(\epsilon \frac{e_2^2 - 2e_1e_3}{e_1^2}\right)$$

• $\mathfrak{s}_{6,92}$

	e_1	e_2	e_3	e_4	e_5
e_6	0	0	0	e_3	e_5

US = [1, 2, 3, 4]
CS = [6, 4, 3, 2, 1]
DS = [6, 4, 0]

Casimir invariants:

$$e_1, \quad 2e_1e_3 - e_2^2$$

• $\mathfrak{s}_{6,93}$

	e_1	e_2	e_3	e_4	e_5
e_6	0	0	ϵe_1	e_3	e_5

US = [1, 2, 3, 4]
CS = [6, 4, 3, 2, 1]
DS = [6, 4, 0]

where $\epsilon = 1$ over the field $\mathbb{C}$, $\epsilon = \pm 1$ over the field $\mathbb{R}$.

Casimir invariants:

$$e_1, \quad e_5^{2\epsilon} \exp\left(\frac{e_2^2 - 2e_1e_3}{e_1^2}\right)$$

• $\mathfrak{s}_{6,94}$

	e_1	e_2	e_3	e_4	e_5
e_6	0	$-e_2$	$-2e_3$	e_4	e_1

US = [1, 2]
CS = [6, 4]
DS = [6, 4, 2, 0]

Casimir invariants:

$$e_1, \quad (2e_1e_3 - e_2^2)\exp\left(2\frac{e_5}{e_1}\right)$$

- $\mathfrak{s}_{6,95}$

	e_1	e_2	e_3	e_4	e_5
e_6	0	$-e_2$	$-2e_3$	$e_4 + e_5$	e_5

US = [1]
CS = [6, 5]
DS = [6, 5, 2, 0]

Casimir invariants:

$$e_1, \quad (2e_1e_3 - e_2^2)e_5^2$$

- $\mathfrak{s}_{6,96}$

	e_1	e_2	e_3	e_4	e_5
e_6	0	$-e_2$	$-2e_3$	e_4	ae_5

US = [1]
CS = [6, 5]
DS = [6, 5, 2, 0]

where $a \neq 0$.
Casimir invariants:

$$e_1, \quad (2e_1e_3 - e_2^2)^a e_5^2$$

- $\mathfrak{s}_{6,97}$

	e_1	e_2	e_3	e_4	e_5
e_6	0	$-e_2$	$-2e_3 + e_5$	e_4	$-2e_5$

US = [1]
CS = [6, 5]
DS = [6, 5, 2, 0]

Casimir invariants:

$$e_1, \quad e_5\exp\left(\frac{2e_1e_3 - e_2^2}{e_1e_5}\right)$$

- $\mathfrak{s}_{6,98}$

	e_1	e_2	e_3	e_4	e_5
e_6	$2e_1$	e_2	e_5	e_4	0

US = [1]
CS = [6, 4, 3]
DS = [6, 4, 1, 0]

Casimir invariants:

$$e_5, \quad e_1\exp\left(\frac{e_2^2 - 2e_1e_3}{e_1e_5}\right)$$

- $\mathfrak{s}_{6,99}$

	e_1	e_2	e_3	e_4	e_5
e_6	e_1	e_2	e_3	e_5	0

US = [1]
CS = [6, 4, 3]
DS = [6, 4, 0]

Casimir invariants:

$$e_5, \quad \frac{2e_1e_3 - e_2^2}{e_1^2}$$

- $\mathfrak{s}_{6,100}$

	e_1	e_2	e_3	e_4	e_5
e_6	e_1	e_2	$\epsilon e_1 + e_3$	e_5	0

US = [1]
CS = [6, 4, 3]
DS = [6, 4, 0]

where $\epsilon = 1$ over the field $\mathbb{C}$, $\epsilon = \pm 1$ over the field $\mathbb{R}$.

Casimir invariants:

$$e_5, \quad e_1^{2\epsilon} \exp\left(\frac{e_2^2 - 2e_1e_3}{e_1^2}\right)$$

- $\mathfrak{s}_{6,101}$

	e_1	e_2	e_3	e_4	e_5
e_6	e_1	0	$-e_3$	$e_4 + e_5$	$e_1 + e_5$

US = [0]
CS = [6, 5]
DS = [6, 5, 2, 0]

Casimir invariants:

$$2e_1e_3 - e_2^2, \quad e_1 \exp\left(-\frac{e_5}{e_1}\right)$$

- $\mathfrak{s}_{6,102}$

	e_1	e_2	e_3	e_4	e_5
e_6	$3e_1$	$2e_2$	e_3	$e_3 + e_4$	ae_5

US = [0]
CS = [6, 5]
DS = [6, 5, 2, 0]

where $a \neq 0$.

Casimir invariants:

$$\frac{(2e_1e_3 - e_2^2)^3}{e_1^4}, \quad \frac{e_5^3}{e_1^a}$$

- $\mathfrak{s}_{6,103}$

	e_1	e_2	e_3	e_4	e_5
e_6	$3e_1$	$2e_2$	$e_3 + e_5$	$e_3 + e_4$	e_5

US = [0]
CS = [6, 5]
DS = [6, 5, 2, 0]

Casimir invariants:

$$\frac{e_5^3}{e_1}, \quad e_1^2 \exp\left(3\frac{e_2^2 - 2e_1e_3}{e_1e_5}\right)$$

- $\mathfrak{s}_{6,104}$

	e_1	e_2	e_3	e_4	e_5
e_6	$3e_1$	$2e_2$	e_3	$e_3 + e_4$	$e_1 + 3e_5$

US = [0]
CS = [6, 5]
DS = [6, 5, 2, 0]

Casimir invariants:

$$\frac{(2e_1e_3 - e_2^2)^3}{e_1^4}, \quad e_1 \exp\left(-3\frac{e_5}{e_1}\right)$$

- $\mathfrak{s}_{6,105}$

	e_1	e_2	e_3	e_4	e_5
e_6	$(a+2)e_1$	$(a+1)e_2$	ae_3	e_4	be_5

US = [0]
CS = [6, 5]
DS = [6, 5, 2, 0]

where $a \neq 0, -2$, $b \neq 0$.

Casimir invariants:

$$\frac{e_5^{a+2}}{e_1^b}, \quad \frac{(2e_1e_3 - e_2^2)^{a+2}}{e_1^{2(1+a)}}$$

- $\mathfrak{s}_{6,106}$

	e_1	e_2	e_3	e_4	e_5
e_6	$(a+2)e_1$	$(a+1)e_2$	ae_3	e_4+e_5	e_5

US = [0]
CS = [6, 5]
DS = [6, 5, 2, 0]

where $a \neq -2, 0$.

Casimir invariants:

$$\frac{e_5^{a+2}}{e_1}, \quad \frac{(2e_1e_3 - e_2^2)^{a+2}}{e_1^{2(a+1)}}$$

- $\mathfrak{s}_{6,107}$

	e_1	e_2	e_3	e_4	e_5
e_6	$(a+2)e_1$	$(a+1)e_2$	ae_3+e_5	e_4	ae_5

US = [0]
CS = [6, 5]
DS = [6, 5, 2, 0]

where $a \neq -2, 0$.

Casimir invariants:

$$\frac{e_5^{a+2}}{e_1^a}, \quad e_1^2 \exp\left(-(a+2)\frac{2e_1e_3 - e_2^2}{e_1e_5}\right)$$

- $\mathfrak{s}_{6,108}$

	e_1	e_2	e_3	e_4	e_5
e_6	$(a+2)e_1$	$(a+1)e_2$	ae_3	e_4	$e_1+(a+2)e_5$

US = [0]
CS = [6, 5]
DS = [6, 5, 2, 0]

where $a \neq -2, 0$.

Casimir invariants:

$$\frac{(2e_1e_3 - e_2^2)^{a+2}}{e_1^{2(1+a)}}, \quad e_1 \exp\left(-(a+2)\frac{e_5}{e_1}\right)$$

- $\mathfrak{s}_{6,109}$

	e_1	e_2	e_3	e_4	e_5
e_6	$2e_1$	e_2	0	e_4+e_5	e_5

US = [0]
CS = [6, 4]
DS = [6, 4, 1, 0]

Casimir invariants:

$$\frac{2e_1e_3 - e_2^2}{e_1}, \quad \frac{e_5^2}{e_1}$$

- $\mathfrak{s}_{6,110}$

	e_1	e_2	e_3	e_4	e_5
e_6	$2e_1$	e_2	0	e_4	ae_5

US = [0]
CS = [6, 4]
DS = [6, 4, 1, 0]

where $a \neq 0$.

Casimir invariants:

$$\frac{2e_1e_3 - e_2^2}{e_1}, \quad \frac{e_5^2}{e_1^a}$$

- $\mathfrak{s}_{6,111}$

	e_1	e_2	e_3	e_4	e_5
e_6	$2e_1$	e_2	0	e_4	$e_1 + 2e_5$

US = [0]
CS = [6, 4]
DS = [6, 4, 1, 0]

Casimir invariants:

$$\frac{2e_1e_3 - e_2^2}{e_1}, \quad e_1 \exp\left(-2\frac{e_5}{e_1}\right)$$

- $\mathfrak{s}_{6,112}$

	e_1	e_2	e_3	e_4	e_5
e_6	e_1	e_2	e_3	0	ae_5

US = [0]
CS = [6, 4]
DS = [6, 4, 0]

where $a \neq 0$.

Casimir invariants:

$$\frac{2e_1e_3 - e_2^2}{e_1^2}, \quad \frac{e_1^a}{e_5}$$

- $\mathfrak{s}_{6,113}$

	e_1	e_2	e_3	e_4	e_5
e_6	e_1	e_2	$e_3 + e_5$	0	e_5

US = [0]
CS = [6, 4]
DS = [6, 4, 0]

Casimir invariants:

$$\frac{e_5}{e_1}, \quad e_1^2 \exp\left(\frac{e_2^2 - 2e_1e_3}{e_1e_5}\right)$$

- $\mathfrak{s}_{6,114}$

	e_1	e_2	e_3	e_4	e_5
e_6	e_1	e_2	e_3	0	$e_1 + e_5$

US = [0]
CS = [6, 4]
DS = [6, 4, 0]

Casimir invariants:

$$\frac{2e_1e_3 - e_2^2}{e_1^2}, \quad e_1 \exp\left(-\frac{e_5}{e_1}\right)$$

- $\mathfrak{s}_{6,115}$

	e_1	e_2	e_3	e_4	e_5
e_6	e_1	e_2	$e_3 + e_5$	0	$\epsilon e_1 + e_5$

US = [0]
CS = [6, 4]
DS = [6, 4, 0]

where $\epsilon = 1$ over the field $\mathbb{C}$, $\epsilon = \pm 1$ over the field $\mathbb{R}$.

Casimir invariants:

$$\frac{2e_1e_3 - e_2^2 - \epsilon e_5^2}{e_1^2}, \quad e_1 \exp\left(-\epsilon\frac{e_5}{e_1}\right)$$

- $\mathfrak{s}_{6,116}$

	e_1	e_2	e_3	e_4	e_5
e_6	e_1	e_2	$\epsilon e_1 + e_3$	0	ae_5

US = [0]
CS = [6, 4]
DS = [6, 4, 0]

where $a \neq 0$ and $\epsilon = 1$ over the field $\mathbb{C}$, $\epsilon = \pm 1$ over the field $\mathbb{R}$.

Casimir invariants:

$$\frac{e_1^a}{e_5}, \quad e_1^2 \exp\left(\epsilon \frac{e_2^2 - 2e_1e_3}{e_1^2}\right)$$

19.5. Solvable six-dimensional Lie algebras with the nilradical $\mathfrak{n}_{5,1}$

	e_1	e_2	e_3	e_4	e_5
e_1	0	0	0	0	0
e_2		0	0	0	0
e_3			0	0	e_1
e_4				0	e_2

- $\mathfrak{s}_{6,117}$

	e_1	e_2	e_3	e_4	e_5
e_6	0	0	e_3	e_4	$-e_5$

US = [2]
CS = [6, 5]
DS = [6, 5, 2, 0]

Casimir invariants:

$$e_1, \quad e_2$$

- $\mathfrak{s}_{6,118}$

	e_1	e_2	e_3	e_4	e_5
e_6	0	e_1	e_3	$e_3 + e_4$	$-e_5$

US = [1, 2]
CS = [6, 5]
DS = [6, 5, 2, 0]

Casimir invariants:

$$e_1, \quad (e_1e_4 - e_2e_3)\exp\left(-\frac{e_2}{e_1}\right)$$

- $\mathfrak{s}_{6,119}$

	e_1	e_2	e_3	e_4	e_5
e_6	0	e_2	0	e_4	e_3

US = [1, 2]
CS = [6, 4, 3, 2]
DS = [6, 4, 0]

Casimir invariants:

$$e_1, \quad \frac{e_1e_4 - e_2e_3}{e_2}$$

- $\mathfrak{s}_{6,120}$

	e_1	e_2	e_3	e_4	e_5
e_6	0	e_2	$-e_1$	$e_2 + e_4$	e_3

US = [1, 2]
CS = [6, 4, 3, 2]
DS = [6, 4, 0]

Casimir invariants:

$$e_1, \quad e_2^2 \exp\left(\frac{e_2e_3 - e_1e_4}{e_1e_2}\right)$$

- $\mathfrak{s}_{6,121}$

	e_1	e_2	e_3	e_4	e_5
e_6	0	e_2	0	e_4	0

US = [1, 2]
CS = [6, 3, 2]
DS = [6, 3, 0]

Casimir invariants:

$$e_1, \quad \frac{e_1e_4 - e_2e_3}{e_2}$$

- $\mathfrak{s}_{6,122}$

	e_1	e_2	e_3	e_4	e_5
e_6	0	e_2	$-e_1$	$e_2 + e_4$	0

US = [1, 2]
CS = [6, 3, 2]
DS = [6, 3, 0]

Casimir invariants:

$$e_1, \quad e_2^2 \exp\left(\frac{e_2e_3 - e_1e_4}{e_1e_2}\right)$$

- $\mathfrak{s}_{6,123}$

	e_1	e_2	e_3	e_4	e_5
e_6	0	$2e_2$	$-e_3$	e_4	$e_4 + e_5$

US = [1]
CS = [6, 5]
DS = [6, 5, 2, 0]

Casimir invariants:

$$e_1, \quad \frac{(e_1e_4 - e_2e_3)^2}{e_2}$$

- $\mathfrak{s}_{6,124}$

	e_1	e_2	e_3	e_4	e_5
e_6	0	e_2	$-ae_3$	$(1-a)e_4$	ae_5

US = [1]
CS = [6, 5]
DS = [6, 5, 2, 0]

where $a \neq 0, 1$.

Casimir invariants:

$$e_1, \quad (e_1e_4 - e_2e_3)e_2^{a-1}$$

- $\mathfrak{s}_{6,125}$

	e_1	e_2	e_3	e_4	e_5
e_6	0	e_2	$e_2 + e_3$	$2e_4$	$-e_5$

US = [1]
CS = [6, 5]
DS = [6, 5, 2, 0]

Casimir invariants:

$$e_1, \quad e_2 \exp\left(\frac{e_1e_4 - e_2e_3}{e_2^2}\right)$$

- $\mathfrak{s}_{6,126}$

	e_1	e_2	e_3	e_4	e_5
e_6	0	e_2	$-e_3$	0	e_5

US = [1]
CS = [6, 4]
DS = [6, 4, 1, 0]

Casimir invariants:

$$e_1, \quad e_1e_4 - e_2e_3$$

- $\mathfrak{s}_{6,127}$

	e_1	e_2	e_3	e_4	e_5
e_6	0	e_2	$-e_3$	e_1	e_5

US = [1]
CS = [6, 4]
DS = [6, 4, 1, 0]

Casimir invariants:

$$e_1, \quad e_2 \exp\left(\frac{e_2e_3 - e_1e_4}{e_1^2}\right)$$

- $\mathfrak{s}_{6,128}$

	e_1	e_2	e_3	e_4	e_5
e_6	$2e_1$	ae_2	e_3	$(a-1)e_4$	$e_3 + e_5$

US = [0]
CS = [6, 5]
DS = [6, 5, 2, 0]

where $a \neq 0, 1$.

Casimir invariants:

$$\frac{(e_1e_4 - e_2e_3)^2}{e_1e_2^2}, \quad \frac{e_1^a}{e_2^2}$$

- $\mathfrak{s}_{6,129}$

	e_1	e_2	e_3	e_4	e_5
e_6	$3e_1$	$2e_2$	$e_2 + 2e_3$	e_4	$e_4 + e_5$

US = [0]
CS = [6, 5]
DS = [6, 5, 2, 0]

Casimir invariants:

$$\frac{e_2^3}{e_1^2}, \quad e_1 \exp\left(3\frac{e_1e_4 - e_2e_3}{e_2^2}\right)$$

- $\mathfrak{s}_{6,130}$

	e_1	e_2	e_3	e_4	e_5
e_6	$2e_1$	$e_1 + 2e_2$	e_3	$e_3 + e_4$	$e_4 + e_5$

US = [0]
CS = [6, 5]
DS = [6, 5, 2, 0]

Casimir invariants:

$$\frac{(e_1e_4 - e_2e_3)^2}{e_1^3}, \quad e_1 \exp\left(-2\frac{e_2}{e_1}\right)$$

- $\mathfrak{s}_{6,131}$

	e_1	e_2	e_3	e_4	e_5
e_6	e_1	ae_2	$(1-b)e_3$	$(a-b)e_4$	be_5

US = [0]
CS = [6, 5]
DS = [6, 5, 2, 0]

where the values of the parameters a, b are $0 < |a| \leq 1$, $b \neq 0, 1, a$. If $|a| = 1$ then $\arg(a) \leq \pi$.

Casimir invariants:

$$\frac{e_1^a}{e_2}, \quad (e_1e_4 - e_2e_3)e_1^{b-a-1}$$

- $\mathfrak{s}_{6,132}$

	e_1	e_2	e_3	e_4	e_5
e_6	e_1	ae_2	$(2-a)e_3$	e_1+e_4	$(a-1)e_5$

US = [0]
CS = [6, 5]
DS = [6, 5, 2, 0]

where $a \neq 0, 1, 2$.

Casimir invariants:

$$\frac{e_1^a}{e_2}, \quad e_1 \exp\left(\frac{e_2e_3 - e_1e_4}{e_1^2}\right)$$

- $\mathfrak{s}_{6,133}$

	e_1	e_2	e_3	e_4	e_5
e_6	e_1	e_1+e_2	$(1-a)e_3$	$e_3+(1-a)e_4$	ae_5

US = [0]
CS = [6, 5]
DS = [6, 5, 2, 0]

where $a \neq 0, 1$.

Casimir invariants:

$$(e_1e_4 - e_2e_3)e_1^{a-2}, \quad e_1 \exp\left(-\frac{e_2}{e_1}\right)$$

- $\mathfrak{s}_{6,134}$ (over the field $\mathbb{C}$: isomorphic to $\mathfrak{s}_{6,131}$)

	e_1	e_2	e_3	e_4	e_5
e_6	$\alpha e_1 - e_2$	$e_1 + \alpha e_2$	$(\alpha-\beta)e_3 - e_4$	$e_3 + (\alpha-\beta)e_4$	βe_5

US = [0]
CS = [6, 5]
DS = [6, 5, 2, 0]

where $0 < \beta$.

Casimir invariants:

$$(e_1e_4 - e_2e_3)^{2\alpha}(e_1^2+e_2^2)^{\beta-2\alpha}, \quad \frac{e_1e_4 - e_2e_3}{e_1^2+e_2^2}\exp\left(-\beta \arctan\frac{e_1}{e_2}\right)$$

- $\mathfrak{s}_{6,135}$

	e_1	e_2	e_3	e_4	e_5
e_6	e_1	$2e_2$	0	e_4	e_4+e_5

US = [0]
CS = [6, 4]
DS = [6, 4, 1, 0]

Casimir invariants:

$$\frac{e_1^2}{e_2}, \quad \frac{e_1e_4 - e_2e_3}{e_1^2}$$

- $\mathfrak{s}_{6,136}$

	e_1	e_2	e_3	e_4	e_5
e_6	$2e_1$	e_2	e_3	0	e_3+e_5

US = [0]
CS = [6, 4]
DS = [6, 4, 1, 0]

Casimir invariants:

$$\frac{e_2^2}{e_1}, \quad \frac{e_1e_4 - e_2e_3}{e_1}$$

- $\mathfrak{s}_{6,137}$

	e_1	e_2	e_3	e_4	e_5
e_6	$2e_1$	e_2	e_2+e_3	0	e_3+e_5

US = [0]
CS = [6, 4]
DS = [6, 4, 1, 0]

Casimir invariants:

$$\frac{e_2^2}{e_1},\quad e_1\exp\left(2\frac{e_1e_4-e_2e_3}{e_2^2}\right)$$

- $\mathfrak{s}_{6,138}$

	e_1	e_2	e_3	e_4	e_5
e_6	e_1	$2e_2$	0	e_1+e_4	e_5

US = [0]
CS = [6, 4]
DS = [6, 4, 1, 0]

Casimir invariants:

$$\frac{e_1^2}{e_2},\quad e_1\exp\left(\frac{e_2e_3-e_1e_4}{e_1^2}\right)$$

- $\mathfrak{s}_{6,139}$

	e_1	e_2	e_3	e_4	e_5
e_6	e_1	ae_2	0	$(a-1)e_4$	e_5

US = [0]
CS = [6, 4]
DS = [6, 4, 1, 0]

where $a \neq 0, 1$.
Casimir invariants:

$$\frac{e_1e_4-e_2e_3}{e_2},\quad \frac{e_1^a}{e_2}$$

- $\mathfrak{s}_{6,140}$

	e_1	e_2	e_3	e_4	e_5
e_6	e_1	ae_2	e_3	ae_4	0

US = [0]
CS = [6, 4]
DS = [6, 4, 0]

where $0 < |a| \leq 1$. If $|a| = 1$ then $\arg(a) \leq \pi$.
Casimir invariants:

$$\frac{e_1e_4-e_2e_3}{e_1e_2},\quad \frac{e_1^a}{e_2}$$

- $\mathfrak{s}_{6,141}$

	e_1	e_2	e_3	e_4	e_5
e_6	e_1	e_2	e_3	e_1+e_4	0

US = [0]
CS = [6, 4]
DS = [6, 4, 0]

Casimir invariants:

$$\frac{e_2}{e_1},\quad e_1\exp\left(\frac{e_2e_3-e_1e_4}{e_1^2}\right)$$

- $\mathfrak{s}_{6,142}$ (over the field $\mathbb{C}$: isomorphic to $\mathfrak{s}_{6,146}$ with $a = 1$)

	e_1	e_2	e_3	e_4	e_5
e_6	e_1	e_2	$-e_2+e_3$	e_1+e_4	0

US = [0]
CS = [6, 4]
DS = [6, 4, 0]

Casimir invariants:

$$\frac{e_2}{e_1}, \quad e_1 \exp\left(\frac{e_2e_3 - e_1e_4}{e_1^2+e_2^2}\right)$$

- $\mathfrak{s}_{6,143}$

	e_1	e_2	e_3	e_4	e_5
e_6	e_1	e_1+e_2	e_3	e_3+e_4	0

US $=[0]$
CS $=[6,4]$
DS $=[6,4,0]$

Casimir invariants:

$$\frac{e_1e_4 - e_2e_3}{e_1^2}, \quad e_1 \exp\left(-\frac{e_2}{e_1}\right)$$

- $\mathfrak{s}_{6,144}$

	e_1	e_2	e_3	e_4	e_5
e_6	e_1	e_1+e_2	e_2+e_3	e_3+e_4	0

US $=[0]$
CS $=[6,4]$
DS $=[6,4,0]$

Casimir invariants:

$$\frac{3e_4e_1^2 - 3e_1e_2e_3 + e_2^3}{e_1^3}, \quad e_1 \exp\left(-\frac{e_2}{e_1}\right)$$

- $\mathfrak{s}_{6,145}$ (over the field $\mathbb{C}$: isomorphic to $\mathfrak{s}_{6,140}$)

	e_1	e_2	e_3	e_4	e_5
e_6	$\alpha e_1 - e_2$	$e_1+\alpha e_2$	$\alpha e_3 - e_4$	$e_3+\alpha e_4$	0

US $=[0]$
CS $=[6,4]$
DS $=[6,4,0]$

where $0 \leq \alpha$.
Casimir invariants:

$$\frac{e_1e_4 - e_2e_3}{e_1^2+e_2^2}, \quad (e_1^2+e_2^2)\exp\left(2\alpha \arctan\frac{e_1}{e_2}\right)$$

- $\mathfrak{s}_{6,146}$

	e_1	e_2	e_3	e_4	e_5
e_6	e_1	ae_2	e_1+e_3	$-e_2+ae_4$	0

US $=[0]$
CS $=[6,4]$
DS $=[6,4,0]$

where $0<|a|\leq 1$. If $|a|=1$ then $\arg(a)\leq\pi$.
Casimir invariants:

$$\frac{e_1^a}{e_2}, \quad e_1^2 \exp\left(\frac{e_1e_4 - e_2e_3}{e_1e_2}\right)$$

- $\mathfrak{s}_{6,147}$ (over the field $\mathbb{C}$: isomorphic to $\mathfrak{s}_{6,146}$)

	e_1	e_2	e_3	e_4	e_5
e_6	$\alpha e_1 - e_2$	$e_1+\alpha e_2$	$e_2+\alpha e_3 - e_4$	$e_3+\alpha e_4$	0

US $=[0]$
CS $=[6,4]$
DS $=[6,4,0]$

Casimir invariants:

$$(e_1^2+e_2^2)\exp\left(2\alpha\frac{2e_1e_4 - 2e_2e_3 - e_1e_2}{e_1^2+e_2^2}\right), \quad \frac{2e_1e_4 - 2e_2e_3 - e_1e_2}{e_1^2+e_2^2} + \arctan\frac{e_2}{e_1}$$

- $\mathfrak{s}_{6,148}$

	e_1	e_2	e_3	e_4	e_5
e_6	e_1	$e_1 + e_2$	0	e_3	e_5

US = [0]
CS = [6, 4, 3]
DS = [6, 4, 1, 0]

Casimir invariants:

$$\frac{e_1e_4 - e_2e_3}{e_1}, \quad e_1 \exp\left(-\frac{e_2}{e_1}\right)$$

- $\mathfrak{s}_{6,149}$

	e_1	e_2	e_3	e_4	e_5
e_6	e_1	e_2	0	0	e_5

US = [0]
CS = [6, 3]
DS = [6, 3, 0]

Casimir invariants:

$$\frac{e_2}{e_1}, \quad \frac{e_1e_4 - e_2e_3}{e_1}$$

19.6. Solvable six-dimensional Lie algebras with the nilradical $\mathfrak{n}_{5,2}$

	e_1	e_2	e_3	e_4	e_5
e_1	0	0	0	0	0
e_2		0	0	0	0
e_3			0	e_2	e_1
e_4				0	e_3

- $\mathfrak{s}_{6,150}$

	e_1	e_2	e_3	e_4	e_5
e_6	0	$3e_2$	e_3	$2e_4$	$-e_5$

US = [1]
CS = [6, 5]
DS = [6, 5, 3, 0]

Casimir invariants:

$$e_1, \quad \frac{(2e_1e_4 - 2e_2e_5 - e_3^2)^3}{e_2^2}$$

- $\mathfrak{s}_{6,151}$

	e_1	e_2	e_3	e_4	e_5
e_6	e_1	$-e_2$	0	$e_2 - e_4$	e_5

US = [0]
CS = [6, 5]
DS = [6, 5, 3, 0]

Casimir invariants:

$$e_1e_2, \quad e_1^2 \exp\left(\frac{-2e_1e_4 + 2e_2e_5 + e_3^2}{e_1e_2}\right)$$

- $\mathfrak{s}_{6,152}$ (over the field $\mathbb{C}$: isomorphic to $\mathfrak{s}_{6,151}$)

	e_1	e_2	e_3	e_4	e_5
e_6	e_2	$-e_1$	0	$-e_5$	$\epsilon e_2 + e_4$

US = [0]
CS = [6, 5]
DS = [6, 5, 3, 0]

where $\epsilon = \pm 1$.

Casimir invariants:

$$e_1^2+e_2^2,\quad \epsilon e_1e_2+2e_1e_4-2e_2e_5-e_3^2-\epsilon(e_1^2+e_2^2)\arctan\frac{e_2}{e_1}$$

- $\mathfrak{s}_{6,153}$

	e_1	e_2	e_3	e_4	e_5
e_6	$3e_1$	e_1+3e_2	$2e_3$	e_4+e_5	e_5

US = [0]
CS = [6, 5]
DS = [6, 5, 3, 0]

Casimir invariants:

$$\frac{(2e_1e_4-2e_2e_5-e_3^2)^3}{e_1^4},\quad e_1\exp\left(-3\frac{e_2}{e_1}\right)$$

- $\mathfrak{s}_{6,154}$ (over the field $\mathbb{C}$: isomorphic to $\mathfrak{s}_{6,155}$)

	e_1	e_2	e_3	e_4	e_5
e_6	$3\alpha e_1+e_2$	$-e_1+3\alpha e_2$	$2\alpha e_3$	αe_4-e_5	$e_4+\alpha e_5$

US = [0]
CS = [6, 5]
DS = [6, 5, 3, 0]

Casimir invariants:

$$\frac{(2e_1e_4-2e_2e_5-e_3^2)^3}{(e_1^2+e_2^2)^2},\quad (e_1^2+e_2^2)\exp\left(6\alpha\arctan\frac{e_2}{e_1}\right)$$

- $\mathfrak{s}_{6,155}$

	e_1	e_2	e_3	e_4	e_5
e_6	$(2a+1)e_1$	$(a+2)e_2$	$(a+1)e_3$	e_4	ae_5

US = [0]
CS = [6, 5]
DS = [6, 5, 3, 0]

where $0<|a|\le 1$, $a\ne -\frac{1}{2}$. If $|a|=1$ then $\arg(a)\le\pi$.

Casimir invariants:

$$\frac{e_2^{2a+1}}{e_1^{a+2}},\quad \frac{(2e_1e_4-2e_2e_5-e_3^2)^{2a+1}}{e_1^{2a+2}}$$

- $\mathfrak{s}_{6,156}$

	e_1	e_2	e_3	e_4	e_5
e_6	e_1	$2e_2$	e_3	e_4	0

US = [0]
CS = [6, 4]
DS = [6, 4, 1, 0]

Casimir invariants:

$$\frac{e_1^2}{e_2},\quad \frac{2e_1e_4-2e_2e_5-e_3^2}{e_2}$$

- $\mathfrak{s}_{6,157}$

	e_1	e_2	e_3	e_4	e_5
e_6	e_1	$2e_2$	e_3	ϵe_1+e_4	0

US = [0]
CS = [6, 4]
DS = [6, 4, 1, 0]

where the value of the parameter ϵ is
over the field $\mathbb{C}$: $\epsilon=1$;
over the field $\mathbb{R}$: $\epsilon=\pm1$.

Casimir invariants:

$$\frac{e_1^2}{e_2}, \quad e_1^2 \exp\left(-\epsilon \frac{2e_1e_4 - 2e_2e_5 - e_3^2}{e_1^2}\right)$$

19.7. Solvable six-dimensional Lie algebras with the nilradical $\mathfrak{n}_{5,3}$

	e_1	e_2	e_3	e_4	e_5
e_1	0	0	0	0	0
e_2		0	0	e_1	0
e_3			0	0	e_1
e_4				0	0

- $\mathfrak{s}_{6,158}$

	e_1	e_2	e_3	e_4	e_5
e_6	0	0	e_3	0	$-e_5$

US = $[1,3]$
CS = $[6,3]$
DS = $[6,3,1,0]$

Casimir invariants:

$$e_1, \quad e_1e_6 + e_3e_5$$

- $\mathfrak{s}_{6,159}$ (over the field $\mathbb{C}$: isomorphic to $\mathfrak{s}_{6,158}$)

	e_1	e_2	e_3	e_4	e_5
e_6	0	0	e_5	0	$-e_3$

US = $[1,3]$
CS = $[6,3]$
DS = $[6,3,1,0]$

Casimir invariants:

$$e_1, \quad 2e_1e_6 + e_3^2 + e_5^2$$

- $\mathfrak{s}_{6,160}$

	e_1	e_2	e_3	e_4	e_5
e_6	0	0	e_3	e_2	$-e_5$

US = $[1,2,3]$
CS = $[6,4,3]$
DS = $[6,4,1,0]$

Casimir invariants:

$$e_1, \quad 2e_1e_6 + 2e_3e_5 - e_2^2$$

- $\mathfrak{s}_{6,161}$ (over the field $\mathbb{C}$: isomorphic to $\mathfrak{s}_{6,160}$)

	e_1	e_2	e_3	e_4	e_5
e_6	0	0	e_5	ϵe_2	$-e_3$

US = $[1,2,3]$
CS = $[6,4,3]$
DS = $[6,4,1,0]$

Casimir invariants:

$$e_1, \quad 2e_1e_6 - \epsilon e_2^2 + e_3^2 + e_5^2$$

- $\mathfrak{s}_{6,162}$

	e_1	e_2	e_3	e_4	e_5
e_6	0	e_2	ae_3	$-e_4$	$-ae_5$

US = $[1]$
CS = $[6,5]$
DS = $[6,5,1,0]$

where $0 < |a| \leq 1$ and $\arg(a) < \pi$. If $|a| = 1$ then $\arg(a) < \frac{\pi}{2}$.

Casimir invariants:

$$e_1, \quad e_1e_6 + e_2e_4 + ae_3e_5$$

- $\mathfrak{s}_{6,163}$

	e_1	e_2	e_3	e_4	e_5
e_6	0	$e_2 + e_3$	e_3	$-e_4$	$-e_4 - e_5$

US = [1]
CS = [6, 5]
DS = [6, 5, 1, 0]

Casimir invariants:

$$e_1, \quad e_1e_6 + e_2e_4 + e_3e_4 + e_3e_5$$

- $\mathfrak{s}_{6,164}$ (over the field $\mathbb{C}$: isomorphic to $\mathfrak{s}_{6,162}$)

	e_1	e_2	e_3	e_4	e_5
e_6	0	αe_2	e_5	$-\alpha e_4$	$-e_3$

US = [1]
CS = [6, 5]
DS = [6, 5, 1, 0]

where $0 < \alpha$.

Casimir invariants:

$$e_1, \quad 2e_1e_6 + 2\alpha e_2e_4 + e_3^2 + e_5^2$$

- $\mathfrak{s}_{6,165}$ (over the field $\mathbb{C}$: isomorphic to $\mathfrak{s}_{6,162}$)

	e_1	e_2	e_3	e_4	e_5
e_6	0	$\alpha e_2 + e_3$	$-e_2 + \alpha e_3$	$-\alpha e_4 + e_5$	$-e_4 - \alpha e_5$

US = [1]
CS = [6, 5]
DS = [6, 5, 1, 0]

where $0 < \alpha$.

Casimir invariants:

$$e_1, \quad e_1e_6 + \alpha e_2e_4 - e_2e_5 + e_3e_4 + \alpha e_3e_5$$

- $\mathfrak{s}_{6,166}$ (over the field $\mathbb{C}$: isomorphic to $\mathfrak{s}_{6,162}$)

	e_1	e_2	e_3	e_4	e_5
e_6	0	e_4	αe_5	$-e_2$	$-\alpha e_3$

US = [1]
CS = [6, 5]
DS = [6, 5, 1, 0]

where $0 < |\alpha| \leq 1$.

Casimir invariants:

$$e_1, \quad 2e_1e_6 + e_2^2 + \alpha e_3^2 + e_4^2 + \alpha e_5^2$$

- $\mathfrak{s}_{6,167}$ (over the field $\mathbb{C}$: isomorphic to $\mathfrak{s}_{6,163}$)

	e_1	e_2	e_3	e_4	e_5
e_6	0	$e_3 + e_4$	$-e_2 + e_5$	e_5	$-e_4$

US = [1]
CS = [6, 5]
DS = [6, 5, 1, 0]

Casimir invariants:

$$e_1, \quad 2e_1e_6 - 2e_2e_5 + 2e_3e_4 + e_4^2 + e_5^2$$

- $\mathfrak{s}_{6,168}$

	e_1	e_2	e_3	e_4	e_5
e_6	$2e_1$	$(1+a)e_2$	$(1+b)e_3$	$(1-a)e_4$	$(1-b)e_5$

US = [0]
CS = [6, 5]
DS = [6, 5, 1, 0]

where $|b| \leq |a|$, $\arg(a), \arg(b) < \pi$, $a, b \neq 1$. If $|a| = |b|$ then $\arg(b) \leq \arg(a)$.
Casimir invariant:

none

- $\mathfrak{s}_{6,169}$

	e_1	e_2	e_3	e_4	e_5
e_6	$2e_1$	$(1+a)e_2$	e_3+e_5	$(1-a)e_4$	e_5

US = [0]
CS = [6, 5]
DS = [6, 5, 1, 0]

where $\arg(a) < \pi$, $a \neq 1$.
Casimir invariant:

none

- $\mathfrak{s}_{6,170}$

	e_1	e_2	e_3	e_4	e_5
e_6	$2e_1$	$(1+a)e_2+e_3$	$(1+a)e_3$	$(1-a)e_4$	$-e_4+(1-a)e_5$

US = [0]
CS = [6, 5]
DS = [6, 5, 1, 0]

where $\arg(a) < \pi$, $a \neq 1$.
Casimir invariant:

none

- $\mathfrak{s}_{6,171}$

	e_1	e_2	e_3	e_4	e_5
e_6	$2e_1$	e_2+e_3	e_3-e_5	e_4	$-e_4+e_5$

US = [0]
CS = [6, 5]
DS = [6, 5, 1, 0]

where $\epsilon = 1$ over the field $\mathbb{C}$, $\epsilon = \pm 1$ over the field $\mathbb{R}$.
Casimir invariant:

none

- $\mathfrak{s}_{6,172}$ (over the field $\mathbb{C}$: isomorphic to $\mathfrak{s}_{6,170}$ with $a = 0$)

	e_1	e_2	e_3	e_4	e_5
e_6	$2e_1$	e_2+e_4	e_3+e_5	e_4	e_5

US = [0]
CS = [6, 5]
DS = [6, 5, 1, 0]

Casimir invariant:

none

- $\mathfrak{s}_{6,173}$ (over the field $\mathbb{C}$: isomorphic to $\mathfrak{s}_{6,168}$)

	e_1	e_2	e_3	e_4	e_5
e_6	$2e_1$	$(1+\alpha)e_2$	$e_3+\beta e_5$	$(1-\alpha)e_4$	$-\beta e_3+e_5$

US = [0]
CS = [6, 5]
DS = [6, 5, 1, 0]

where $0 \leq \alpha$, $\alpha \neq 1$, $0 < \beta$.

Casimir invariant:

none

- $\mathfrak{s}_{6,174}$ (over the field $\mathbb{C}$: isomorphic to $\mathfrak{s}_{6,168}$)

	e_1	e_2	e_3	e_4	e_5
e_6	$2e_1$	$e_2+\alpha e_4$	$e_3+\beta e_5$	$-\alpha e_2+e_4$	$-\beta e_3+e_5$

US $=[0]$
CS $=[6,5]$
DS $=[6,5,1,0]$

where $0<|\beta|\le\alpha$.
Casimir invariant:

none

- $\mathfrak{s}_{6,175}$ (over the field $\mathbb{C}$: isomorphic to $\mathfrak{s}_{6,168}$)

	e_1	e_2	e_3	e_4	e_5
e_6	$2e_1$	$(1+\alpha)e_2+\beta e_3$	$-\beta e_2+(1+\alpha)e_3$	$(1-\alpha)e_4+\beta e_5$	$-\beta e_4+(1-\alpha)e_5$

US $=[0]$ CS $=[6,5]$ DS $=[6,5,1,0]$

where $0<\beta$, $0<\alpha$.
Casimir invariant:

none

- $\mathfrak{s}_{6,176}$ (over the field $\mathbb{C}$: isomorphic to $\mathfrak{s}_{6,169}$)

	e_1	e_2	e_3	e_4	e_5
e_6	$2e_1$	e_2+e_4	$e_3+\alpha e_5$	e_4	$-\alpha e_3+e_5$

US $=[0]$
CS $=[6,5]$
DS $=[6,5,1,0]$

where $\alpha\neq 0$.
Casimir invariant:

none

- $\mathfrak{s}_{6,177}$ (over the field $\mathbb{C}$: isomorphic to $\mathfrak{s}_{6,170}$)

	e_1	e_2	e_3	e_4	e_5
e_6	$2e_1$	$e_2+\alpha e_3+e_4$	$-\alpha e_2+e_3+e_5$	$e_4+\alpha e_5$	$-\alpha e_4+e_5$

US $=[0]$
CS $=[6,5]$
DS $=[6,5,1,0]$

where $0<\alpha$.
Casimir invariant:

none

- $\mathfrak{s}_{6,178}$

	e_1	e_2	e_3	e_4	e_5
e_6	$2e_1$	$(1+a)e_2$	$2e_3$	$(1-a)e_4$	0

US $=[0]$
CS $=[6,4]$
DS $=[6,4,1,0]$

where $a\neq 1$, $\arg(a)<\pi$.
Casimir invariant:

none

- $\mathfrak{s}_{6,179}$

	e_1	e_2	e_3	e_4	e_5
e_6	$2e_1$	$2e_2$	e_3+e_5	0	e_5

US = [0]
CS = [6, 4]
DS = [6, 4, 1, 0]

Casimir invariant:

none

- $\mathfrak{s}_{6,180}$ (over the field $\mathbb{C}$: isomorphic to $\mathfrak{s}_{6,178}$)

	e_1	e_2	e_3	e_4	e_5
e_6	$2e_1$	$2e_2$	$e_3+\alpha e_5$	0	$-\alpha e_3+e_5$

US = [0]
CS = [6, 4]
DS = [6, 4, 1, 0]

where $\alpha \neq 0$.

Casimir invariant:

none

- $\mathfrak{s}_{6,181}$

	e_1	e_2	e_3	e_4	e_5
e_6	$2e_1$	$2e_2+e_3$	$2e_3$	0	$-e_4$

US = [0]
CS = [6, 4, 3]
DS = [6, 4, 1, 0]

Casimir invariant:

none

- $\mathfrak{s}_{6,182}$

	e_1	e_2	e_3	e_4	e_5
e_6	e_1	e_2	e_3	0	0

US = [0]
CS = [6, 3]
DS = [6, 3, 0]

Casimir invariant:

none

19.8. Solvable six-dimensional Lie algebras with the nilradical $\mathfrak{n}_{5,4}$

	e_1	e_2	e_3	e_4	e_5
e_1	0	0	0	0	0
e_2		0	0	0	e_1
e_3			0	e_1	0
e_4				0	e_2

- $\mathfrak{s}_{6,183}$

	e_1	e_2	e_3	e_4	e_5
e_6	0	$-e_2$	$2e_3$	$-2e_4$	e_5

US = [1]
CS = [6, 5]
DS = [6, 5, 2, 0]

Casimir invariants:

$$e_1, \quad e_1^2e_6 + 2e_1e_3e_4 - e_1e_2e_5 - e_2^2e_3$$

- $\mathfrak{s}_{6,184}$

	e_1	e_2	e_3	e_4	e_5
e_6	$(a+2)e_1$	$(a+1)e_2$	$2e_3$	ae_4	e_5

US = [0]
CS = [6, 5]
DS = [6, 5, 2, 0]

where $a \neq 0, -2$.

Casimir invariant:

none

- $\mathfrak{s}_{6,185}$

	e_1	e_2	e_3	e_4	e_5
e_6	$3e_1$	$2e_2$	e_2+2e_3	e_4	$-e_4+e_5$

US = [0]
CS = [6, 5]
DS = [6, 5, 2, 0]

Casimir invariant:

none

- $\mathfrak{s}_{6,186}$

	e_1	e_2	e_3	e_4	e_5
e_6	$4e_1$	$3e_2$	$2e_3$	e_3+2e_4	e_5

US = [0]
CS = [6, 5]
DS = [6, 5, 2, 0]

Casimir invariant:

none

- $\mathfrak{s}_{6,187}$

	e_1	e_2	e_3	e_4	e_5
e_6	$2e_1$	e_2	σe_1+2e_3	0	e_5

US = [0]
CS = [6, 4]
DS = [6, 4, 1, 0]

where $\sigma = 0, 1$.

Casimir invariant:

none

- $\mathfrak{s}_{6,188}$

	e_1	e_2	e_3	e_4	e_5
e_6	e_1	e_1+e_2	0	e_4	$-e_3$

US = [0]
CS = [6, 4, 3]
DS = [6, 4, 1, 0]

Casimir invariant:

none

- $\mathfrak{s}_{6,189}$

	e_1	e_2	e_3	e_4	e_5
e_6	e_1	e_2	0	e_4	0

US = [0]
CS = [6, 3]
DS = [6, 3, 0]

Casimir invariant:

none

19.9. Solvable six-dimensional Lie algebras with the nilradical $\mathfrak{n}_{5,5}$

	e_1	e_2	e_3	e_4	e_5
e_1	0	0	0	0	0
e_2		0	0	0	e_1
e_3			0	0	e_2
e_4				0	e_3

- $\mathfrak{s}_{6,190}$

	e_1	e_2	e_3	e_4	e_5
e_6	0	$-e_2$	$-2e_3$	$-3e_4$	e_5

US = [1]
CS = [6, 5]
DS = [6, 5, 3, 0]

Casimir invariants:

$$e_1, \quad \frac{(3e_1^2e_4 - 3e_1e_2e_3 + e_2^3)^2}{(2e_1e_3 - e_2^2)^3}$$

- $\mathfrak{s}_{6,191}$

	e_1	e_2	e_3	e_4	e_5
e_6	$4e_1$	$3e_2$	$2e_3$	e_4	$e_4 + e_5$

US = [0]
CS = [6, 5]
DS = [6, 5, 3, 0]

Casimir invariants:

$$\frac{(2e_1e_3 - e_2^2)^2}{e_1^3}, \quad \frac{(3e_1^2e_4 - 3e_1e_2e_3 + e_2^3)^4}{e_1^9}$$

- $\mathfrak{s}_{6,192}$

	e_1	e_2	e_3	e_4	e_5
e_6	$(3+a)e_1$	$(2+a)e_2$	$(1+a)e_3$	ae_4	e_5

US = [0]
CS = [6, 5]
DS = [6, 5, 3, 0]

where $a \neq 0, -3$.

Casimir invariants:

$$\frac{(3e_1^2e_4 - 3e_1e_2e_3 + e_2^3)^2}{(2e_1e_3 - e_2^2)^3}, \quad \frac{(2e_1e_3 - e_2^2)^{3+a}}{e_1^{2(2+a)}}$$

- $\mathfrak{s}_{6,193}$

	e_1	e_2	e_3	e_4	e_5
e_6	$3e_1$	$2e_2$	e_3	0	e_5

US = [0]
CS = [6, 4]
DS = [6, 4, 2, 0]

Casimir invariants:

$$\frac{(2e_1e_3 - e_2^2)^3}{e_1^4}, \quad \frac{3e_1^2e_4 - 3e_1e_2e_3 + e_2^3}{e_1^2}$$

- $\mathfrak{s}_{6,194}$

	e_1	e_2	e_3	e_4	e_5
e_6	e_1	e_2	e_3	e_4	0

US = [0]
CS = [6, 4]
DS = [6, 4, 0]

Casimir invariants:

$$\frac{2e_1e_3 - e_2^2}{e_1^2}, \quad \frac{3e_1^2e_4 - 3e_1e_2e_3 + e_2^3}{e_1^3}$$

- $\mathfrak{s}_{6,195}$

	e_1	e_2	e_3	e_4	e_5
e_6	e_1	e_2	$\epsilon e_1 + e_3$	$\epsilon e_2 + e_4$	0

US $= [0]$
CS $= [6, 4]$
DS $= [6, 4, 0]$

where the values of the parameter ϵ are
over the field $\mathbb{C}$: $\epsilon = 1$;
over the field $\mathbb{R}$: $\epsilon = \pm 1$.
Casimir invariants:

$$\frac{3e_1^2e_4 - 3e_1e_2e_3 + e_2^3}{e_1^3}, \quad e_1^2 \exp\left(\epsilon \frac{e_2^2 - 2e_1e_3}{e_1^2}\right)$$

- $\mathfrak{s}_{6,196}$

	e_1	e_2	e_3	e_4	e_5
e_6	e_1	e_2	$ae_1 + e_3$	$e_1 + ae_2 + e_4$	0

US $= [0]$
CS $= [6, 4]$
DS $= [6, 4, 0]$

Casimir invariants:

$$\frac{3e_1e_2^2 - 6e_1^2e_3 + 6ae_1^2e_4 - 6ae_1e_2e_3 + 2ae_2^3}{e_1^3}, \quad e_1^{2a} \exp\left(\frac{e_2^2 - 2e_1e_3}{e_1^2}\right)$$

19.10. Solvable six-dimensional Lie algebra with the nilradical $\mathfrak{n}_{5,6}$

	e_1	e_2	e_3	e_4	e_5
e_1	0	0	0	0	0
e_2		0	0	0	e_1
e_3			0	e_1	e_2
e_4				0	e_3

- $\mathfrak{s}_{6,197}$

	e_1	e_2	e_3	e_4	e_5
e_6	$5e_1$	$4e_2$	$3e_3$	$2e_4$	e_5

US $= [0]$
CS $= [6, 5]$
DS $= [6, 5, 3, 0]$

Casimir invariant:

none

19.11. Solvable six-dimensional Lie algebras with the nilradical $4\mathfrak{n}_{1,1}$

	e_1	e_2	e_3	e_4
e_1	0	0	0	0
e_2		0	0	0
e_3			0	0

- $\mathfrak{s}_{6,198}$

	e_1	e_2	e_3	e_4	e_5
e_5	0	e_1	0	e_4	0
e_6	0	0	e_3	ae_4	e_2

US = [1, 2]
CS = [6, 4, 3, 2]
DS = [6, 4, 0]

where $0 < |a| \leq 1$. If $|a| = 1$ then $\arg(a) \leq \pi$.

Casimir invariants:

$$e_1, \quad \frac{e_3^a}{e_4} \exp\left(\frac{e_2}{e_1}\right)$$

- $\mathfrak{s}_{6,199}$ (over the field $\mathbb{C}$: isomorphic to $\mathfrak{s}_{6,198}$)

	e_1	e_2	e_3	e_4	e_5
e_5	0	e_1	$-e_4$	e_3	0
e_6	0	0	e_3	e_4	e_2

US = [1, 2]
CS = [6, 4, 3, 2]
DS = [6, 4, 0]

Casimir invariants:

$$e_1, \quad e_2 + e_1 \arctan \frac{e_3}{e_4}$$

- $\mathfrak{s}_{6,200}$ (over the field $\mathbb{C}$: isomorphic to $\mathfrak{s}_{6,198}$)

	e_1	e_2	e_3	e_4	e_5
e_5	0	e_1	e_3	e_4	0
e_6	0	0	$\alpha e_3 - e_4$	$e_3 + \alpha e_4$	e_2

US = [1, 2]
CS = [6, 4, 3, 2]
DS = [6, 4, 0]

where $\alpha \geq 0$.

Casimir invariants:

$$e_1, \quad (e_3^2 + e_4^2) \exp\left(-2\frac{e_2}{e_1} + 2\alpha \arctan \frac{e_3}{e_4}\right)$$

- $\mathfrak{s}_{6,201}$

	e_1	e_2	e_3	e_4	e_5
e_5	0	e_1	0	e_4	0
e_6	0	0	e_3	ae_4	0

US = [1, 2]
CS = [6, 3, 2]
DS = [6, 3, 0]

where $0 < |a| \leq 1$. If $|a| = 1$ then $\arg(a) \leq \pi$.

Casimir invariants:

$$e_1, \quad \frac{e_3^a}{e_4} \exp\left(\frac{e_2}{e_1}\right)$$

- $\mathfrak{s}_{6,202}$ (over the field $\mathbb{C}$: isomorphic to $\mathfrak{s}_{6,201}$)

	e_1	e_2	e_3	e_4	e_5
e_5	0	e_1	$-e_4$	e_3	0
e_6	0	0	e_3	e_4	0

US = [1, 2]
CS = [6, 3, 2]
DS = [6, 3, 0]

Casimir invariants:

$$e_1, \quad e_2 + e_1 \arctan \frac{e_3}{e_4}$$

- $\mathfrak{s}_{6,203}$ (over the field $\mathbb{C}$: isomorphic to $\mathfrak{s}_{6,201}$)

	e_1	e_2	e_3	e_4	e_5
e_5	0	e_1	e_3	e_4	0
e_6	0	0	$\alpha e_3 - e_4$	$e_3 + \alpha e_4$	0

US $= [1, 2]$
CS $= [6, 3, 2]$
DS $= [6, 3, 0]$

where $\alpha \geq 0$.

Casimir invariants:

$$e_1, \quad (e_3^2 + e_4^2) \exp\left(-2\frac{e_2}{e_1} + 2\alpha \arctan \frac{e_3}{e_4}\right)$$

- $\mathfrak{s}_{6,204}$

	e_1	e_2	e_3	e_4	e_5
e_5	0	ae_2	0	e_4	0
e_6	0	be_2	e_3	0	e_1

US $= [1]$
CS $= [6, 4, 3]$
DS $= [6, 4, 0]$

where $0 \leq |b| \leq |a| \leq 1$, $a \neq 0$.

Casimir invariants:

$$e_1, \quad \frac{e_3^b e_4^a}{e_2}$$

- $\mathfrak{s}_{6,205}$

	e_1	e_2	e_3	e_4	e_5
e_5	0	e_2	$e_2 + e_3$	0	0
e_6	0	0	0	e_4	e_1

US $= [1]$
CS $= [6, 4, 3]$
DS $= [6, 4, 0]$

Casimir invariants:

$$e_1, \quad e_2 \exp\left(-\frac{e_3}{e_2}\right)$$

- $\mathfrak{s}_{6,206}$

	e_1	e_2	e_3	e_4	e_5
e_5	0	e_2	e_3	ae_4	0
e_6	0	0	e_2	e_4	e_1

US $= [1]$
CS $= [6, 4, 3]$
DS $= [6, 4, 0]$

Casimir invariants:

$$e_1, \quad \frac{e_2^a}{e_4} \exp\left(\frac{e_3}{e_2}\right)$$

- $\mathfrak{s}_{6,207}$ (over the field $\mathbb{C}$: isomorphic to $\mathfrak{s}_{6,204}$)

	e_1	e_2	e_3	e_4	e_5
e_5	0	$\alpha e_2 - e_3$	$e_2 + \alpha e_3$	0	0
e_6	0	0	0	e_4	e_1

US $= [1]$
CS $= [6, 4, 3]$
DS $= [6, 4, 0]$

where $\alpha \geq 0$.

Casimir invariants:

$$e_1, \quad (e_2^2 + e_3^2) \exp\left(2\alpha \arctan \frac{e_2}{e_3}\right)$$

- $\mathfrak{s}_{6,208}$ (over the field $\mathbb{C}$: isomorphic to $\mathfrak{s}_{6,204}$)

	e_1	e_2	e_3	e_4	e_5
e_5	0	e_3	$-e_2$	αe_4	0
e_6	0	e_2	e_3	βe_4	e_1

US $= [1]$
CS $= [6, 4, 3]$
DS $= [6, 4, 0]$

where $\alpha^2 + \beta^2 \neq 0$, $\alpha \geq 0$.
Casimir invariants:

$$e_1, \quad \frac{e_4^2}{(e_2^2 + e_3^2)^\beta} \exp\left(2\alpha \arctan \frac{e_3}{e_2}\right)$$

- $\mathfrak{s}_{6,209}$

	e_1	e_2	e_3	e_4	e_5
e_5	e_1	e_2	e_3	0	0
e_6	0	e_1	e_2	e_4	0

US $= [0]$
CS $= [6, 4]$
DS $= [6, 4, 0]$

Casimir invariants:

$$\frac{2e_1e_3 - e_2^2}{e_1^2}, \quad e_4 \exp\left(-\frac{e_2}{e_1}\right)$$

- $\mathfrak{s}_{6,210}$

	e_1	e_2	e_3	e_4	e_5
e_5	e_1	e_2	ae_3	ae_4	0
e_6	0	e_1	e_3	$e_3 + e_4$	0

US $= [0]$
CS $= [6, 4]$
DS $= [6, 4, 0]$

where $0 \leq |a| \leq 1$. If $|a| = 1$ then $\arg(a) \leq \pi$.
Casimir invariants:

$$\frac{e_1e_4 - e_2e_3}{e_1e_3}, \quad \frac{e_1^a}{e_3} \exp\left(\frac{e_2}{e_1}\right)$$

- $\mathfrak{s}_{6,211}$ (over the field $\mathbb{C}$: isomorphic to $\mathfrak{s}_{6,210}$)

	e_1	e_2	e_3	e_4	e_5
e_5	$-e_2$	e_1	$-e_4$	e_3	0
e_6	e_1	e_2	$e_1 + e_3$	$e_2 + e_4$	0

US $= [0]$
CS $= [6, 4]$
DS $= [6, 4, 0]$

Casimir invariants:

$$\frac{e_1e_4 - e_2e_3}{e_1^2 + e_2^2}, \quad (e_1^2 + e_2^2) \exp\left(-2\frac{e_1e_3 + e_2e_4}{e_1^2 + e_2^2}\right)$$

- $\mathfrak{s}_{6,212}$

	e_1	e_2	e_3	e_4	e_5
e_5	e_1	e_2	$e_2 + e_3$	0	0
e_6	0	0	e_1	e_4	0

US $= [0]$
CS $= [6, 4]$
DS $= [6, 4, 0]$

Casimir invariants:

$$\frac{e_2}{e_1}, \quad e_4 \exp\left(-\frac{e_3}{e_1}\right) e_1^{\frac{e_2}{e_1}}$$

- $\mathfrak{s}_{6,213}$

	e_1	e_2	e_3	e_4	e_5
e_5	ae_1	be_2	0	e_4	0
e_6	ce_1	de_2	e_3	0	0

US = [0]
CS = [6, 4]
DS = [6, 4, 0]

where $a \neq 0$, $0 \leq |b| \leq |a|$. If one or both equalities hold further restrictions on the parameters are necessary in order to avoid redundancies, as discussed in Section 10.4.

Casimir invariants:

$$\frac{e_1^b e_3^{ad-cb}}{e_2^a}, \quad \frac{e_2^c e_4^{ad-cb}}{e_1^d}$$

- $\mathfrak{s}_{6,214}$

	e_1	e_2	e_3	e_4	e_5
e_5	e_1	e_2	ae_3	be_4	0
e_6	0	e_1	e_3	ce_4	0

US = [0]
CS = [6, 4]
DS = [6, 4, 0]

where $|c| \leq 1$, $|b| + |c| \neq 0$. If $|c| = 1$ then $\arg(c) \leq \pi$. If $c = 1$ then $|b| \leq |a|$. If $c = 1$ and $|b| = |a|$ then $\arg(a) \leq \arg(b)$.

Casimir invariants:

$$\frac{e_1^{b-ac} e_3^c}{e_4}, \quad \frac{e_3}{e_1^a} \exp\left(-\frac{e_2}{e_1}\right)$$

- $\mathfrak{s}_{6,215}$

	e_1	e_2	e_3	e_4	e_5
e_5	e_1	$e_1 + e_2$	0	ae_4	0
e_6	0	0	e_3	be_4	0

US = [0]
CS = [6, 4]
DS = [6, 4, 0]

where $a \neq 0$, $0 < |b| \leq 1$. If $|b| = 1$ then $\arg(b) \leq \pi$. If $b = 1$ then $\arg(a) < \pi$.

Casimir invariants:

$$\frac{e_4}{e_1^a e_3^b}, \quad e_1 \exp\left(-\frac{e_2}{e_1}\right)$$

- $\mathfrak{s}_{6,216}$ (over the field $\mathbb{C}$: isomorphic to $\mathfrak{s}_{6,213}$)

	e_1	e_2	e_3	e_4	e_5
e_5	$\alpha e_1 - e_2$	$e_1 + \alpha e_2$	0	βe_4	0
e_6	0	0	e_3	γe_4	0

US = [0]
CS = [6, 4]
DS = [6, 4, 0]

where $0 < \beta$, $0 < |\gamma| \leq 1$.

Casimir invariants:

$$\frac{e_3^{2\alpha\gamma}(e_1^2 + e_2^2)^\beta}{e_4^{2\alpha}}, \quad \frac{e_4}{e_3^\gamma} \exp\left(\beta \arctan \frac{e_1}{e_2}\right)$$

- $\mathfrak{s}_{6,217}$ (over the field $\mathbb{C}$: isomorphic to $\mathfrak{s}_{6,213}$)

	e_1	e_2	e_3	e_4	e_5
e_5	e_2	$-e_1$	αe_3	βe_4	0
e_6	e_1	e_2	γe_3	δe_4	0

US = [0]
CS = [6, 4]
DS = [6, 4, 0]

where $\alpha^2 + \gamma^2 \neq 0$, $\beta^2 + \delta^2 \neq 0$, $\alpha \geq 0$, $\alpha \geq |\beta|$. If $\alpha = \beta$ then $\gamma \geq \delta$.

Casimir invariants:

$$\frac{e_4^{2\alpha}(e_1^2+e_2^2)^{\beta\gamma-\alpha\delta}}{e_3^{2\beta}}, \qquad \frac{e_3^2}{(e_1^2+e_2^2)^{\gamma}}\exp\left(2\alpha\arctan\frac{e_2}{e_1}\right).$$

When $\alpha\to 0$ these invariants become dependent and one of them should be replaced by

$$\frac{e_4^2}{(e_1^2+e_2^2)^{\delta}}\exp\left(2\beta\arctan\frac{e_2}{e_1}\right)$$

- $\mathfrak{s}_{6,218}$

	e_1	e_2	e_3	e_4	e_5
e_5	e_1	e_1+e_2	e_3	0	0
e_6	0	0	e_1	e_4	0

US = [0]
CS = [6, 4]
DS = [6, 4, 0]

Casimir invariants:

$$e_1\exp\left(-\frac{e_2}{e_1}\right), \qquad e_4\exp\left(-\frac{e_3}{e_1}\right)$$

- $\mathfrak{s}_{6,219}$

	e_1	e_2	e_3	e_4	e_5
e_5	e_1	e_2	0	e_3	0
e_6	ae_1	e_1+ae_2	e_3	e_4	0

US = [0]
CS = [6, 4]
DS = [6, 4, 0]

Casimir invariants:

$$e_3\exp\left(-\frac{e_2}{e_1}\right), \qquad e_1\exp\left(-\frac{e_1e_4+ae_2e_3}{e_1e_3}\right)$$

- $\mathfrak{s}_{6,220}$

	e_1	e_2	e_3	e_4	e_5
e_5	e_1	e_1+e_2	ae_3	ae_4	0
e_6	be_1	be_2	e_3	e_3+e_4	0

US = [0]
CS = [6, 4]
DS = [6, 4, 0]

where $a\neq 0$, $|b|\leq|a|$, $ab\neq 1$. If $|a|=|b|$ then $\arg(a)\leq\arg(b)$.
Casimir invariants:

$$\frac{e_3^b}{e_1}\exp\left((1-ab)\frac{e_2}{e_1}\right), \qquad e_1\exp\left(-\frac{be_1e_4+e_2e_3}{e_1e_3}\right)$$

- $\mathfrak{s}_{6,221}$

	e_1	e_2	e_3	e_4	e_5
e_5	e_1	e_1+e_2	e_2+e_3	0	0
e_6	0	ae_1	be_1+ae_2	e_4	0

US = [0]
CS = [6, 4]
DS = [6, 4, 0]

where $|a|+|b|\neq 0$.
Casimir invariants:

$$e_1e_4^a\exp\left(-\frac{e_2}{e_1}\right), \qquad e_1^{2b}\exp\left(\frac{2ae_1e_3-2be_1e_2-ae_2^2}{e_1^2}\right)$$

- $\mathfrak{s}_{6,222}$ (over the field $\mathbb{C}$: isomorphic to $\mathfrak{s}_{6,214}$)

	e_1	e_2	e_3	e_4	e_5
e_5	$\alpha e_1 - e_2$	$e_1 + \alpha e_2$	0	e_3	0
e_6	0	0	e_3	e_4	0

US = [0]
CS = [6, 4]
DS = [6, 4, 0]

where $\alpha \geq 0$.

Casimir invariants:

$$(e_1^2 + e_2^2)\exp\left(-2\alpha\frac{e_4}{e_3}\right), \quad \frac{e_4}{e_3} + \arctan\frac{e_1}{e_2}$$

- $\mathfrak{s}_{6,223}$ (over the field $\mathbb{C}$: isomorphic to $\mathfrak{s}_{6,214}$)

	e_1	e_2	e_3	e_4	e_5
e_5	$\alpha e_1 - e_2$	$e_1 + \alpha e_2$	βe_3	βe_4	0
e_6	e_1	e_2	0	e_3	0

US = [0]
CS = [6, 4]
DS = [6, 4, 0]

where $\beta > 0$.

Casimir invariants:

$$\frac{(e_1^2 + e_2^2)^\beta}{e_3^{2\alpha}}\exp\left(-2\beta\frac{e_4}{e_3}\right), \quad e_3\exp\left(\beta\arctan\frac{e_1}{e_2}\right)$$

- $\mathfrak{s}_{6,224}$ (over the field $\mathbb{C}$: isomorphic to $\mathfrak{s}_{6,214}$, resp. $\mathfrak{s}_{6,215}$ when $\beta = 0$)

	e_1	e_2	e_3	e_4	e_5
e_5	$\alpha e_1 - e_2$	$e_1 + \alpha e_2$	βe_3	βe_4	0
e_6	γe_1	γe_2	e_3	$e_3 + e_4$	0

US = [0]
CS = [6, 4]
DS = [6, 4, 0]

where $\alpha \geq 0$, $\beta^2 + \gamma^2 \neq 0$. If $\alpha = 0$ then $\beta \geq 0$.

Casimir invariants:

$$\frac{e_3^{2\alpha}}{(e_1^2 + e_2^2)^\beta}\exp\left(2(\beta\gamma - \alpha)\frac{e_4}{e_3}\right), \quad e_3\exp\left(-\frac{e_4}{e_3} + \beta\arctan\frac{e_1}{e_2}\right)$$

When $\beta \to 0$ these invariants become dependent and one of them should be replaced by

$$\frac{e_3^{2\gamma}}{e_1^2 + e_2^2}\exp\left(2(\beta\gamma - \alpha)\arctan\frac{e_1}{e_2}\right)$$

- $\mathfrak{s}_{6,225}$ (over the field $\mathbb{C}$: isomorphic to $\mathfrak{s}_{6,220}$)

	e_1	e_2	e_3	e_4	e_5
e_5	$-e_2$	e_1	$e_1 - e_4$	$e_2 + e_3$	0
e_6	e_1	e_2	$\alpha e_1 - \beta e_2 + e_3$	$\beta e_1 + \alpha e_2 + e_4$	0

US = [0]
CS = [6, 4]
DS = [6, 4, 0]

where $\alpha \geq 0$.

Casimir invariants:

$$(e_1^2 + e_2^2)^\beta\exp\left(2\frac{e_2e_3 - e_1e_4}{e_1^2 + e_2^2}\right), \quad (e_1^2 + e_2^2)^\alpha\exp\left(-2\left(\frac{e_1e_3 + e_2e_4}{e_1^2 + e_2^2} + \arctan\frac{e_1}{e_2}\right)\right)$$

- $\mathfrak{s}_{6,226}$ (over the field $\mathbb{C}$: isomorphic to $\mathfrak{s}_{6,213}$)

	e_1	e_2	e_3	e_4	e_5
e_5	$-e_2$	e_1	$\alpha e_3 - \beta e_4$	$\beta e_3 + \alpha e_4$	0
e_6	e_1	e_2	γe_3	γe_4	0

US = [0]
CS = [6, 4]
DS = [6, 4, 0]

where $0 < \beta \leq 1$, $\alpha \geq 0$. If $\beta = 1$ then $|\gamma| \leq 1$.

Casimir invariants:

$$\arctan \frac{e_4}{e_3} - \beta \arctan \frac{e_2}{e_1}, \quad \frac{e_3^2 + e_4^2}{(e_1^2 + e_2^2)^\gamma} \exp\left(-2\alpha \arctan \frac{e_2}{e_1}\right)$$

- $\mathfrak{s}_{6,227}$ (over the field $\mathbb{C}$: isomorphic to $\mathfrak{s}_{6,214}$, resp. $\mathfrak{s}_{6,215}$ when $\beta = 0$)

	e_1	e_2	e_3	e_4	e_5
e_5	$-e_2$	e_1	αe_3	$e_3 + \alpha e_4$	0
e_6	e_1	e_2	βe_3	βe_4	0

US = [0]
CS = [6, 4]
DS = [6, 4, 0]

where $\alpha \geq 0$, $\alpha^2 + \beta^2 \neq 0$.

Casimir invariants:

$$\frac{e_4}{e_3} + \arctan \frac{e_1}{e_2}, \quad \frac{e_3^2}{(e_1^2 + e_2^2)^\beta} \exp\left(2\alpha \arctan \frac{e_1}{e_2}\right)$$

- $\mathfrak{s}_{6,228}$ (over the field $\mathbb{C}$: isomorphic to $\mathfrak{s}_{6,213}$)

	e_1	e_2	e_3	e_4	e_5
e_5	$\alpha e_1 - e_2$	$e_1 + \alpha e_2$	βe_3	βe_4	0
e_6	γe_1	γe_2	$\delta e_3 - e_4$	$e_3 + \delta e_4$	0

US = [0]
CS = [6, 4]
DS = [6, 4, 0]

where $0 \leq \beta \leq \gamma$, $0 < \gamma$. If $\beta = 0$ then $\alpha \geq 0$. If $\beta = \gamma$ then $\alpha \leq \delta$.

Casimir invariants:

$$(e_1^2 + e_2^2) \exp\left(-2\left(\alpha \arctan \frac{e_2}{e_1} + \gamma \arctan \frac{e_4}{e_3}\right)\right),$$
$$\frac{\exp(2\beta \arctan(e_2/e_1) + 2\delta \arctan(e_4/e_3))}{e_3^2 + e_4^2}$$

19.12. Solvable six-dimensional Lie algebras with the nilradical $\mathfrak{n}_{3,1} \oplus \mathfrak{n}_{1,1}$

	e_1	e_2	e_3	e_4
e_1	0	0	0	0
e_2		0	e_1	0
e_3			0	0

- $\mathfrak{s}_{6,229}$

	e_1	e_2	e_3	e_4	e_5
e_5	0	e_2	$-e_3$	0	0
e_6	0	0	0	e_4	$-e_1$

US = [1]
CS = [6, 4]
DS = [6, 4, 1, 0]

Casimir invariants:

$$e_1, \quad e_4 \exp\left(\frac{e_1 e_5 + e_2 e_3}{e_1^2}\right)$$

- $\mathfrak{s}_{6,230}$ (over the field $\mathbb{C}$: isomorphic to $\mathfrak{s}_{6,229}$)

	e_1	e_2	e_3	e_4	e_5
e_5	0	e_3	$-e_2$	0	0
e_6	0	0	0	e_4	ϵe_1

US = [1]
CS = [6, 4]
DS = [6, 4, 1, 0]

Casimir invariants:

$$e_1, \quad e_4^2 \exp\left(-\frac{2e_1e_5 + e_2^2 + e_3^2}{\epsilon e_1^2}\right)$$

- $\mathfrak{s}_{6,231}$

	e_1	e_2	e_3	e_4	e_5
e_5	e_1	e_2	0	0	0
e_6	e_1	0	e_3	0	e_4

US = [1]
CS = [6, 4, 3]
DS = [6, 4, 1, 0]

Casimir invariants:

$$e_4, \quad e_1 \exp\left(\frac{e_1e_6 - e_1e_5 - e_2e_3}{e_1e_4}\right)$$

- $\mathfrak{s}_{6,232}$ (over the field $\mathbb{C}$: isomorphic to $\mathfrak{s}_{6,231}$)

	e_1	e_2	e_3	e_4	e_5
e_5	$2e_1$	e_2	e_3	0	0
e_6	0	e_3	$-e_2$	0	e_4

US = [1]
CS = [6, 4, 3]
DS = [6, 4, 1, 0]

Casimir invariants:

$$e_4, \quad e_1 \exp\left(\frac{2e_1e_6 + e_2^2 + e_3^2}{e_1e_4}\right)$$

- $\mathfrak{s}_{6,233}$

	e_1	e_2	e_3	e_4	e_5
e_5	0	e_2	$-e_3$	0	0
e_6	e_1	0	e_3	$e_1 + e_4$	0

US = [0]
CS = [6, 4]
DS = [6, 4, 1, 0]

Casimir invariants:

$$\frac{e_1e_5 + e_2e_3}{e_1}, \quad e_1 \exp\left(-\frac{e_4}{e_1}\right)$$

- $\mathfrak{s}_{6,234}$

	e_1	e_2	e_3	e_4	e_5
e_5	e_1	e_2	0	ae_4	0
e_6	e_1	0	e_3	be_4	0

US = [0]
CS = [6, 4]
DS = [6, 4, 1, 0]

where $|a| \leq |b|$, $b \neq 0$. If $|a| = |b|$ then $\arg(a) \leq \arg(b)$.

Casimir invariant:

none

- $\mathfrak{s}_{6,235}$

	e_1	e_2	e_3	e_4	e_5
e_5	$2e_1$	e_2	e_3	ae_4	0
e_6	0	e_3	0	e_4	0

US = [0]
CS = [6, 4]
DS = [6, 4, 1, 0]

Casimir invariant:

none

- $\mathfrak{s}_{6,236}$

	e_1	e_2	e_3	e_4	e_5
e_5	e_1	ae_2	$(1-a)e_3$	e_4	0
e_6	0	e_2	$-e_3$	e_1	0

US = [0]
CS = [6, 4]
DS = [6, 4, 1, 0]

where $\operatorname{Im} a \geq 0$. If $a \in \mathbb{R}$ then $a \geq 1$.
Casimir invariant:

none

- $\mathfrak{s}_{6,237}$

	e_1	e_2	e_3	e_4	e_5
e_5	$(1+a)e_1$	ae_2	e_3	e_4	0
e_6	e_1	e_2	e_4	0	0

US = [0]
CS = [6, 4]
DS = [6, 4, 1, 0]

Casimir invariant:

none

- $\mathfrak{s}_{6,238}$

	e_1	e_2	e_3	e_4	e_5
e_5	e_1	e_2	0	0	0
e_6	e_1	0	e_3+e_4	e_4	0

US = [0]
CS = [6, 4]
DS = [6, 4, 1, 0]

Casimir invariant:

none

- $\mathfrak{s}_{6,239}$ (over the field $\mathbb{C}$: isomorphic to $\mathfrak{s}_{6,234}$)

	e_1	e_2	e_3	e_4	e_5
e_5	$2e_1$	e_2	e_3	αe_4	0
e_6	0	e_3	$-e_2$	βe_4	0

US = [0]
CS = [6, 4]
DS = [6, 4, 1, 0]

where $\alpha^2+\beta^2 \neq 0$, $\beta \geq 0$.
Casimir invariant:

none

- $\mathfrak{s}_{6,240}$ (over the field $\mathbb{C}$: isomorphic to $\mathfrak{s}_{6,236}$)

	e_1	e_2	e_3	e_4	e_5
e_5	$2e_1$	e_2	e_3	$2e_4$	0
e_6	0	e_3	$-e_2$	e_1	0

US = [0]
CS = [6, 4]
DS = [6, 4, 1, 0]

Casimir invariant:

none

- $\mathfrak{s}_{6,241}$ (over the field $\mathbb{C}$: isomorphic to $\mathfrak{s}_{6,236}$)

	e_1	e_2	e_3	e_4	e_5
e_5	$2\alpha e_1$	$\alpha e_2 - e_3$	$e_2+\alpha e_3$	$2\alpha e_4$	0
e_6	$2e_1$	e_2	e_3	e_1+2e_4	0

US = [0]
CS = [6, 4]
DS = [6, 4, 1, 0]

where $\alpha \geq 0$.
Casimir invariant:

none

19.13. Solvable six-dimensional Lie algebra with the nilradical $\mathfrak{n}_{4,1}$

	e_1	e_2	e_3	e_4
e_1	0	0	0	0
e_2		0	0	e_1
e_3			0	e_2

- $\mathfrak{s}_{6,242}$

	e_1	e_2	e_3	e_4	e_5
e_5	$2e_1$	e_2	0	e_4	0
e_6	e_1	e_2	e_3	0	0

US = [0]
CS = [6, 4]
DS = [6, 4, 2, 0]

Casimir invariant:

none

19.14. Simple six-dimensional Lie algebra

- $\mathfrak{so}(1,3)$ (over the field $\mathbb{C}$: isomorphic to $\mathfrak{sl}(2,\mathbb{C}) \oplus \mathfrak{sl}(2,\mathbb{C})$)

	e_1	e_2	e_3	e_4	e_5	e_6
e_1	0	e_3	$-e_2$	0	e_6	$-e_5$
e_2		0	e_1	$-e_6$	0	e_4
e_3			0	e_5	$-e_4$	0
e_4				0	$-e_3$	e_2
e_5					0	$-e_1$

US = [0]
CS = [6]
DS = [6]

Casimir invariants:

$$e_1e_4 + e_2e_5 + e_3e_6, \quad e_1^2 + e_2^2 + e_3^2 - e_4^2 - e_5^2 - e_6^2$$

19.15. Six-dimensional Levi decomposable Lie algebras

- $\mathfrak{sl}(2,\mathbb{F}) \niplus \mathfrak{n}_{3,1}$

	e_1	e_2	e_3	e_4	e_5	e_6
e_1	0	$2e_1$	$-e_2$	0	e_6	0
e_2		0	$2e_3$	0	e_5	$-e_6$
e_3			0	0	0	e_5
e_4				0	0	0
e_5					0	e_4

US = [1]
CS = [6]
DS = [6]

Casimir invariants:

$$e_4, \quad 4e_1e_3e_4 - 2e_1e_5^2 + e_2^2e_4 + 2e_2e_5e_6 + 2e_3e_6^2$$

- $\mathfrak{sl}(2,\mathbb{F}) \niplus 3\mathfrak{n}_{1,1}$

	e_1	e_2	e_3	e_4	e_5	e_6
e_1	0	$2e_1$	$-e_2$	0	$2e_4$	$-e_5$
e_2		0	$2e_3$	$-2e_4$	0	$2e_6$
e_3			0	e_5	$-2e_6$	0
e_4				0	0	0
e_5					0	0

US = [0]
CS = [6]
DS = [6]

Casimir invariants:

$$2e_1e_6 + e_2e_5 + 2e_3e_4, \quad 4e_6e_4 + e_5^2$$

- $\mathfrak{so}(3,\mathbb{R}) \inplus 3\mathfrak{n}_{1,1}$ (over the field $\mathbb{C}$: isomorphic to $\mathfrak{sl}(2,\mathbb{C}) \inplus 3\mathfrak{n}_{1,1}$)

	e_1	e_2	e_3	e_4	e_5	e_6
e_1	0	e_3	$-e_2$	0	e_6	$-e_5$
e_2		0	e_1	$-e_6$	0	e_4
e_3			0	e_5	$-e_4$	0
e_4				0	0	0
e_5					0	0

US = [0]
CS = [6]
DS = [6]

Casimir invariants:

$$e_1e_4 + e_2e_5 + e_3e_6, \quad e_4^2 + e_5^2 + e_6^2$$

- $\mathfrak{sl}(2,\mathbb{F}) \inplus \mathfrak{s}_{3,1,a=1}$

	e_1	e_2	e_3	e_4	e_5	e_6
e_1	0	$2e_1$	$-e_2$	e_5	0	0
e_2		0	$2e_3$	e_4	$-e_5$	0
e_3			0	0	e_4	0
e_4				0	0	e_4
e_5					0	e_5

US = [0]
CS = [6, 5]
DS = [6, 5]

Casimir invariant:

none

Bibliography

1. L. Abellanas and L. Martínez Alonso, *A general setting for Casimir invariants*, J. Mathematical Phys. **16** (1975), 1580–1584.
2. ______, *Invariants in enveloping algebras under the action of Lie algebras of derivations*, J. Math. Phys. **20** (1979), no. 3, 437–440.
3. J. M. Ancochea Bermúdez, R. Campoamor-Stursberg, and L. García Vergnolle, *Solvable Lie algebras with naturally graded nilradicals and their invariants*, J. Phys. A **39** (2006), no. 6, 1339–1355.
4. ______, *Indecomposable Lie algebras with nontrivial Levi decomposition cannot have filiform radical*, Int. Math. Forum **1** (2006), no. 5-8, 309–316.
5. R. E. Beck, B. Kolman, and I. N. Stewart, *Computing the structure of a Lie algebra*, Computers in Nonassociative Rings and Algebras (San Antonio, TX, 1976) (R. E. Beck and B. Kolman, eds.), Academic Press, New York, 1977, pp. 167–188.
6. L. Bianchi, *Sugli spazii a tre dimensioni che ammettono un gruppo continuo di movimenti*, Soc. Ital. Sci. Mem. di Mat. **11** (1898), 267–352.
7. V. Boyko, J. Patera, and R. Popovych, *Computation of invariants of Lie algebras by means of moving frames*, J. Phys. A **39** (2006), no. 20, 5749–5762.
8. ______, *Invariants of Lie algebras with fixed structure of nilradicals*, J. Phys. A **40** (2007), no. 1, 113–130.
9. ______, *Invariants of triangular Lie algebras*, J. Phys. A **40** (2007), no. 27, 7557–7572.
10. ______, *Invariants of triangular Lie algebras with one nil-independent diagonal element*, J. Phys. A **40** (2007), no. 32, 9783–9792.
11. ______, *Invariants of solvable Lie algebras with triangular nilradicals and diagonal nilindependent elements*, Linear Algebra Appl. **428** (2008), no. 4, 834–854.
12. N. Burgoyne and R. Cushman, *Conjugacy classes in linear groups*, J. Algebra **44** (1977), no. 2, 339–362.
13. R. Campoamor-Stursberg, *Invariants of solvable rigid Lie algebras up to dimension* 8, J. Phys. A **35** (2002), no. 30, 6293–6306.
14. ______, *Non-semisimple Lie algebras with Levi factor* $\mathfrak{so}(3)$, $\mathfrak{sl}(2,\mathbb{R})$ *and their invariants*, J. Phys. A **36** (2003), no. 5, 1357–1369.
15. ______, *On the invariants of some solvable rigid Lie algebras*, J. Math. Phys. **44** (2003), no. 2, 771–784.
16. ______, *Some remarks concerning the invariants of rank one solvable real Lie algebras*, Algebra Colloq. **12** (2005), no. 3, 497–518.
17. ______, *A note on the classification of nine dimensional Lie algebras with nontrivial Levi decomposition*, Int. Math. Forum **2** (2007), no. 25-28, 1341–1344.
18. ______, *Structural data and invariants of nine dimensional real Lie algebras with nontrivial Levi decomposition*, Nova Science Publishers Inc., New York, 2009.
19. ______, *Solvable Lie algebras with an* $\mathbb{N}$*-graded nilradical of maximal nilpotency degree and their invariants*, J. Phys. A **43** (2010), no. 14, 145202.
20. É. Cartan, *Les groupes réels simples, finis et continus*, Ann. Sci. École Norm. Sup. (3) **31** (1914), 263–355.
21. ______, *Groupes simples clos et ouverts et géométrie riemannienne*, J. Math. Pures Appl. **8** (1929), 1–34.
22. ______, *Sur la structure des groupes des transformations finis et continus*, 2nd ed., Vuibert, Paris, 1933.
23. ______, *La méthode du repère mobile, la théorie des groupes continus et les espaces généralisés*, Actualites Sci. Indust., vol. 194, Hermann, Paris, 1935.

24. ______, *La théorie des groupes finis et continus et la géométrie différentielle traitées par la méthode du repère mobile. Leçons professées à la Sorbonne. Rédigées par Jean Leray*, Cahiers Sci., vol. 18, Gauthier-Villars, Paris, 1937.
25. H. B. G. Casimir, *Über die Konstruktion einer zu den irreduziblen Darstellungen halbeinfacher kontinuierlicher Gruppen gehörigen Differentialgleichung*, Proc. Kon. Akad. Wetensch. **34** (1931), 844–846.
26. H. B. G. Casimir and B. L. van der Waerden, *Algebraischer Beweis der vollständigen Reduzibilität der Darstellungen halbeinfacher Liescher Gruppen*, Math. Ann. **111** (1935), no. 1, 1–12.
27. A. A. Chesnokov, *Symmetries and exact solutions of the rotating shallow-water equations*, European J. Appl. Math. **20** (2009), no. 5, 461–477.
28. C. Chevalley, *Théorie des groupes de Lie*. Tome II: *Groupes algébriques*, Actualités Sci. Ind., vol. 1152, Hermann, Paris, 1951.
29. C. Chevalley and S. Eilenberg, *Cohomology theory of Lie groups and Lie algebras*, Trans. Amer. Math. Soc. **63** (1948), 85–124.
30. A. M. Cohen and W. A. de Graaf, *Lie algebraic computation*, Comput. Phys. Comm. **97** (1996), no. 1-2, 53–62.
31. A. M. Cohen, W. A. de Graaf, and L. Rónyai, *Computations in finite-dimensional Lie algebras*, Discrete Math. Theor. Comput. Sci. **1** (1997), no. 1, 129–138.
32. W. A. de Graaf, *An algorithm for the decomposition of semisimple Lie algebras*, Theoret. Comput. Sci. **187** (1997), no. 1-2, 117–122.
33. ______, *Calculating the structure of a semisimple Lie algebra*, J. Pure Appl. Algebra **117/118** (1997), 319–329.
34. ______, *Lie algebras: theory and algorithms*, North-Holland Math. Library, vol. 56, North-Holland, Amsterdam, 2000.
35. W. A. de Graaf, G. Ivanyos, A. Küronya, and L. Rónyai, *Computing Levi decompositions in Lie algebras*, Appl. Algebra Engrg. Comm. Comput. **8** (1997), no. 4, 291–303.
36. W. A. de Graaf, G. Ivanyos, and L. Rónyai, *Computing Cartan subalgebras of Lie algebras*, Appl. Algebra Engrg. Comm. Comput. **7** (1996), no. 5, 339–349.
37. M. A. del Olmo, M. A. Rodríguez, P. Winternitz, and H. Zassenhaus, *Maximal abelian subalgebras of pseudounitary Lie algebras*, Linear Algebra Appl. **135** (1990), 79–151.
38. A. Di Bucchianico and D. Loeb, *Umbral calculus*, Electron. J. Combin. **Dynamic Surveys** (2000), DS3.
39. L. E. Dickson, *Linear algebras*, Hafner Publishing Co., New York, 1960. Reprinting of Cambridge Tracts in Math. Math. Phys., vol. 16.
40. A. Dimakis, F. Müller-Hoissen, and T. Striker, *Umbral calculus, discretization, and quantum mechanics on a lattice*, J. Phys. A **29** (1996), no. 21, 6861–6876.
41. J. Dixmier, *Enveloping algebras*, Grad. Stud. Math., vol. 11, Amer. Math. Soc., Providence, RI, 1996. Revised reprint of the 1977 translation.
42. D. Ž. Djoković, J. Patera, P. Winternitz, and H. Zassenhaus, *Normal forms of elements of classical real and complex Lie and Jordan algebras*, J. Math. Phys. **24** (1983), no. 6, 1363–1374.
43. E. B. Dynkin, *Classification of the simple Lie groups*, Rec. Math. [Mat. Sbornik] N. S. **18(60)** (1946), no. 3, 347–352 (Russian).
44. ______, *The structure of semi-simple algebras*, Uspehi Matem. Nauk (N.S.) **2** (1947), no. 4(20), 59–127 (Russian); English transl. in Amer. Math. Soc. Translation **1950** (1950), no. 17.
45. F. J. Echarte, J. R. Gómez, and J. Núñez, *Les algèbres de Lie filiformes complexes dérivées d'autres algèbres de Lie*, Lois d'algèbres et variétés algébriques (Colmar, 1991) (M. Goze, ed.), Travaux en Cours, vol. 50, Hermann, Paris, 1996, pp. 45–55.
46. K. Erdmann and M. J. Wildon, *Introduction to Lie algebras*, Springer Undergrad. Math. Ser., Springer, London, 2006.
47. G. Favre, *Une algèbre de Lie caractéristiquement nilpotente de dimension* 7, C. R. Acad. Sci. Paris Sér. A-B **274** (1972), A1338–A1339.
48. M. Fels and P. J. Olver, *Moving coframes*. I: *A practical algorithm*, Acta Appl. Math. **51** (1998), no. 2, 161–213; II: *Regularization and theoretical foundations* **55** (1999), no. 2, 127–208.

49. L. Gagnon and P. Winternitz, *Lie symmetries of a generalised nonlinear Schrödinger equation. I: The symmetry group and its subgroups*, J. Phys. A **21** (1988), no. 7, 1493–1511; II: *Exact solutions* **22** (1989), no. 5, 469–497.
50. F. Gantmacher, *On the classification of real simple Lie groups*, Rec. Math. [Mat. Sbornik] N.S. **5 (47)** (1939), 217–250.
51. M.-P. Gong, *Classification of nilpotent Lie algebras of dimension* 7 *(over algebraically closed fields and* $\mathbb{R}$*)*, ProQuest LLC, Ann Arbor, MI, 1998. Ph.D. Thesis, University of Waterloo.
52. M. Goze and Yu. Khakimdjanov, *Nilpotent Lie algebras*, Mathematics Appl., vol. 361, Kluwer, Dordrecht, 1996.
53. ———, *Nilpotent and solvable Lie algebras*, Handbook of Algebra, Vol. 2 (M. Hazewinkel, ed.), North-Holland, Amsterdam, 2000, pp. 615–663.
54. Harish-Chandra, *The characters of semisimple Lie groups*, Trans. Amer. Math. Soc. **83** (1956), 98–163.
55. J. Hietarinta, *Direct methods for the search of the second invariant*, Phys. Rep. **147** (1987), no. 2, 87–154.
56. J. Hrivnák and P. Novotný, *Twisted cocycles of Lie algebras and corresponding invariant functions*, Linear Algebra Appl. **430** (2009), no. 4, 1384–1403.
57. V. Hussin, P. Winternitz, and H. Zassenhaus, *Maximal abelian subalgebras of complex orthogonal Lie algebras*, Linear Algebra Appl. **141** (1990), 183–220.
58. N. Jacobson, *Basic algebra*. I, W. H. Freeman and Co., San Francisco, CA, 1974.
59. ———, *Lie algebras*, Dover Publications Inc., New York, 1979. Republication of the 1962 original.
60. W. Killing, *Die Zusammensetzung der stetigen endlichen Transformationsgruppen.*, Math. Ann. **31** (1888), no. 2, 252–290; **33** (1888), no. 1, 1–48; **34** (1889), no. 1, 57–122; **36** (1890), no. 2, 161–189.
61. A. A. Kirillov, *Elements of the theory of representations*, 2nd ed., "Nauka", Moscow, 1978 (Russian).
62. A. W. Knapp, *Lie groups beyond an introduction*, 2nd ed., Progr. Math., vol. 140, Birkhäuser, Boston, MA, 2002.
63. B. Kostant, *On the conjugacy of real Cartan subalgebras*. I, Proc. Nat. Acad. Sci. U. S. A. **41** (1955), 967–970.
64. M. Krawtchouk, *Über vertauschbare Matrizen*, Rend. Circ. Mat. Palermo (1) **51** (1927), no. 1, 126–130.
65. G. I. Kruchkovich, *Classification of three-dimensional Riemannian spaces according to groups of motions*, Uspehi Matem. Nauk (N.S.) **9** (1954), no. 1(59), 3–40 (Russian).
66. J. M. Lee, *Manifolds and differential geometry*, Grad. Stud. Math., vol. 107, Amer. Math. Soc., Providence, RI, 2009.
67. G. F. Leger and E. M. Luks, *Cohomology of nilradicals of Borel subalgebras*, Trans. Amer. Math. Soc. **195** (1974), 305–316.
68. D. Levi, M. C. Nucci, C. Rogers, and P. Winternitz, *Group theoretical analysis of a rotating shallow liquid in a rigid container*, J. Phys. A **22** (1989), no. 22, 4743–4767.
69. D. Levi, P. Tempesta, and P. Winternitz, *Lorentz and Galilei invariance on lattices*, Phys. Rev. D **69** (2004), no. 10, 105011.
70. ———, *Umbral calculus, difference equations and the discrete Schrödinger equation*, J. Math. Phys. **45** (2004), no. 11, 4077–4105.
71. E. E. Levi, *Sulla struttura dei gruppi finiti e continui*, Atti Accad. Sci. Torino **40** (1905), 551–565.
72. S. Lie, *Theorie der Transformationsgruppen*. Vol. I, B. G. Teubner, Leipzig, 1888; Vol. II, 1890; Vol. III, 1893.
73. A. I. Mal′cev, *On the representation of an algebra as a direct sum of the radical and a semi-simple subalgebra*, C. R. (Doklady) Acad. Sci. URSS (N.S.) **36** (1942), 42–45.
74. ———, *Commutative subalgebras of semi-simple Lie algebras*, Bull. Acad. Sci. URSS. Sér. Math. [Izvestia Akad. Nauk SSSR] **9** (1945), 291–300 (Russian); English transl. in Amer. Math. Soc. Translation **1951** (1951), no. 40.
75. ———, *Foundations of linear algebra* (J. B. Roberts, ed.), translated by T. C. Brown, W. H. Freeman, San Francisco, CA – London, 1963.
76. L. Martina, G. Soliani, and P. Winternitz, *Partially invariant solutions of a class of nonlinear Schrödinger equations*, J. Phys. A **25** (1992), no. 16, 4425–4435.

77. V. V. Morozov, *Classification of nilpotent Lie algebras of sixth order*, Izv. Vysš. Učebn. Zaved. Matematika **1958** (1958), no. 4 (5), 161–171 (Russian).
78. G. M. Mubarakzjanov, *On solvable Lie algebras*, Izv. Vysš. Učehn. Zaved. Matematika **1963** (1963), no. no 1 (32), 114–123 (Russian).
79. ______, *Classification of real structures of Lie algebras of fifth order*, Izv. Vysš. Učebn. Zaved. Matematika **1963** (1963), no. 3 (34), 99–106 (Russian).
80. ______, *Classification of solvable Lie algebras of sixth order with a non-nilpotent basis element*, Izv. Vysš. Učebn. Zaved. Matematika **1963** (1963), no. 4 (35), 104–116 (Russian).
81. ______, *Certain theorems on solvable Lie algebras*, Izv. Vysš. Učebn. Zaved. Matematika **1966** (1966), no. 6 (55), 95–98 (Russian).
82. J.-C. Ndogmo, *Invariants of solvable Lie algebras of dimension six*, J. Phys. A **33** (2000), no. 11, 2273–2287.
83. J.-C. Ndogmo and P. Winternitz, *Generalized Casimir operators of solvable Lie algebras with abelian nilradicals*, J. Phys. A **27** (1994), no. 8, 2787–2800.
84. ______, *Solvable Lie algebras with abelian nilradicals*, J. Phys. A **27** (1994), no. 2, 405–423.
85. P. Novotný and J. Hrivnák, *On (α, β, γ)-derivations of Lie algebras and corresponding invariant functions*, J. Geom. Phys. **58** (2008), no. 2, 208–217.
86. P. J. Olver, *Applications of Lie groups to differential equations*, Grad. Texts in Math., vol. 107, Springer, New York, 1986.
87. J. Patera, R. T. Sharp, P. Winternitz, and H. Zassenhaus, *Invariants of real low dimension Lie algebras*, J. Mathematical Phys. **17** (1976), no. 6, 986–994.
88. J. Patera, P. Winternitz, and H. Zassenhaus, *The maximal solvable subgroups of the* $\mathrm{SU}(p,q)$ *groups and all subgroups of* $\mathrm{SU}(2,1)$, J. Mathematical Phys. **15** (1974), 1378–1393.
89. ______, *The maximal solvable subgroups of* $\mathrm{SO}(p,q)$ *groups*, J. Mathematical Phys. **15** (1974), 1932–1938.
90. ______, *Maximal abelian subalgebras of real and complex symplectic Lie algebras*, J. Math. Phys. **24** (1983), no. 8, 1973–1985.
91. J. N. Pecina-Cruz and Y. Ne'eman, *On the calculation of invariants of Lie algebras*, Canad. J. Phys. **72** (1994), no. 7-8, 466–496 (English, with English and French summaries).
92. J. Pedlosky, *Geophysical Fluid Dynamics*, Springer, Providence, RI, 1990.
93. M. Perroud, *The maximal solvable subalgebras of the real classical Lie algebras*, J. Mathematical Phys. **17** (1976), no. 6, 1028–1033.
94. ______, *The fundamental invariants of inhomogeneous classical groups*, J. Math. Phys. **24** (1983), no. 6, 1381–1391.
95. A. Z. Petrov, *Einstein spaces* (J. Woodrow, ed.), translated by R. F. Kelleher, Pergamon Press, Oxford, 1969.
96. C. Quesne, *Giant dipole transitions in the nuclear* $\mathrm{WSp}(6,\mathbb{R})$ *model*, Phys. Lett. B **188** (1987), no. 1, 1–5.
97. ______, *The nuclear collective* $\mathrm{wsp}(6,\mathbb{R})$ *model*, Ann. Physics **185** (1988), no. 1, 46–85.
98. G. Racah, *Group theory and spectroscopy*, Ergeb. Exakt. Naturwiss., vol. 37, Springer, Berlin, 1965, pp. 28–84.
99. A. Ramani, B. Grammaticos, and T. Bountis, *The Painlevé property and singularity analysis of integrable and nonintegrable systems*, Phys. Rep. **180** (1989), no. 3, 159–245.
100. D. W. Rand, *PASCAL programs for identification of Lie algebras.* I: *RADICAL—a program to calculate the radical and nil radical of parameter-free and parameter-dependent Lie algebras*, Comput. Phys. Comm. **41** (1986), no. 1, 105–125; III: *Levi decomposition and canonical basis* **46** (1987), no. 2, 311–322.
101. D. W. Rand, P. Winternitz, and H. Zassenhaus, *PASCAL programs for the identification of Lie algebras.* II: *SPLIT—a program to decompose parameter-free and parameter-dependent Lie algebras into direct sums*, Comput. Phys. Comm. **46** (1987), no. 2, 297–309.
102. ______, *On the identification of a Lie algebra given by its structure constants.* I: *Direct decompositions, Levi decompositions, and nilradicals*, Linear Algebra Appl. **109** (1988), 197–246.
103. S. M. Roman and G.-C. Rota, *The umbral calculus*, Advances in Math. **27** (1978), no. 2, 95–188.
104. G. Rosensteel and D. J. Rowe, *Nuclear* $\mathrm{Sp}(3,\mathbb{R})$ *model*, Phys. Rev. Lett. **38** (1977), no. 1, 10–14.

105. ______, *On the algebraic formulation of collective models. III: The symplectic shell model of collective motion*, Ann. Physics **126** (1980), no. 2, 343–370.
106. J. L. Rubin and P. Winternitz, *Solvable Lie algebras with Heisenberg ideals*, J. Phys. A **26** (1993), no. 5, 1123–1138.
107. È. N. Safiullina, *Classification of nilpotent Lie algebras of order* 7, Candidates' Works (1964), Math., Mech., Phys., Izdat. Kazan. Univ., Kazan, 1964, pp. 66–69 (Russian).
108. H. Samelson, *Notes on Lie algebras*, 2nd ed., Universitext, Springer, New York, 1990.
109. D. H. Sattinger and O. L. Weaver, *Lie groups and algebras with applications to physics, geometry, and mechanics*, Appl. Math. Sci., vol. 61, Springer, New York, 1986.
110. C. Seeley, 7*-dimensional nilpotent Lie algebras*, Trans. Amer. Math. Soc. **335** (1993), no. 2, 479–496.
111. A. Shabanskaya and G. Thompson, *Six-dimensional Lie algebras with a five-dimensional nilradical*, J. Lie Theory **23** (2013), no. 2, 313–355.
112. T. Skjelbred and T. Sund, *Sur la classification des algèbres de Lie nilpotentes*, C. R. Acad. Sci. Paris Sér. A-B **286** (1978), no. 5, A241–A242.
113. L. Šnobl, *On the structure of maximal solvable extensions and of Levi extensions of nilpotent Lie algebras*, J. Phys. A **43** (2010), no. 50, 505202.
114. L. Šnobl, *Maximal solvable extensions of filiform algebras*, Arch. Math. (Brno) **47** (2011), no. 5, 405–414.
115. L. Šnobl and D. Karásek, *Classification of solvable Lie algebras with a given nilradical by means of solvable extensions of its subalgebras*, Linear Algebra Appl. **432** (2010), no. 7, 1836–1850.
116. L. Šnobl and P. Winternitz, *A class of solvable Lie algebras and their Casimir invariants*, J. Phys. A **38** (2005), no. 12, 2687–2700.
117. ______, *All solvable extensions of a class of nilpotent Lie algebras of dimension n and degree of nilpotency $n-1$*, J. Phys. A **42** (2009), no. 10, 105201.
118. ______, *Solvable Lie algebras with Borel nilradicals*, J. Phys. A **45** (2012), no. 9, 095202.
119. J. Sonn and H. Zassenhaus, *On the theorem on the primitive element*, Amer. Math. Monthly **74** (1967), 407–410.
120. H. Stephani, D. Kramer, M. MacCallum, C. Hoenselaers, and E. Herlt, *Exact solutions of Einstein's field equations*, 2nd ed., Cambridge Monogr. Math. Phys., Cambridge Univ. Press, Cambridge, 2003.
121. M. Sugiura, *Conjugate classes of Cartan subalgebras in real semi-simple Lie algebras*, J. Math. Soc. Japan **11** (1959), 374–434.
122. D. A. Suprunenko and R. I. Tyshkevich, *Commutative matrices*, Academic Press, New York, 1968.
123. S. Tremblay and P. Winternitz, *Solvable Lie algebras with triangular nilradicals*, J. Phys. A **31** (1998), no. 2, 789–806.
124. ______, *Invariants of the nilpotent and solvable triangular Lie algebras*, J. Phys. A **34** (2001), no. 42, 9085–9099.
125. Gr. Tsagas, *Classification of nilpotent Lie algebras of dimension eight*, J. Inst. Math. Comput. Sci. Math. Ser. **12** (1999), no. 3, 179–183.
126. Gr. Tsagas and A. Kobotis, *Nilpotent Lie algebras of dimension nine*, J. Inst. Math. Comput. Sci. Math. Ser. **4** (1991), no. 1, 21–28.
127. Gr. Tsagas, A. Kobotis, and T. Koukouvinos, *Classification of nilpotent Lie algebras of dimension nine whose maximum abelian ideal is of dimension seven*, Int. J. Comput. Math. **74** (2000), no. 1, 5–28.
128. P. Turkowski, *Low-dimensional real Lie algebras*, J. Math. Phys. **29** (1988), no. 10, 2139–2144.
129. ______, *Solvable Lie algebras of dimension six*, J. Math. Phys. **31** (1990), no. 6, 1344–1350.
130. ______, *Structure of real Lie algebras*, Linear Algebra Appl. **171** (1992), 197–212.
131. M. Vergne, *Cohomologie des algèbres de Lie nilpotentes: Application à l'étude de la variété des algèbres de Lie nilpotentes*, C. R. Acad. Sci. Paris Sér. A-B **267** (1968), A867–A870.
132. J. Voisin, *On some unitary representations of the Galilei group.* I: *Irreducible representations*, J. Mathematical Phys. **6** (1965), 1519–1529; II: *Two-particle systems*, 1822–1832.
133. Y. Wang, J. Lin, and S. Deng, *Solvable Lie algebras with quasifiliform nilradicals*, Comm. Algebra **36** (2008), no. 11, 4052–4067.
134. J. H. M. Wedderburn, *Lectures on matrices*, Dover Publications Inc., New York, 1964.

135. G. B. Whitham, *Linear and nonlinear waves*, Pure Appl. Math., Wiley, New York, 1974.

136. P. Winternitz, *Subalgebras of Lie algebras: Example of* $\mathrm{sl}(3,\mathbb{R})$, Symmetry in Physics (Montréal, QC, 2002) (P. Winternitz, J. Harnad, C. S. Lam, and J. Patera, eds.), CRM Proc. Lecture Notes, vol. 34, Amer. Math. Soc., Providence, RI, 2004, pp. 215–227.

137. P. Winternitz and H. Zassenhaus, *Decomposition theorems for maximal abelian subalgebras of the classical algebras*, Technical Report CRMA-1199, Centre de recherches en mathématiques appliquées, Université de Montréal, Montréal, QC, 1984.

138. H. Zassenhaus, *Über eine Verallgemeinerung des Henselschen Lemmas*, Arch. Math. (Basel) **5** (1954), 317–325.

139. ______, *Lie groups, Lie algebras and representation theory*, Sem. Math. Sup., vol. 75, Presses Univ. Montréal, Montréal, QC, 1981.

140. D. Zwillinger, *Handbook of differential equations*, 2nd ed., Academic Press, Boston, MA, 1992.

Index